T0314151

Wood Deterioration, Protection and Maintenance

Wood Deterioration, Protection and Maintenance

Ladislav Reinprecht
Faculty of Wood Sciences and Technology,
Technical University of Zvolen, Slovakia

WILEY Blackwell

This edition first published 2016

Registered office
John Wiley & Sons, Ltd, The Atrium, Southern Gate, Chichester, West Sussex, PO19 8SQ, United Kingdom

Editorial offices
9600 Garsington Road, Oxford, OX4 2DQ, United Kingdom
The Atrium, Southern Gate, Chichester, West Sussex, PO19 8SQ, United Kingdom

For details of our global editorial offices, for customer services and for information about how to apply for permission to reuse the copyright material in this book please see our website at www.wiley.com/wiley-blackwell.

Library of Congress Cataloging-in-Publication Data applied for.

A catalogue record for this book is available from the British Library.

ISBN: 9781119106531

Wiley also publishes its books in a variety of electronic formats. Some content that appears in print may not be available in electronic books.

Cover image: Gettyimages/Katsumi Murouchi

Set in 10/12pt MinionPro by Aptara Inc., New Delhi, India
Printed and bound in Malaysia by Vivar Printing Sdn Bhd

1 2016

Contents

Preface ix
About the Author xi

1 Wood Durability and Lifetime of Wooden Products 1
1.1 Basic information about wood structure and its properties 1
1.1.1 Wood structure 3
1.1.2 Wood properties 10
1.2 Types and principles of wood degradation 12
1.3 Natural durability of wood 14
1.4 Methods of wood protection for improvement its durability 17
1.5 Service life prediction of wooden products 18
1.5.1 Lifetime of wooden products 20
1.5.2 Service life prediction of wooden products by factor method 21
1.5.3 Life cycle assessment of wooden products 22
References 25
Standards 27

2 Abiotic Degradation of Wood 28
2.1 Wood damaged by weather factors 28
2.2 Wood damaged thermally and by fire 34
2.2.1 Thermal wood decomposition 34
2.2.2 Wood burning: fire 36
2.3 Wood damaged by aggressive chemicals 45
2.3.1 Corrosion of wood by chemicals under aerobic conditions 45
2.3.2 Corrosion of wood by chemicals under anaerobic conditions: wood fossilization 49
2.4 Properties of abiotically damaged wood 50
2.4.1 Properties of weathered wood 50
2.4.2 Impact of increased temperature and fire on wood properties 52
2.4.3 Impact of water and other chemicals on wood properties 53
References 57
Standards 61

3 Biological Degradation of Wood 62
3.1 Wood damaged by bacteria 62

3.2 Wood damaged by fungi 65
3.2.1 Reproduction, classification and physiology of wood-damaging fungi 66
3.2.2 Wood-decaying fungi 76
3.2.3 Wood-staining fungi and moulds 88
3.3 Wood damaged by insects 91
3.3.1 Reproduction, classification and physiology of wood-damaging insects 91
3.3.2 Wood-damaging insects 97
3.4 Wood damaged by marine organisms 106
3.4.1 Shipworms 106
3.4.2 Limnoria 107
3.5 Mechanisms of wood biodegradation 108
3.5.1 Biodegradation of cellulose 110
3.5.2 Biodegradation of hemicelluloses 113
3.5.3 Biodegradation of lignin 114
3.6 Properties of biologically damaged wood 117
3.6.1 Properties of rotten wood 117
3.6.2 Properties of wood having galleries 118
References 120

4 Structural Protection of Wood **126**
4.1 Methodology of structural protection of wood 126
4.2 Selection of suitable wood materials 126
4.3 Design proposals for permanently low moisture of wood 129
4.3.1 Estimated moisture of wood 129
4.3.2 Shape optimizations for wood moisture reduction 131
4.3.3 Waterproofing and other isolations of wood and wooden composites from water sources 137
4.3.4 Structural design to prevent condensed water generation 140
4.3.5 Regulation of climatic conditions in interiors 141
4.4 Fire sections and other fire-safety measures 142
References 143
Standards 144

5 Chemical Protection of Wood **145**
5.1 Methodology, ecology and regulation of chemical protection of wood 145
5.1.1 Methodology and legislation of chemical protection of wood 146
5.1.2 Toxicological and ecotoxicological standpoints of chemical protection of wood 149
5.1.3 Regulation of chemical protection of wood 151
5.2 Preservatives for wood protection 152
5.2.1 Bactericides 152

5.2.2 Fungicides: for decay, sap-stain and mould control 153
5.2.3 Insecticides 163
5.2.4 Fire retardants 167
5.2.5 Protective coatings against weather impacts 170
5.2.6 Evaluation of new preservatives 172
5.3 Technologies of chemical protection of wood 173
5.3.1 Improvement of permeability and impregnability of wood 174
5.3.2 Application properties of preservatives 176
5.3.3 Flow and diffusion transport of preservatives in wood 177
5.3.4 Fixation of preservatives in wood 182
5.3.5 Non-autoclave technologies of chemical protection of wood 182
5.3.6 Autoclave technologies of chemical protection of wood 186
5.3.7 Nanotechnologies and nano-compounds for chemical protection of wood 191
5.3.8 Quality control of chemically protected wood 193
5.4 Chemical protection of wooden composites 196
5.4.1 Wooden composites and their susceptibility to damage 196
5.4.2 Principles and technologies of chemical protection of wooden composites 199
References 206
Standards 215
Directives 217

6 Modifying Protection of Wood 218
6.1 Methodology, ecology and effectiveness of wood modification 218
6.1.1 Methods of wood modification: mechanical, physical, chemical and biological 219
6.1.2 Ecology of wood modification 221
6.1.3 Effectiveness of wood modification 221
6.2 Thermally modified wood 223
6.2.1 Principles, methods and technology of thermal wood modification 223
6.2.2 Durability and other properties of thermally modified wood 226
6.2.3 Applications of thermally modified wood 230
6.3 Chemically modified wood 231
6.3.1 Principles, methods and technology of chemical wood modification 231
6.3.2 Substances intentionally or randomly reacting with wood components 233

6.3.3 Durability and other properties of chemically modified wood 242
6.3.4 Applications of chemically modified wood 247
6.4 Biologically modified wood 247
6.4.1 Microorganisms suppressing the activity of wood-damaging fungi and insects 247
6.4.2 Gene engineering for increasing durability of wood and decreasing the activity of fungal enzymes 249
References 250
Standards 259

7 Maintenance of Wood and Restoration of Damaged Wood 260
7.1 Aims and enforcement of the maintenance and the restoration of damaged wood 260
7.2 Wood maintenance 260
7.2.1 Principles of wood maintenance in exteriors and interiors 260
7.2.2 Principles of the fight against the active stages of wood pests 261
7.3 Diagnosis of damaged wood 264
7.3.1 Sensory diagnostic methods 265
7.3.2 Instrumental diagnostic methods 266
7.3.3 Diagnosing the age of wood 278
7.4 Sterilization of biologically damaged wood 279
7.4.1 Physical sterilization of wood 280
7.4.2 Chemical sterilization of wood 286
7.5 Conservation of damaged wood 289
7.5.1 Natural and synthetic agents for wood conservation 289
7.5.2 Methods and technologies for the conservation of air-dried damaged wood 301
7.5.3 Methods and technologies for the conservation of waterlogged wood 306
7.6 Renovation of damaged wood 308
7.6.1 General requirements for the renovation of wooden objects 309
7.6.2 Techniques for strengthening of individual wooden elements 313
7.6.3 Techniques for strengthening of whole wooden structural units 324
References 327
Standards 337

Index 339

Preface

It is well known that wood has been used since ancient times as a structural material for buildings, as the main or auxiliary material in agriculture and later in industrial products, and as a material for furniture and various artistic products.

The service life of wooden products depends first of all on the natural durability of the wood species and wooden composites used, but also very significantly on their design, their methods of chemical and modifying protection, their exposure and their maintenance.

People are able to prolong the lifetime of wooden products based on practical knowledge related to wood-damaging agents (e.g. solar radiation, water, fire, aggressive chemicals, wood-decaying fungi, moulds, wood-destroying insects or marine borers), and also from theoretical studies related to the mechanisms of their action on wood at the molecular, anatomical, morphological and geometry levels.

The structural protection of wooden products is based first of all on the application of durable wood species and other high-quality materials. Simultaneously, the presence of wood-damaging agents has to be limited by using suitable designs with the aim to reduce contact of wood with rain and other sources of water, to reduce the creation of water condensate, and to reduce the impact of ultraviolet (UV) light and fire. So, for this purpose, suitable atmospheric, moisture-impermeable, UV and fire-retardant insulations are applied.

Chemical protection of wood is performed with preservatives; that is, mainly with fungicides, toxic and hormonal insecticides, fire retardants and UV-protective finishes that are applied on the wood's surface and also into its depth. Currently, for wood preservatives not only is their efficiency important, but also their effects on human health and the environment. The optimization of wood pretreatment (e.g. debarking, drying, improving permeability) and its chemical preservation technology (e.g. time of dipping, vacuum, and/or pressure) derive from the theoretical principles of flow and diffusion of preservative substances in the capillary structure of wood. Plywood, particleboards and other wooden composites can be chemically treated during their production or subsequently.

The modifying protection of wood is a prospective mode for improving its resistance against biological agents and dimensional changes. Using active chemical modification, the —OH groups of the lignin–saccharide wood matrix react with molecules of a suitable chemical. This results in a decrease in wood hygroscopicity, and fungi, insects or marine borers then have less interest in this treated wood (e.g. acetylated wood). Thermally modified wood also leads to good resistance to atmospheric factors and biological damage, mainly where there is no contact with the ground.

Wooden products have to be regularly maintained with the aim to increase their lifetime. However, when they became damaged, a thorough inspection of their actual state is important; that is, the diagnosis of the cause, type, degree and range of their damage. Biologically damaged wood should be sterilized. Subsequently, for restoration of smaller wooden elements are used conservation methods, working with natural and synthetic substances. Load-bearing elements of wooden houses, roofs, ceilings and other constructions can be reinforced with prostheses, splicing, special bracing or other methods.

Acknowledgements

The writing of this book was inspired by Mr Gervais Sawyer, editor of the *International Wood Products Journal*. At the same time I would like to thank the professional members of the Wiley publishing house, personally to the publisher Mr Paul Sayer for finding external examiners and for his important comments, to all the anonymous reviewers for ideas on improving some sections of this book, to the editorial assistant Ms Viktora Vida for painstaking collection and control of individual chapters, to Mr Peter Lewis for careful copyediting work and preparing the manuscript for typesetting, and to the production editor Ms Audrey Koh and the project manager from Aptara Ms Baljinder Kaur for creation of the proofs and stylish press quality files. Finally, I would like to thank my daughter Mrs Judita Quiňones for her help in the first grammar correction of this book.

Ladislav Reinprecht
Zvolen, December 2015

About the Author

Ladislav Reinprecht is a professor at the Faculty of Wood Sciences and Technology, Technical University in Zvolen, Slovakia. He obtained an MSc degree in organic chemistry, a postgraduate degree in mycology and a PhD degree in wood technology. For students and specialists he has written many books and monographs, both in Slovak (e.g. *Wood Protection, ThermoWood, SilanWood, Processes of Wood Deterioration, Reconstruction of Damaged Wood Structures, Wooden Buildings – Constructions, Protection and Maintenance, Wooden Ceilings and Trusses – Types, Failures, Inspections and Reconstructions*), and in English (e.g. *Strength of Deteriorated Wood in Relation to its Structure, TCMTB and Organotin Fungicides for Wood Preservation*). His primary research interest lies in the analysis of abiotic and biological defects in wood structure – the conditions for their creation and the methods of their detection, inhibition and prevention. He has published the results of his experimental work in many articles in scientific journals and presented results at various international and domestic conferences.

1 Wood Durability and Lifetime of Wooden Products

Wooden products (furniture, flooring, doors, etc.) and constructions (log cabins, bridges, ceilings, trusses, etc.) produced from various species of wood and types of wooden composites are in practice exposed to different environments, where they can be subjected to more forms of degradation (see Chapters 2 and 3).

With the aim to suppress the degradation processes in the wood, and also in glues, paints and other materials used for wooden products and constructions, it is desirable to use suitable forms of their structural, chemical and modifying protection so that their lifetime can be suitably increased (see Chapters 4, 5 and 6).

The service life of wooden products and constructions can be increased by their regular maintenance. However, when degradation processes in wood and/or in additional materials occur and cause damage, appropriate restoration methods should be used (see Chapter 7).

1.1 Basic information about wood structure and its properties

The structure of wood and wooden composites (Figures 1.1 and 1.2) and their exposure in conditions suitable for the action of abiotic factors and/or the activity of biological pests (Figure 1.3) are the basic prerequisites for potential damage of wooden products and constructions.

Wood is a biopolymer, created by a genetically encoded system of photosynthetic and subsequent biochemical reactions in the cambial initials of trees (Figure 1.1). Trees consist of approximately 70–93 vol.% of wood, with the rest being bast, bark and needles or leaves. Wood is the internal, lignified part of the stem, branches and roots. The characteristics of wood include: (1) anisotropy, typical in three anatomical directions – longitudinal, radial and tangential;

Wood Deterioration, Protection and Maintenance, First Edition. Ladislav Reinprecht.

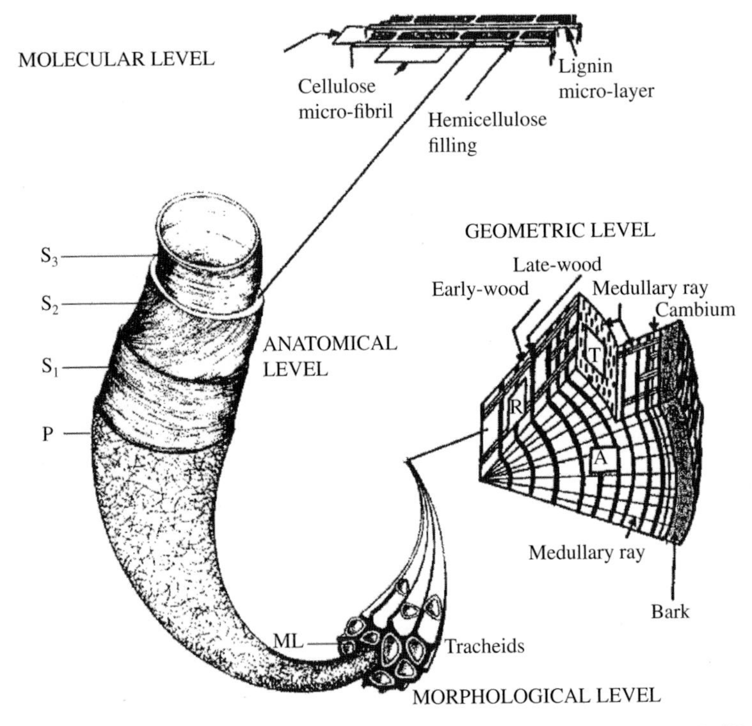

Figure 1.1 Structural levels of wood (modified from Eriksson et al. (1990) and Reinprecht (2008))

Source: Eriksson, K-E., Blanchette, R. A. and Ander, P. (1990) Microbial and enzymatic degradation of wood and wood components. Springer Verlag – Berlin Heidelberg, 407 p. Reproduced by permission of Springer

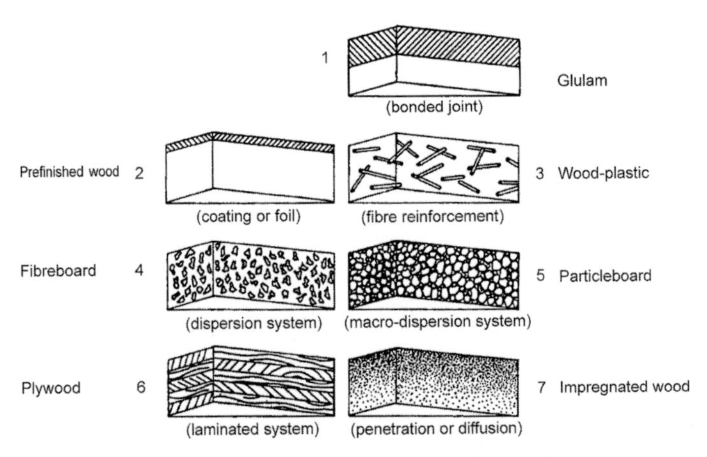

Figure 1.2 The basic types of wooden composites: (1) glulam (glued joints); (2) prefinished wood (coatings or foils); (3) wood–plastic (fibre reinforcement); (4) fibreboard (dispersion systems); (5) particleboard (macro-dispersion systems); (6) plywood (laminated systems); (7) impregnated wood (penetrations or diffusions). (Note: composite is a multicomponent system of materials consisting of at least two macroscopically distinguishable phases, of which at least one is solid)

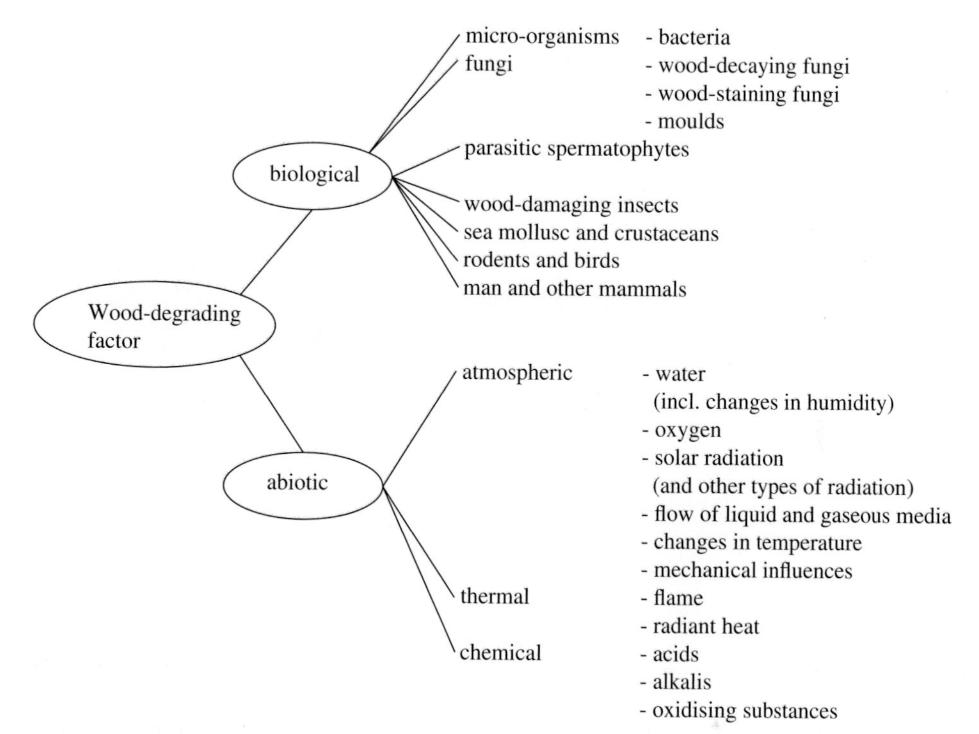

Figure 1.3 Biological and abiotic wood-degrading factors

Source: R., L. (2013) *Wood Protection*, Handbook, TU Zvolen, Slovakia, 134 p. Reproduced by permission of TU Zvolen

(2) inhomogeneity, influenced by the sapwood and heartwood, the early wood and late wood, and so on; (3) specificity, given by the wood species; and (4) variability, given by the growth conditions of the tree of a given wood species.

Wood is a traditional material, used for producing wooden buildings, furniture, work and sport tools, as well as art works. It is currently an irreplaceable raw material for the production of bio-based composites with the targeted combination of wood particles in various stages of disintegration and pretreatment with a complementary system of adhesives, waxes and other additives (Figure 1.2).

1.1.1 Wood structure

The *structure of wood* (Figure 1.1, Boxes 1.1, 1.2, 1.3 and 1.4) and *wooden composites* (Figure 1.2) is defined at four levels:

- primary (i.e. molecular/chemical structure);
- secondary (i.e. anatomical/submicroscopic structure);
- tertiary (i.e. morphological/microscopic structure);
- quaternary (i.e. geometric/macroscopic structure).

Box 1.1 A basic preview of the geometric structure of wood

The geometric structure of wood

Defines

The external appearance – shape, volume, colour, the ratio of tangential, radial and facial areas, the proportion of sapwood, heartwood and/or mature wood, the proportion of early and late wood in annual rings, the roughness and overall quality of the surfaces, and so on.

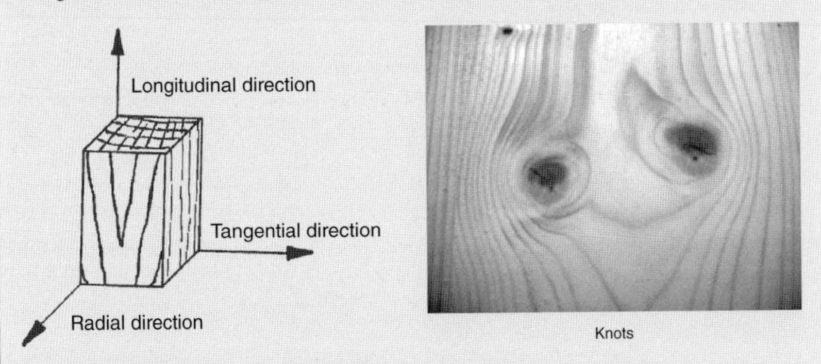

Knots

The macroscopic inhomogeneities – knots, compression or tension wood, juvenile wood, false heart, resin chanals, and so on, together with their type, frequency and state of health (e.g. damage by rot).

Depends on

- the morphological structural level (i.e. the proportional and spatial distribution of various types of cell elements in the wood);
- the growth defects and anomalies in the wood;
- the mechanical and other loads/treatments of the wood.

Source: R., L. (2008) *Ochrana Dreva* (*Wood Protection*), Handbook, TU Zvolen, Slovakia, 453 p. Reproduced by permission of TU Zvolen.

Box 1.2 A basic preview of the morphological structure of wood

The morphological structure of wood

Defines

The individual cells – type, shape, dimensions, slenderness factor, orientation to the pith (longitudinal, radial), thickness of the cell wall, thinning in the cell wall (type, frequency, location), and so on.

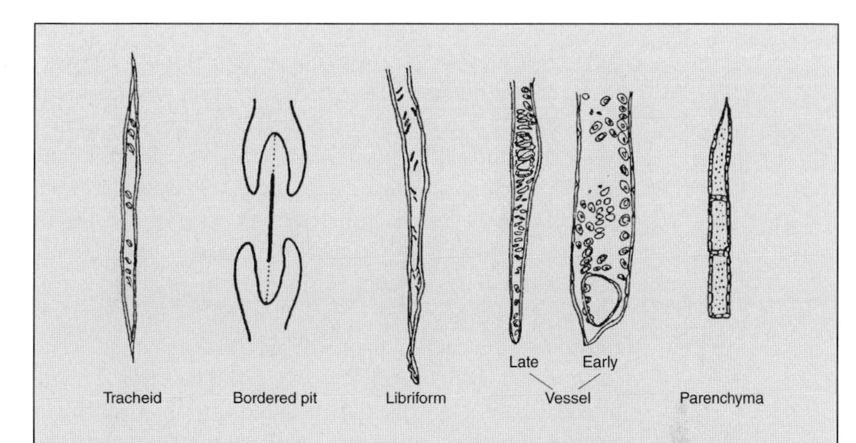

The grouping of cells – proportion and location of parenchymatic, libriform, vessel, tracheid and other cell-types in the wood tissues.

Depends on

The wood species (Fengel & Wegener, 2003; Wagenführ, 2007; Wiedenhoeft, 2010; Wiemann, 2010):

- *Wood of coniferous species* has a simple and fairly regular morphological structure. Approximately 90–95% of wood volume is formed of early and late tracheids. Tracheids have a conductive and strengthening function. They are 2–5 mm long (late are approximately 10% longer) and 0.015–0.045 mm wide. Their cell walls, with a thickness of 0.002–0.008 mm, contain a fairly high number of pit pairs, usually 60–100 in early tracheids and 5–25 in late tracheids. Pit-pairs with a diameter of 0.008–0.03 mm are mainly at the end of tracheids on their radial walls. Opened pit-pairs provide interconnection between tracheids, which is used in the transport of liquids into the wood at its chemical protection and modification. Parenchymatic, thin-walled cells form stock tissue with living protoplasm. They are located in radially oriented pith beams and in longitudinally oriented parenchymatic fibres and resin channels. Resin chanals are lacking in some coniferous species (i.e. they are not present in fir or yew wood).
- *Wood of broadleaved species* has a more complicated morphological structure compared with coniferous wood. Libriform fibres, present in a volume of 36–76%, have a strengthening function. They are relatively short, from 0.3 to 2.2 mm, with a width from 0.005 to 0.03 mm. They have a weak connection with other types of cells due to the small number of simple pit or half-pit thinned areas. Vessels, present in a volume of 20–40%, have a conductive function. Their conductive function is important for the transport of nutrients during a tree's growth, as well as for transport of preservatives and modifying substances into wood.

In ring-porous species (ash, elm, hickory, oak), large vessels in early wood have a diameter from 0.2 to 0.5 mm, whilst small vessels in late wood are from 0.016 to 0.1 mm. The length of vascular systems are usually up to 0.1 m, but in some wood species this can even be several metres (e.g. as long as 7 m in oak). They are created from a long, vertical line of vessels connected via openings – simple, reticular or ranking perforations. Cell walls of vessels have circular and spiral thickenings. The conductive function of vessels decreases under the influence of tyloses (i.e. when blocked by outgrowth from the surrounding paratracheal parenchyma). Parenchymatic cells, present in a volume of 2–15%, mainly have a storage function. Longitudinal, paratracheal parenchymata (single-sided, group, vasicentric, etc.) group around the vessels and vessel tracheids and connect to them via single-sided pit pairs. Longitudinal, apotracheal parenchymata do not come into contact with the vessels. In radially oriented pith beams, several parenchymatic cells are combined with a rectangular shape, horizontal or vertical, either in morphological unity (homogeneous beam) or in morphological diversity (heterogenic beam).

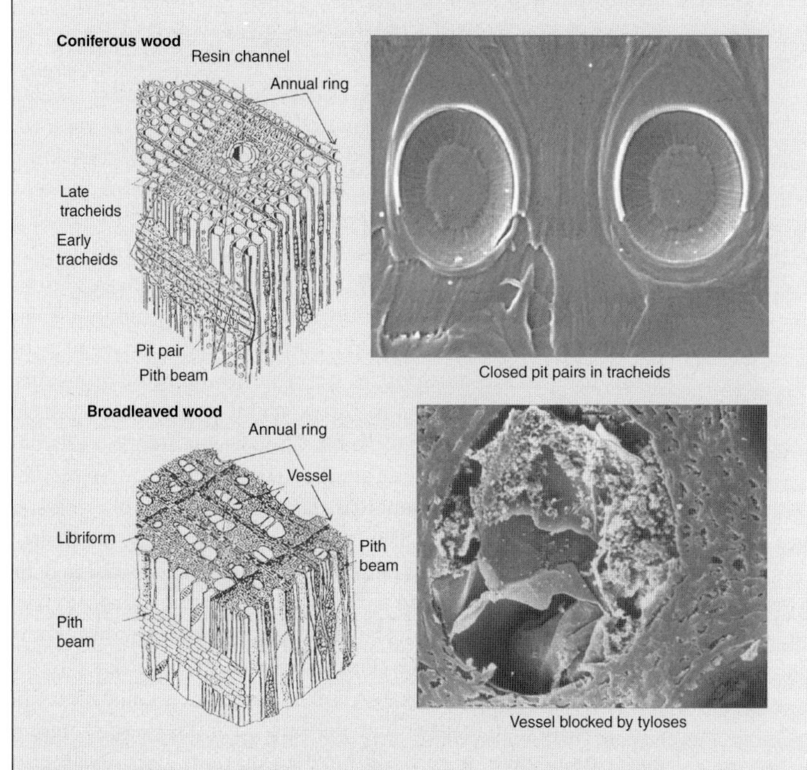

Closed pit pairs in tracheids

Vessel blocked by tyloses

Source: R., L. (2008) *Ochrana Dreva* (*Wood Protection*), Handbook, TU Zvolen, Slovakia, 453 p. Reproduced by permission of TU Zvolen.

Box 1.3 A basic preview of the anatomical structure of wood

The anatomical structure of wood

Defines

The structure of the cell walls of wood's cells:

- layering (i.e. the individual layers ML, P, S_1, S_2, S_3 – see Figure 1.1);
- proportion and localization of the structural polymers (cellulose, hemicelluloses and lignin) and extractives in the individual layers of the cell wall.

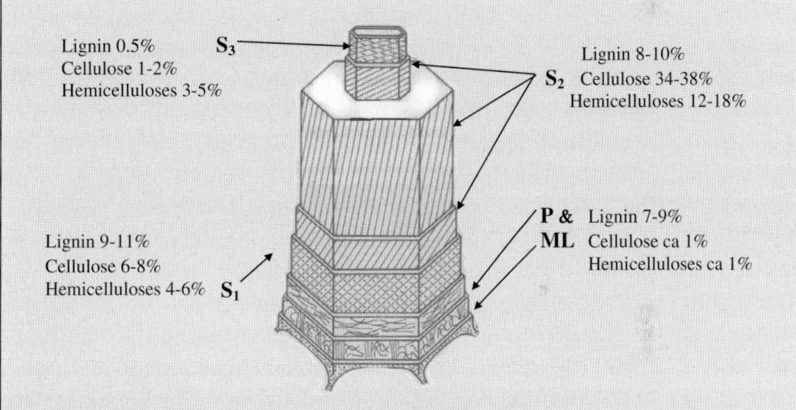

Depends on

The wood species and the type of cell (Fengel & Wegener, 2003; Wiedenhoeft, 2010):

- Elementary fibrils, formed usually of 40 macromolecules of cellulose, are the basic elements of the cells' walls with a cross-section of ca 3.4 nm × 3.8 nm. Microfibrils consist of 20–60 elementary fibrils. Macrofibrils consist of cellulose microfibrils as well as of hemicellulose fillings and lignin microlayers.
- Microfibrils and macrofibrils form substantial lamellae that are the structural base for individual layers of a cell wall (i.e. the ML, P, S_1, S_2 and S_3):
  - ML → middle lamella, mainly formed of lignin granules;
  - P → primary wall, with a thickness of 0.06–0.09 μm, formed of a high proportion of lignin and cellulose fibrils orientated randomly into a multilayered network;

- S → secondary wall, with a thickness of 1–6 μm, formed of three separate layers, S_1, S_2 and S_3; these layers differ in thickness, orientation of fibrils and the proportion and structure of lignin and polysaccharides; for example, in the tracheids of conifers the ratio of layers $S_1/S_2/S_3$ is around 12/78/10.

Affects

The permeability of the wood:

- Cell walls of wood are able to transmit gases and polar liquids. This is due to their microporous structure with vacant pores of size 1–80 nm, as well as due to hydroxyl (—OH), carbonyl (C=O) and other polar groups of lignin and polysaccharides. Macromolecules of polysaccharides in the cell walls repel each other in the presence of polar liquid molecules (e.g. water), which also continuously increase the porosity of the cell wall to a maximum size of ~80 nm. Therefore, its permeability for diffusion and capillary transports continually increases.
- Micropores in the cell walls of wood – gaps in elementary fibrils (~1 nm), capillaries in microfibrils (~10 nm), capillaries in macrofibrils (<80 nm), pores in a pit membrane (<150 nm).

The mechanical properties of the wood:

- For example, cell walls with a greater proportion of the S_2 layer, and also therefore cellulose, provide wood with a greater tensile strength along the fibres.

Source: R., L. (2008) *Ochrana Dreva* (*Wood Protection*), Handbook, TU Zvolen, Slovakia, 453 p. Reproduced by permission of TU Zvolen.

Box 1.4 A basic preview of the molecular structure of wood

The molecular structure of wood

Defines

The types and chemical structure of wood components – cellulose, hemicelluloses and lignin located in the cell walls, and extractives (accompanying substances) located in the cell walls or also in the lumens.

The physical-chemical status of wood components – the degree of polymerization, conformation and configuration structures (spatial grouping into globules, rods, helixes), the supramolecular status (crystalline, amorphous), the physical status (glassy, plastic, viscoelastic), the ability to form intramolecular bonds (hydrogen bonds, van der Waals interactions).

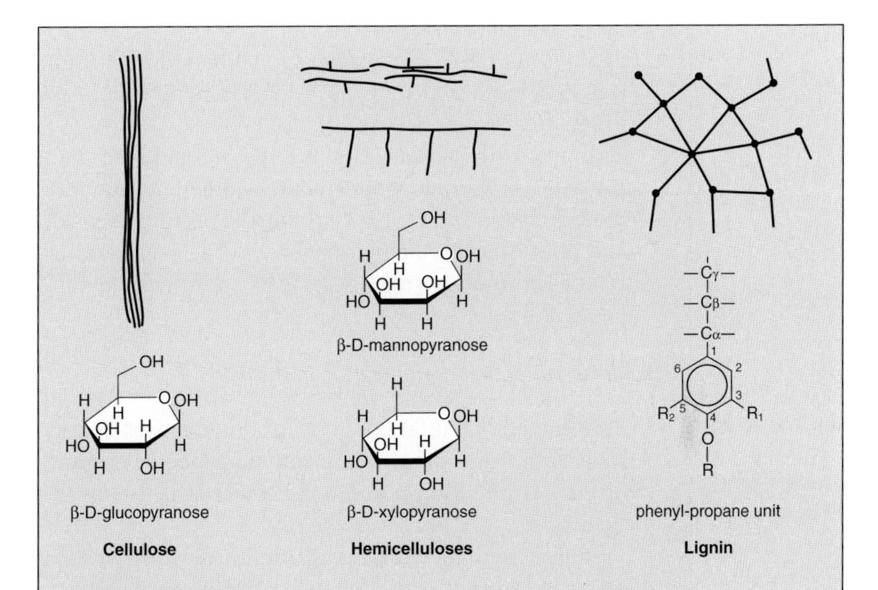

Depends on

The wood species, the type of cell, and the specifics of its composition (Eriksson et al., 1990; Fengel & Wegener 2003):

- Cellulose is a linear polymer consisting of 1,4-β-D-glucopyranose units. These are either arranged into crystalline units (elementary fibrils) or are in an amorphous state.
- Hemicelluloses form branched macromolecular systems of mannanes (in coniferous wood), xylanes (in broadleaved wood) and other polysaccharides.
- Lignin in coniferous woods is a guaiacyl-type based on coniferyl phenyl-propane units (i.e. 15% $-OCH_3$ groups/C_9). Lignin in broadleaved woods is a mixture of guaiacyl-type and syringyl-type, at which lignin of syringyl-type is based on synapyl phenyl-propane units (i.e. 20–21% $-OCH_3$ groups/C_9).
- Terpenes are accompanying biologically effective substances in the wood of more durable coniferous species. Tannins, flavonoids and some other substances play this role in the wood of more durable broadleaved species.

Affects

The durability of the wood:

- hemicelluloses are the overall most unstable component of the wood, mainly against high temperatures and hydrolysis in the presence of acids in the environment or enzymes produced by wood-decaying fungi;

- lignin is not stable when facing oxidation induced by ultraviolet (UV) radiation in exterior, or by peroxidases and other enzymes of white-rot fungi;
- accompanying substances affect the resistance of wood to biological damage; various woods contain different amounts of (1) easily biodegradable substances (e.g. starch, pectin, glycosides and lipids) and (2) substances biologically effective against wood-decaying fungi, moulds or wood-boring insects (e.g. tannins, flavonoids, stilbenes, terpenes and resin acids).

The preservation and modification of the wood:

- diffusion and fixation processes of preservatives and modifying substances in wood depend not only upon their physical and chemical properties but also upon the molecular structure of the cell walls in the wood;
- modification of the molecular structure of wood (acetylation, etherification, etc.) can increase resistance to biological pests, and the dimensional stability and strength properties can be improved.

Source: R., L. (2008) *Ochrana Dreva* (*Wood Protection*), Handbook, TU Zvolen, Slovakia, 453 p. Reproduced by permission of TU Zvolen.

The structure of wood significantly determines its natural durability, defined as its resistance to abiotic and biological damages (see Chapters 2 and 3). In this view, the structure of wood also affects the conditions for storing of cut logs and produced timber, the methods and technologies for the structural, chemical and modifying protection of wooden products (see Chapters 4, 5 and 6), as well as the methods and technologies for their maintenance and restoration (see Chapter 7).

1.1.2 Wood properties

The properties of wood (Box 1.5) usually worsen due to its damage (see Chapters 2 and 3). The restoration of damaged wood returns its original properties – strength, dimensional stability, aesthetics, and so on (see Chapter 7).

Box 1.5 A basic preview of the properties of wood

See also Sections 2.4 and 3.6.

Density

The wood of broadleaved species commonly used in Europe for products and constructions (Table 1.3: beech, birch, black locust, elm, hornbeam, linden, maple, oak, poplar, etc., 440–800 kg/m^3) is usually more dense than the wood of commonly used coniferous species (Table 1.3: cedar, Douglas fir, fir, larch, pine, spruce, 370–530 kg/m^3). The density of wood is decreased after being damaged by fire, fungal rot or insect galleries. However, the opposite trend (i.e. an increase of density) is not uncommon in subfossil wood or wood attacked by alkalis.

Strength properties

Wood has a relatively high strength in relation to its density when compared with other materials used in construction. The strength properties of wood (compression, tension, bending, hardness, etc.) depend upon its density and structure, which assist us in selecting a suitable type of wood for a particular use. Depolymerization of polysaccharides in decayed or otherwise damaged wood decreases its strength, mainly in its wet state, where the support strengthening effect of hydrogen bonds and van der Waals interactions already do not apply.

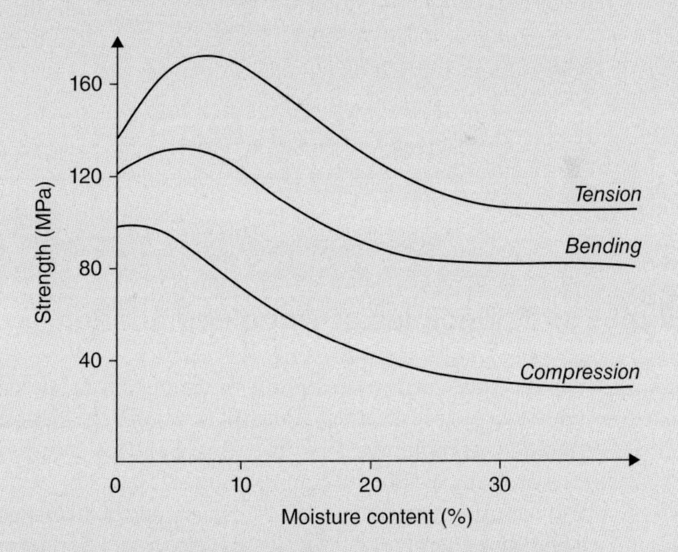

Moisture properties

The humidity of wood adapts to climatic conditions of its exposure. During long-term exposure to air with a relative humidity of 95–99%, wood greatly humidifies and its equilibrium moisture content settles around 28–30%; that is, the fibre saturation point (FSP). The wood also easily receives liquid water via capillary forces, and its maximum moisture depends upon its porosity/density; for example, beech with a density of 600 kg/m^3 has a maximum moisture of ~120%, whereas for spruce with density of 400 kg/m^3 it is as much as ~200%.

The swelling and shrinkage of wood are processes connected with receiving of bound water until it reaches the FSP (wood swells) and vice versa with its drying when water is released (wood shrinkages). The dimensions of wood change as well as when its moisture changes from FSP to 0%; for example, for common wood species the maximum shrinkages are in a longitudinal direction $\alpha_L = 0.15–0.65\%$, a radial direction $\alpha_R = 2.5–6.7\%$ and in a tangential direction $\alpha_T = 8.3–14.7\%$.

The moisture properties of wood also affect its strength, durability and use; that is, (1) strength of wood usually decreases with increased moisture within the range from 0% to the FSP; (2) resistance of wood to biological damage is usually lower at higher moistures; and (3) frequent and marked changes in the moisture content of wood lead to shape deformation and the creation of cracks.

Thermal properties

Wood has relatively good thermal insulation properties. However, it does not resist temperatures over 150 °C for a long time and may ignite. Despite the fact that it is flammable, during fires it is often more stable in terms of shape and strength than metals or plastics are.

Source: R., L. (2008) *Ochrana Dreva* (*Wood Protection*), Handbook, TU Zvolen, Slovakia, 453 p. Reproduced by permission of TU Zvolen.

1.2 Types and principles of wood degradation

Wood is more or less susceptible to various forms of degradation (Figure 1.3, Table 1.1; and see Chapters 2 and 3). In wood degradation, the dominant role is played by its molecular structure (Box 1.4). However, its higher structural levels – anatomical (Box 1.3), morphological (Box 1.2), and geometric (Box 1.1) – also have significant roles. Wood can already be damaged during its growth in trees, subsequently at harvesting, during storage and transport of logs and timbers, and also after processing on products and constructions.

Changes in structural levels of wood are caused by abiotic agents or energies and biological pests. Subsequently, its strength, hygroscopic, thermal, aesthetic and other properties are also changed, and usually impaired. The intensity and scope of the structural and property changes of wood depend upon the type and mechanism of the degradation process. In the case of some degradation types (e.g. due to weathering or moulds), just surface damage of wood occurs. In contrast, in the case of fire, fungal decay or feeding by wood-damaging insects, wood is degraded to a greater depth, often in full.

Damage that begins at the molecular structure of wood is the most important for changes in its properties (Table 1.1). All degradation effects in the

Table 1.1 Types of wood degradation related to the deterioration of its polymers (polysaccharides and lignin) or without their deterioration

Type of wood degradation	Wood-degrading factor
With destruction of wood polymers	
Photo-oxidations (mainly in lignin, 0.05–2.5 mm from surface)	UV radiation
Thermo-oxidations, dehydrations (mainly in hemicelluloses)	Thermal effects • temperature of air above ~150 °C • fire (flame)
Hydrolytic reactions, lignin plasticizing	Hydrothermal effects • temperature of water/steam above ~70 °C
Various reactions – hydrolytic, dehydration, oxidation, etc., cellulose de-crystallization	Aggressive chemicals • emissions (e.g. SO_2, NO_x) • acids and alkalis • inorganic fungicides • fire retardants
Biochemical reactions catalysed by enzymes and low molecular weight agents of fungi and bacteria	Wood-decaying fungi • white rot • brown rot • soft rot Bacteria
Mechanical decompositions and then biochemical reactions in digestive tract	Wood-damaging insects Marine organisms
Without destruction of wood polymers	
Mechanical cracks	Humidity and thermal gradients
Mechanical holes, nibbling marks	Some insects, birds and mammals
Colour changes	Wood-staining fungi and moulds
Degradation of bordered pits in tracheids	Bacteria and moulds

Source: R., L. (2013) *Wood Protection*, Handbook, TU Zvolen, Slovakia, 134 p. Reproduced by permission of TU Zvolen.

molecular structure of wood – in its polymers, caused by atmospheric factors, high temperatures, aggressive chemicals and fungal decay – are reflected also at its anatomical and morphological structural levels (e.g. with regard to damage of cell walls and entire tissues), and usually also at its geometric level (e.g. more intensive degradation of sapwood or early wood), and of course in its properties as well.

Damage that begins only at the upper structural levels of wood is usually less important for changes in its properties. For example, small changes in the density and strength of wood can occur as a result of microscopic and macroscopic cracks created by moisture stresses due to badly regulated drying, although without great changes in its hygroscopicity and colour.

1.3 Natural durability of wood

Wood has several implicit advantages in comparison with other materials (e.g. stone, clay, brick, concrete, metals, plastics):

- it is a permanently renewable material source (e.g. it has low impact on the environment and low energetic demands for processing in total);
- it is easily workable;
- it has high strength in relation to density;
- it has low thermal conductivity;
- aesthetically pleasing qualities in products.

In contrast, *wood also has negative properties, reflected by its lower natural durability*:

- weathering
- flammability
- biodegradability.

The *natural durability of wood* is its inherent resistance to various abiotic factors and biological pests (see Chapters 2 and 3). The natural durability of a defined wood species (and also of a defined type of wooden composite) may be only supposed – it may not be exactly defined. The cause is the complementary effects of many variables. The most significant of these are:

- Differences in the structure of the individual wood species; that is, there is a specific dependence on the age of the wood and the presence of juvenile wood, and also on the climatic, soil and other conditions of tree growth (Table 1.2).
- The environment around the wooden product. That is, there is usually a difference between interior and exterior environments, as well as in various exterior climatic zones; there is also an influence of the structural protection of wood used (prEN 16818). For example, weather factors acting on the exterior are more aggressive in a direct contact with the ground or above ground without shelter (Brischke & Rapp, 2008); a northerly orientation is usually more suitable for the activity of wood-damaging fungi, while abiotic degradation of wood surfaces due to UV radiation and temperature changes is stronger in a southerly orientation.

The *natural durability of individual wood species* is known from practice; however, it is permanently studied also on the basis of both laboratory and field tests. Several studies on the natural durability of various wood species were elaborated by, for example, Rapp et al. (2000), Van Acker et al. (2003) and Van Acker and Stevens (2003). The natural durability of woods is now based on practical knowledge and experiments assembled in the form of (1) the percentage ratio

Table 1.2 Natural durability of wood predetermined by its structure

Structural level of wood	Wood durability
Molecular	
• Accompanying substances:	
Tannins (e.g. black locust, chestnut, oak)	higher resistance to fungi and insects
Resins (e.g. Douglas fir, larch, pine)	higher resistance to fungi and insects lower resistance against ignition
Inorganic substances (e.g. containing Na, K, Ca, Mg, P, S) comprised mainly in the wood of fast-growing species (e.g. poplar, alder)	lower resistance against fungi
• Crystalline cellulose	higher resistance against fungi
• Lignin	higher resistance against combustion lower resistance against UV radiation
Anatomical	
• Both polysaccharides and lignin in cell walls	impeded transfer of enzymes of fungi and bacteria in cell walls
Morphological	
• Parenchyma cells	easily attacked by bacteria and fungi (since they comprise nutrients)
• Vessels	easily permeable for fungi hyphae (also for liquids and gases)
• Libriform fibres	easily attacked by some decay-causing fungi (compared with vessels)
• Opened pits in cell walls	easier transfer of enzymes in wood easier changes of wood moisture
Geometric	
• More frontal surfaces	worse durability
• More sap-wood	worse durability
• More early wood	usually worse durability
• Rougher surface	usually worse durability

Source: R., L. (2013) *Wood Protection*, Handbook, TU Zvolen, Slovakia, 134 p. Reproduced by permission of TU Zvolen.

durability, with regard to a well-known wood species (e.g. to oak heartwood), or (2) durability classes (e.g. by EN 350-2), which rank woods on the basis of their resistance to activity of selected biological pests (Table 1.3).

The natural durability of individual wood species is influenced also by the pest's interest about the wood, or by the ability of a specific chemical compound to be attacked in the wood. For example, some species of wood-damaging insects attack only wood of conifers and only in interiors – typically, the house longhorn beetle (*Hylotrupes bajulus*). Woods having tannins or similar extractives (e.g. oak) become black in colour near to a contact with iron nails or screws, but others woods (e.g. beech) are resistant to such colour changes.

Table 1.3 Classes of natural durability of selected wood species in their contact with ground – against rot (modified from EN 350-2)

Durability class	Commercial name	Scientific name	B or C[a]	Density (kg/m^3)	Occurrence
1					
Very durable	Greenheart	*Ocotea rodiaei*	B	1030	South America
	Jarrah	*Eucalyptus marginata*	B	830	Australia
	Mansonia	*Mansonia altissima*	B	620	West Africa
	Okan	*Cylicodiscus gabunensis*	B	920	West Africa
	Padouk	*Pterocarpus soyauxii*	B	740	West Africa
	Teak	*Tectona grandis*	B	680	Asia
	Walaba	*Eperua falcata*	B	900	South America
1–2					
	Black locust	*Robinia pseudoacacia*	B	740	Europe
	Kapur	*Dryobalanops aromatica*	B	700	South-East Asia
2					
Durable	Bubinga	*Guibourtia demeusii*	B	830	West Africa
	Chestnut	*Castanea sativa*	B	590	Europe
	Oak	*Quercus robur (Q. petraea)*	B	710	Europe
	White cedar	*Thuja plicata*	C	370	North America
3					
Medium durable	Douglas fir	*Pseudotsuga menziesii*	C	530	North America
	Turkey oak	*Quercus cerris*	B	770	Europe
	Walnut	*Juglans regia*	B	670	Europe
3–4					
	Larch	*Larix decidua*	C	600	Europe
	Pine	*Pinus sylvestris*	C	520	Europe
4					
Less durable	Elm	*Ulmus* sp.	B	650	Europe
	Fir	*Abies alba*	C	460	Europe
	Spruce	*Picea abies*	C	460	Europe
5					
Non-durable	Ash	*Fraxinus excelsior*	B	700	Europe
	Beech	*Fagus sylvatica*	B	710	Europe

Table 1.3 (*Continued*)

Durability class	Commercial name	Scientific name	B or C[a]	Density (kg/m^3)	Occurrence
	Birch	*Betula pubescens*	B	660	Europe
	Hornbeam	*Carpinus betulus*	B	800	Europe
	Lime tree	*Tilia cordata*	B	540	Europe
	Maple	*Acer pseudo-platanus*	B	640	Europe
	Poplar	*Populus* sp.	B	440	Europe

Durability classes 1 to 5 are relative; that is, usable only for mutual comparison of durability of the individual types of wood.
Durability classes are applicable only to heartwood.
Sapwood of all broadleaved and coniferous species is classified in class 5 (non-durable), unless other data are available.
[a] B: broadleaved; C: coniferous.

From the point of view of the natural durability of wood against pests, it can be generally stated that this does not depend upon the wood density. For example, the mature or heart parts of beech, hornbeam and other woods having a high density but no biologically active agents (tannins, stilbenes, terpenes, etc.) belong to the less durable species like alder, lime tree and poplar woods that have substantially lower densities.

A *decrease in the high natural durability (resistance to pests)* of several exotic wood species (Kazemi, 2003; Yamamoto et al., 2004; Nzokou et al., 2005) or some European woods (e.g. black locust and oak) can occur during their exposure when mainly in a permanently moist environment. Tannins and also other low molecular weight substances, which significantly increase the resistance of woods to biological pests, may be gradually washed out or evaporated from the wood, and thus suffer from a decreased natural durability against fungi or other pests over time. Similar experiences are known for archaeological or subfossil oaks that have lain under a wet ground for several thousand years (Horský & Reinprecht, 1986).

1.4 Methods of wood protection for improvement its durability

Wood protection is defined as the set of measures for securing its quality and increasing its natural durability. Protection of wood is carried out from growing interventions by foresters, through suitable tree harvesting, logs and sawn timbers transportation and storage, compliance with technological principles of production and protection of wooden products up to their use in practice.

Protection of wood can be performed by applying various methods throughout the various stages:

- protection of growing trees (i.e. in the forest) against physiological pests and other factors, which is provided for by foresters (i.e. forest and tree protection);
- protection of harvested wood (i.e. on round-wood yards) and during first-stage processing to sawn timber, veneers, chips and other intermediate products (i.e. the physical – totally wet conditions or quick drying) and possibly also short-term chemical wood protection;
- protection of new wooden products (i.e. use of suitable structural designs and selection of suitable chemical, modifying and other processing) – all these methods of preventive and supplementary wood protection are based on physical, structural, chemical and/or modification technologies (Table 1.4; and see Chapters 4, 5 and 6);
- protection of older wooden products – that is, maintenance of wood, or conservation and reconstruction of markedly degraded wood, typically of historical artefacts and constructions (see Chapter 7).

In general, the objective of wood protection is to create such conditions in its structure and surrounding environment that are unfavourable for the damaging effect of abiotic factors and biological pests. The natural durability of wood can be improved using several methods (Table 1.4). However, it holds true that improvement of the natural durability of wood should be specifically reasonable for every type of wooden product; that is, it is not expedient to unreasonably improve the service life of a wooden product using such technology that evidently burdens the environment and the protected wood becomes harmful to the health of people and animals.

1.5 Service life prediction of wooden products

Wooden products and constructions wear in time both physically (i.e. due to the impact of defects caused by various wood-degrading factors) and morally (i.e. due to the changed demands of humans regarding their functionality, aesthetic aspects, etc.).

The service life means the period of time after installation during which a building or its parts meets or exceeds the performance requirements (ISO 15686). Similarly, the service life of log cabins, trusses, ceilings, windows, doors, furniture and other products made of wood is defined by the time for which they should meet the function, technical and aesthetic requirements under the supposed conditions of application. Their service life can be defined also as the time after which they get to a so-called terminal condition (i.e. they become unusable). The service life is a variable value since it is determined from supposed exposure conditions that are not always fully implemented

Table 1.4 Principles of preventive protection of wooden products in order to increase their durability

Principle of wood protection		Technology of wood protection	Utilization in practice
Natural durability		Application of: (a) more durable wood species and wooden composites; that is, using durable woods with tannins, terpenoids, and so on (e.g. heartwood of teak, black locust, oak, larch) and durable wood species in plywood, oriented strand board, and so on	+++
		(b) durable glues and other agents in wooden composites	+
		Gene engineering applied for cultivation of trees	N
Exposures		Permanently dry conditions (biological pests are inactive); that is, timber drying-up and correct structural protection of wooden products and constructions	+++
		Permanently wet conditions with minimal amount of oxygen (biological pests are inactive, expect for anaerobic bacteria and marine organisms); that is, logs sprayed with water, stored in water pools or wet ground	++
		Atmosphere unsuitable for biological pests; that is, placement of wood into inert gases (nitrogen, argon)	+
		Barriers created on wood surfaces; that is, regulation of air and moisture transport to wood and out from wood	+
Preservatives	Biocides	Toxic biocides against pests; that is, bactericides, fungicides and insecticides (e.g. creosote, boric acid, quaternary ammonium compounds, triazoles, pyrethroids)	+++
		Nontoxic biocides against pests (e.g. growth regulators of insects)	++
		Vegetable extracts; that is, substances with toxic effect against some pests and also with hydrophobic effect	+
		Pheromones, attractants, repellents; that is, regulators of behaviour of insects and other pests during their life	+

(*continued*)

Table 1.4 (*Continued*)

Principle of wood protection		Technology of wood protection	Utilization in practice
	Others	Fire retardants	+++
		Resins and oils with hydrophobic effect	++
		Film-forming paints against weathering (e.g. polymers with UV-stabilizers, water repellents, and other additives)	+++
Modification		Thermal treatment (at ~160–220 °C)	++
		Chemical treatment (e.g. acetylation, furfurylation)	+
		Mineralization (e.g. silicates)	(+)
		Enzymatic treatment	N
		Bio-control (i.e. antagonistic organisms – bacteria, moulds, etc.) against wood-decaying fungi or/and wood-damaging insects	(+)

Source: R., L. (2013) *Wood Protection*, Handbook, TU Zvolen, Slovakia, 134 p. Reproduced by permission of TU Zvolen.
+++: significant application; ++: medium application; +: little (limited, specialized) application; (+): rare application; N: application in practice maybe in future.

(Van Acker et al., 2014). Moisture content of wood is frequently used as an input variable to modelling conditions and the resulting risk of decay, and then also of the service life prediction of wooden products (Brischke & Thelandersson, 2014).

1.5.1 Lifetime of wooden products

The lifetime of materials and structures is defined as the:

- *physical useful lifetime* – reflecting the real technical condition;
- *ethical lifetime* – relating mainly to aesthetic aspects and the satisfaction of functional demands of the present user (e.g. stylishness, spatialization and/or higher demands for thermal and acoustic insulation);
- *economical lifetime* – considering the time within which the costs of maintenance, operation and depreciation are still economical with regard to usability.

The *useful lifetime* is not usually the same for the entire wooden product or wooden structure; it means for all the wooden elements in a log cabin, truss, and so on (see Chapters 4 and 7). For example, in log cabins, the most suitable humidity conditions for action of biological pests are created in the lowest beams, which are more often in contact with rain or capillary water. Shortening of the physical lifetime of products/structures can be caused mainly due to the following effects:

- failures in a project (i.e. errors in static, material composition, structural protection, etc.);
- failures in execution (i.e. technological errors);
- failures during usage (i.e. increased aggressiveness of environment, increased mechanical load, poor or insufficient maintenance);
- unforeseeable events (i.e. fire, storm, etc.);
- amendments in standards and regulations (i.e. innovations in static standards, new safety regulations, etc.).

The useful lifetime, or service life, of wooden buildings or bridges is usually from 40 to 200 years, trusses from 60 to 400 years, windows from 30 to 70 years, untreated sleepers from 3 to 5 years, sleepers preserved with creosote from 25 to 50 years, and so on. In Eurocode, the suggested minimum design service life of a building's elements varies from 10 to 100 years. However, the age of well-maintained wooden historical buildings and their components is often greater; for example, there are log-cabin-churches more than 200 years old (Reinprecht, 2004; Viitanen, 2013), and there are windows greater than 100 years old (Menzies, 2013). Hansson et al. (2012), on the basis of selected climatic data (i.e. the average outdoor temperatures and average daily precipitations from 28 European field trials), proposed decay risk models that should be helpful for service life prediction of wooden constructions in above-ground exposures. Kirker and Winandy (2014) concisely summarized biotic and abiotic factors that impact service life of wood and wood-based materials above ground.

1.5.2 Service life prediction of wooden products by factor method

The *service life prediction or planning* of wooden constructions and wooden products can be performed by the factor method (ISO 15686).

The factor method is used to obtain an estimated service life of a component or a design object by modifying a reference service life by considering the differences between the object-specific and the reference in-use conditions under which the reference service life is valid:

$$\mathrm{ESL_{WP}} = f(A, B, C, D, E, F, G) = \mathrm{RSL_{WP}} \times (A \times B \times C \times D \times E \times F \times G) \tag{1.1}$$

where $\mathrm{ESL_{WP}}$ (years) is the estimated (predicted) service life of the wooden product, $\mathrm{RSL_{WP}}$ (years) is the reference service life of the wooden product (i.e. its lifetime under standard production and application conditions) and A–G are nondimensional factors, the values of which are usually in the range 0.8–1.2 but which in some situations (e.g. an error in design or, in contrast, a perfect maintenance) can also cover a greater range (e.g. 0.5–3). A is a measure of the quality of the components (i.e. of the wood and also of any complementary materials – see Chapters 5 and 6); *B* is a measure of the design (i.e. of the product/structure in its entirety, and its details – see Chapter 4); *C* is a measure of the technology level of works carried out (see Chapters 4, 5 and 6); *D* is a measure of the internal environment (i.e. in joints and other details); *E* is a measure of the external environment (i.e. around the product/structure); *F* is a measure of the user conditions (e.g. unexpected changes in loading, storm); and *G* is a measure of the maintenance level (see Chapter 7).

However, in practice the computation of ESL_{WP} by the factor method is sometimes connected with a certain subjectivity and sometimes also with less professionalism (Viitanen, 2005, 2013):

- it is not a really scientific method;
- need for intensive preparatory work – for example, many-year observations in terrain and/or model experiments, with the aim to reliably define all specific effects influencing the individual factors *A* to *G*;
- there can be a subjective overestimation of the importance of a certain factor and the underestimation of another;
- there can be ambiguities in the mutual interactions of the individual factors *A* to *G*.

An example of the application of the factor method, with the subjective highlighting of the importance of using chemical protection, regular maintenance with needed repairs and reconstruction works is the calculation of the ESL of wooden bridges (Figure 1.4, Table 1.5). The service life of a correctly designed, materially and technologically executed and maintained wooden bridge is usually from 40 to 200 years, sometimes even more, which is in accordance with the results in Table 1.5. However, there are also cases where, when selecting less durable types of wood species and not implementing a suitable protection method (e.g. badly structurally designing details) and when neglecting maintenance, the service life of a wooden bridge can be significantly reduced, sometimes not even reaching 5–10 years.

Similarly, it is possible to calculate the lifetime of wooden ceilings, trusses, pergolas, children's playgrounds, shingle roofs or claddings, windows and other wooden structures and products. However, it is necessary to always bear in mind that the reliability of the calculation depends upon the selected RSL_{WP} to a notable extent and at the same time also upon the selection of specific nondimensional factors *A* to *G* in relation to a particular project, and its implementation into practice.

1.5.3 Life cycle assessment of wooden products

Life cycle assessment (LCA) is a technique for assessing the environmental aspects and potential impacts associated with a product and has been used by

(a)

(b)

Figure 1.4 Estimated service life (ESL_{WP}) of wooden uncovered bridges is ~40–50 years when using suitable design for rainfall drain and at the same time also chemical protection of wood elements, either with creosote (a) or with inorganic biocide fixable in wood (b); examples are from Norway (Reinprecht, 2008)

Source: R., L. (2008) *Ochrana Dreva* (*Wood Protection*), Handbook, TU Zvolen, Slovakia, 453 p. Reproduced by permission of TU Zvolen

Table 1.5 Calculation of estimated service life (ESL) of wooden bridges (Reinprecht, 2005)

Wooden bridge without a roof – good lifetime	
Glued lamellae prepared from Norway spruce (*Picea abies*), which according to EN 350-2 is less durable wood species against wood-decaying fungi (class 4) and is also prone to degradation by wood-damaging insects (class SH).	$A = 3$
The individual lamellae are first treated with stable and highly efficient biocide (fungicide and insecticide) against pests using vacuum-pressure technology, and then bonded with stable water-resistant adhesive.	
The glued lamellae are also treated with a water-repellent paint.	
Finally, biocide cartridges are inserted to the end (frontal) sections of glued lamellae.	
The bridge is not roofed. However, the details are structurally correctly designed, with possible drain of rainwater.	$B = 0.5$
Proper input humidity of wood when producing glued lamellae (moisture content of wood $w = 15\,\%$).	$C = 1$
Microclimate in details of bridge is sometimes (e.g. after rains) suitable for fungal rot.	$D = 0.6$
The bridge is exposed to rainfall, alternation of rainy and sunny weather, thermal differences between day and night, and so on. This all enables the swelling and shrinkage of wood, and gradually creation of cracks. Environment for activity of fungi and insects is also suitable.	$E = 0.6$
The bridge is not exposed to more significant mechanical stresses, emissions, and so on.	$F = 1$
Regular inspection with diagnosis of damage (every third year), and suitable maintenance (exchange of biocide cartridges, new paints, etc.) and improvement interventions (replacement or reinforcement of damaged elements by prostheses, etc.) carried out.	$G = 2$
$ESL_{WP} = RSL_{WP} \times (A \times B \times C \times D \times E \times F \times G) = 40 \times (3 \times 0.5 \times 1 \times 0.6 \times 0.6 \times 1 \times 2) \approx 43.2$ years	
Wooden bridge with a roof – very good lifetime	
Solid larch (*Larix decidua*) heartwood, which according to EN 350-2 is from moderately durable to less durable against wood-decaying fungi (class 3–4) and durable against wood-damaging insects (only sapwood is susceptible to attack – class S).	$A = 2.25$
The individual wooden elements of bridge (note: elements dimensionally greater than lamellae and thus harder to impregnate) treated with stable biocide (fungicide and insecticide) against wood pests using a long-term dipping technology.	
Finally, all surfaces of wood elements painted with a water-repellent coating.	
The bridge is both roofed and structurally correctly designed.	$B = 1$
Proper input humidity of wood for the production of bridge elements ($w = 18\,\%$).	$C = 1$

Table 1.5 (*Continued*)

Wooden bridge with a roof – very good lifetime	
Microclimate in construction details is rarely suitable for fungal rot.	$D = 0.8$
The bridge is not exposed to direct water precipitations.	$E = 1$
The bridge is not exposed to more significant mechanical stresses, emissions, and so on.	$F = 1$
Regular inspection with diagnosis of damage (every third year), and the execution of both suitable maintenance and improvement interventions.	$G = 2$
$ESL_{WP} = RSL_{WP} \times (A \times B \times C \times D \times E \times F \times G) = 40 \times (2.25 \times 1 \times 1 \times 0.8 \times 1 \times 1 \times 2) \approx 144$ years	

Source: R., L. (2005) Investigation of wooden constructions – the basis for their renovation. In: *Restoration and Reconstruction of Constructions*, 27th Conference WTA, pp. 23–30. Brno, Czech Republic. (In Slovak.)

industry in many situations; for example, to help reduce overall environmental burdens across the whole life cycle of goods and services, to improve the competitiveness of a company's products or to communicate with governmental bodies. It can also be used in decision-making, as a tool to improve material composition and design of the product, the selection of optimal technologies, and so on. The benefit of LCA is that it provides a single tool that is able to provide insights into the upstream and downstream trade-offs associated with environmental pressures, human health and the consumption of resources (Ferreira et al., 2015).

The *LCA of wooden products and constructions* is usually better for comparing with those made from steel, aluminium, concrete or other materials (Brischke & Rapp, 2010). For example, the Bolin and Smith (2011) state that the individual factors of LCA are evidently better for borate-treated lumber structural framing in comparison with galvanized steel framing: (1) there are 3.7 times less fossil fuel use and 38 times less water use; (2) there are 1.8 times lower greenhouse gases, 3.5 times lower acid rain, 2.8 times lower smog and 3.3 times lower emissions; and (3) they are candidates for energy recovery as a renewable fuel source. Similarly, in a comparison of wooden and plastic PVC windows, in a mild exposure scenario the service life of plastic windows should be beyond 35 years and wooden windows 60–65 years, at which time the recycling of wooden windows is easier (Menzies, 2013).

References

Bolin, A. & Smith, S. (2011) Life cycle assessment of ACQ-treated lumber with comparison to wood plastic composite decking. *Journal of Cleaner Production* 19(6–7), 620–629.

Brischke, C. & Rapp, A. O. (2008) Influence of wood moisture content and wood temperature on fungal decay in the field: observations in different micro-climates. *Wood Science and Technology* 42(8), 663–677.

Brischke, C. & Rapp, A. O. (2010) Service life prediction of wooden components – part 1: determination of dose–response functions for above ground decay. In *IRG/WP 10*, Biarritz, France. IRG/WP 10-20439.

Brischke, C. & Thelandersson, S. (2014) Modelling the outdoor performance of wood products – a review on existing approaches. *Construction and Building Materials* 66, 384–397.

Eriksson, K-E., Blanchette, R. A. & Ander, P. (1990) *Microbial and Enzymatic Degradation of Wood and Wood Components*. Springer Verlag, Berlin.

Fengel, D. & Wegener, G. (2003) *Wood – Chemistry, Ultrastructure, Reactions*. Verlag Kessel, Remagen.

Ferreira, J., Esteves, B., Nunes, L. & Domingos, I. (2015) Life cycle assessment as a tool to promote sustainable thermowood boards: a Portuguese case study. In *The Eighth European Conference on Wood Modification*, Helsinki, Finland, pp. 289–296.

Hansson, E. F., Brischke, Ch., Meyer, L., et al. (2012) Durability of timber outdoor structures – modelling performance and climate impacts. In *World Conference on Timber Engineering*, Auckland, New Zealand.

Horský, D. & Reinprecht, L. (1986) *Study of Subfossil Oak Wood*. Monograph. VŠLD Zvolen, Slovakia (in Slovak, English abstract).

Kazemi, S. M. (2003) Relationship of wood durability and extractives. In *IRG/WP 03*, Brisbane, Australia. IRG/WP 03-10493.

Kirker, G. and Winandy, J. (2014) Above ground deterioration of wood and wood-based materials. In *Deterioration and Protection of Sustainable Biomaterials*, Schultz, T., Goodell, B. & Nicholas, D. D. (eds). ACS Symposium Series, vol. 1158. American Chemical Society, Washington, DC, pp. 114–129.

Menzies, G. F. (2013) Whole life analysis of timber, modified timber and aluminium-clad timber windows: service life planning (SLP), whole life costing (WLC) and life cycle assessment (LCA). Report for the Wood Window Alliance, Institute for Building and Urban Design, Heriot Watt University.

Nzokou, P., Wehner, K. & Kamdem, D. P. (2005) Natural durability of eight hardwoods species from Africa. In *IRG/WP 05*, Bangalore, India. IRG/WP 05-10563.

Rapp, A., Peek, R-D. & Sailer, M. (2000) Modelling the moisture induced risk of decay for treated and untreated wood above ground. *Holzforschung* 54(2), 111–118.

Reinprecht, L. (2004) Biodegradation and treatment of ancient wood in Slovakia. In *Master Conservation Plan – Wooden Churches*, Part 1. Technical Group Findings, Powter, A. (ed.), Appendix 1H. Prešov, Slovakia, pp. 159–169.

Reinprecht, L. (2005) Investigation of wooden constructions – the basis for their renovation. In *Restoration and Reconstruction of Constructions, 27th Conference WTA*, Brno, Czech Republic, pp. 23–30 (in Slovak).

Reinprecht, L. (2008) *Wood Protection*, Handbook. TU Zvolen, Slovakia (in Slovak).

Reinprecht, L. (2013) *Wood Protection*, Handbook. TU Zvolen, Slovakia.

Van Acker, J. & Stevens, M. (2003) Biological durability of wood in relation to end-use – part 2. The use of an accelerated outdoor L-joint performance test. *Holz als Roh- und Werkstoff* 61, 125–132.

Van Acker, J., Stevens, M., Carey, J., et al. (2003) Biological durability of wood in relation to end-use – part 1. Towards a European standard for laboratory testing of the biological durability of wood. *Holz als Roh- und Werkstoff* 61, 35–45.

Van Acker, J., De Windt, I., Li, W. & Van den Bulcke, J. (2014) Critical parameters on moisture dynamics in relation to time of wetness as factor in service life prediction. In *IRG/WP 14*, St George, UT, USA. IRG/WP 14-20555.

Viitanen, H. (2005) Test of resistance and durability of wooden products – an example of evaluation of service life for wooden façade. In *Sustainability through New Technologies for Enhanced Wood Durability*, COST Action E37, WG2 – Service life prediction tools. Oslo, Norway.

Viitanen, H. (2013) 100 year's service life of wood material in dry conditions. Research Report VTT-R-08243-13, Finnish Wood Research Oy, VTT, Helsinki, Finland.

Wagenführ, R. (2007) *Holzatlas*. Fachbuchverlag, Leipzig.

Wiedenhoeft, A. (2010) Structure and function of wood. In *Wood Handbook – Wood as an Engineering Material*, Forest Product Laboratory, Madison, WI, chapter 3.

Wiemann, M. C. (2010) Characteristics and availability of commercially important woods. In *Wood Handbook – Wood as an Engineering Material*, Forest Product Laboratory, Madison, WI, chapter 2.

Yamamoto, K., Tamura, A. & Nakada, R. (2004) Variation of natural durability of sugi (*Cryptomeria japonica*) wood in 15 clones examined by decay test – preliminary report. In *IRG/WP 04*, Ljubljana, Slovenia. IRG/WP 04-10526.

Standards

EN 350-2 (1994) Durability of wood and wood based products – Natural durability of solid wood – Guide to natural durability and treatability of selected wood species of importance in Europe.

ISO 15686: 1–10 (2007–2008) Building and constructed assets – Service-life planning – Parts 1 to 10.

prEN 16818 (2015) Durability of wood and wood-based products – Moisture dynamics of wood and wood-based products.

2 Abiotic Degradation of Wood

Wood is susceptible to damage by various abiotic influences. For example, sunlight, oxygen and other weather factors erode and discolour wood surfaces; at temperatures above ~200 °C it ignites and burns; cracks and deformations in its geometric structure are created due to moisture stresses during drying or as a result of mechanical impacts; or it fiberizes and often also discolours in the presence of aggressive chemicals.

Abiotic material and energetic potentials have different intensities in interior and exterior environments, which corresponds also with the modes and intensity of wood abiotic degradation (Table 2.1).

2.1 Wood damaged by weather factors

Atmospheric corrosion is a natural ageing process that causes damage to wood, and also other materials, due to the presence of several abiotic weather factors, usually also in the presence of microorganisms (Figure 2.1). Its intensity depends mainly upon UV radiation, humidity and temperature of the air – and oscillation of these parameters in time, so atmospheric corrosion is more significant in exteriors (Williams, 2005). Atmospheric corrosion of wood logically relates to the presence of direct solar radiation, rainfall and more significant changes in climate during a year as well as during a single day. However, photodegradation rate and extent also depend on the wood species (Pandey & Vuorinen, 2008). During weathering, a partial increase in the crystallinity index of cellulose (Lionetto et al., 2012) and change in its colour (Matsuo et al., 2012) occurs as well. Such changes in cellulose are probably as a result of simultaneously thermally activated processes in its structure (Tolvaj et al., 2011).

Wood Deterioration, Protection and Maintenance, First Edition. Ladislav Reinprecht.

Table 2.1 Abiotic degradation of wood in interior and exterior environments

	Mode and intensity of wood degradation			
	Interior		Exterior	
Abiotic factor	**Mode**	**Intensity**	**Mode**	**Intensity**
Thermal				
(a) intense	Fire	Significant	Fire	Significant
(b) moderate	Darkening	Significant	Darkening	Significant
	Strength decrease	Moderate	Strength decrease	Moderate
Radiation (UV, visible, IR)				
		Moderate		Significant
	Photo-oxidation	'On surface'	Photo-oxidation	'On surface'
	Colour changes	'On surface'	Colour changes	'On surface'
Water				
			Leaching out of photo-oxidized lignin	Significant 'on surface'
			Hydrolysis lasting many years	Moderate 'fossilization'
Sand, etc.				
			Fibrillation and roughening	Significant 'on surface'
Drying/ mechanical				
	Cracks	Moderate	Cracks	Moderate
	Surface wear	Moderate	Surface wear	Moderate
Chemical				
	Fibrillation and roughening	—[a]	Fibrillation and roughening	—[a]
	Colour changes	—[a]	Colour changes	—[a]
	Strength changes (decrease)	—[a]	Strength changes (decrease)	—[a]

Source: R., L. (2013) *Wood Protection*, Handbook, TU Zvolen, Slovakia, 134 p. Reproduced by permission of TU Zvolen.

[a] Intensity of the chemical corrosion of wood depends on the type and concentration of the aggressive chemical (emissions, solutions of acids, etc.).

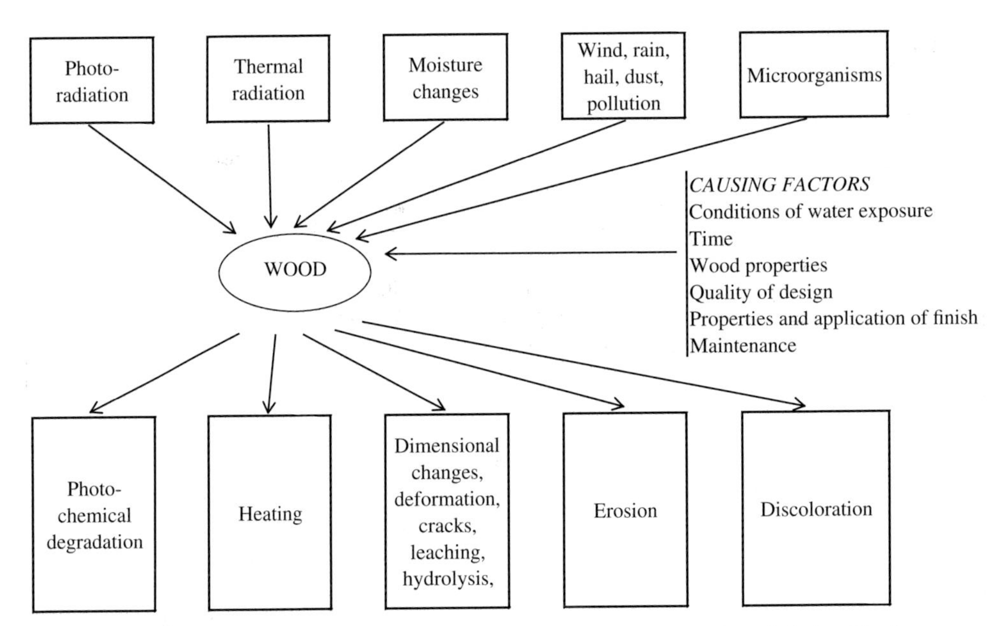

Figure 2.1 Scheme of factors causing weather impact on wood and their effect on changes in wood

Generally, weathering of wood is a result of photolytic, oxidation and hydrolytic reactions, taking place particularly in lignin (photo-oxidation) and also in hemicelluloses of wood (photo-oxidation and hydrolysis). Evans (1989) found that damage to wood surfaces caused by weather factors is more significant under horizontal exposures than under vertical exposures.

The *atmospheric erosion of wood is caused by greater number of abiotic factors*:

- material agents (e.g. water, oxygen, aggressive gases, emissions and liquids, dusts, sands, tars);
- energy impacts (i.e. solar radiation – UV, visible, infrared (IR) and other spectra), heat, wind in form of laminar and turbulent flows, and so on.

These factors usually act in synergy and therefore their degradation effect is multiplied. Solar radiation and water have the decisive impact on weather ageing of wood. Oxygen, and also other factors, mainly emissions, dust, extreme temperatures and strong wind, increase the ageing of wood even more (Figure 2.1). Better conditions for attacking the wood by biological pests are often created by its previous weathering. The cause is the depolymerization of lignin and hemicelluloses to low molecular weight substances more easily attackable by moulds, staining fungi and some wood-boring insect species, or also the washing out from wood of tannins and other extractives that have a biocide effect.

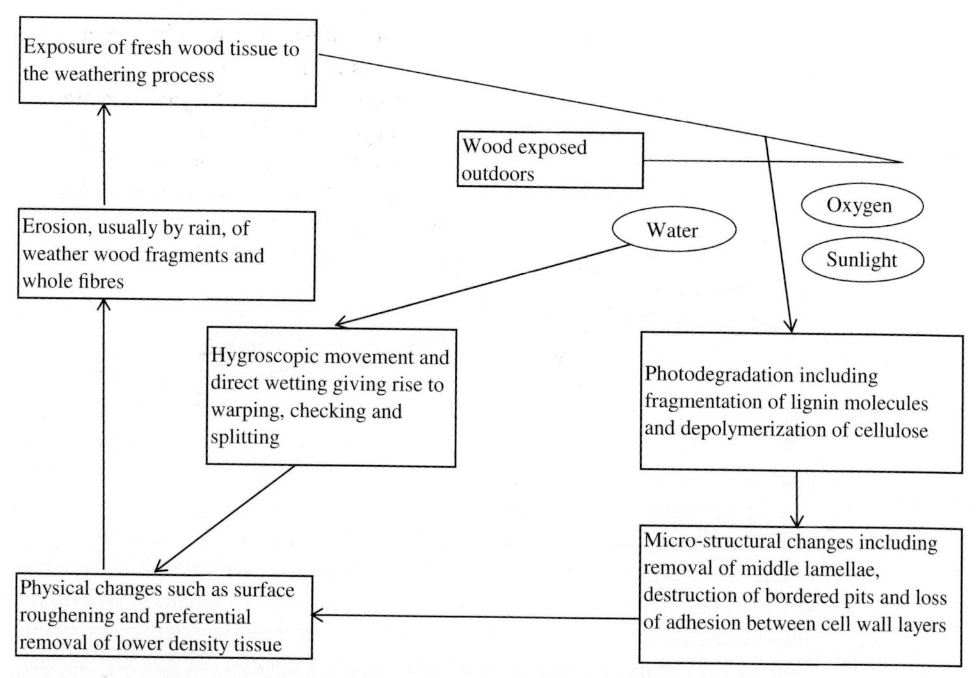

Figure 2.2 Cyclic sequence of damaging processes during atmospheric corrosion of wood in external environment (modified from Feist (1990))

Source: Reproduced by permission of Feist, W. C. (1990) Outdoor wood weathering and protection. In: *Archaeological Wood – Properties, Chemistry, and Preservation* (eds. R. M. Rowell, J. R. Barbour), Adv Chem Ser 225, Am Chem Soc, pp. 263–298. Washington DC, USA. Copyright © 1990, American Chemical Society

The sequence of cyclic atmospheric erosion of wood in the direction from its outside surfaces to a depth of several millimetres is pertinently documented by the scheme in Figure 2.2. In principle, there are three cyclically repeated events:

- photodegradation of lignin and partially also hemicelluloses due to solar radiation;
- washing away of photo-oxidized lignin and hemicelluloses by water;
- mechanical extraction of freed fibrils of cellulose from the wood surface, which takes place due to solid substances (sand, dust, ice) and liquid substances (water rain) in flowing air.

The surfaces of wood exposed to solar radiation gradually change their appearance (Derbyshire & Miller, 1981; Chang et al., 1982; Lindergaard & Morsing, 2003; Tolvaj et al., 2011; Pánek & Reinprecht, 2016). Bright types of wood, such as maple, beech or hornbeam, usually get darker, the cause of which is the chemical reactions in lignin invoked by UV radiation. Should rainfall act concurrently, the wood will become grey in colour, since the photochemically depolymerized yellow–brown polar fractions of lignin are washed away from the wood. The importance of solar radiation, water and oxygen for erosion of wood is decisive. However, we must not forget further factors, such as temperature and air flow. At higher temperatures, the solubility of photochemically disrupted wood components (lignin and hemicelluloses) is increased and they can be better washed away from weathered surfaces of wood by water. At higher speeds of air flow, the intensity of secession of cellulose fibrils from the photochemically weakened surfaces of wood is increased.

The *intensity of atmospheric erosion of wood surfaces* depends upon the wood species, the climatic conditions and the mode of wood exposure in various dry/wet, cold/hot environments; that is, the angle of slope, orientation to cardinal directions, roofed or without a roof, and so on. A wood surface is reduced by 1–13 mm after 100 years in exterior environments, depending on wood species and the environmental conditions (Williams et al., 2001).

Within various wood species, their surface erosion is significantly affected by the different densities of early and late wood and by the thickness of cell walls. For example, Feist (1990) studied accelerated weathering (~18 times quicker than natural weathering) of Norway spruce and Douglas fir woods and determined 4.25 times greater erosion of the three times less dense early wood (300 kg/m^3) than of the late wood (900 kg/m^3); that is, the average accelerated erosion of the early wood was 0.175 μm/h and that of the late wood only 0.041 μm/h.

The erosion of wood grows linearly with reduction of its density in the interval 300–1000 kg/m^3 (Sell & Feist, 1986). However, the wood is also affected by its other structural characteristics, such as:

- higher portion of lignin in coniferous species → faster erosion;
- greater differences in the density of early and late wood of conifers → plastic texture;
- type and quantity of extractives → specific discoloration.

The *plastic texture of wood* is the consequence of nonuniform erosion of early and late wood, typically in conifers. It contributes to more significant erosion of more porous early wood (Figure 2.3). Plastic texture is also created on surfaces of some wooden composites; for example, plywood (Figure 2.4). When proposing the composition of plywood or laminated veneer lumber (LVL) for exterior use, it is desirable to consider more intense reduction of early wood by erosion when compared with late wood; thus, for example, for spruce plywood or LVL the surface veneers should have a tangential rather than radial orientation. In addition to plastic texture, the colour of atmospherically attacked wood is often changed; that is, it usually gets darker from the surface to a depth of several micrometres, and cracks or other defects of predominantly surface character are created in it, whereby the wood is then also better accessible for attack by bacteria and moulds.

Figure 2.3 Plastic texture of the surface of weathered spruce wood
Source: R., L. (2013) *Wood Protection*, Handbook, TU Zvolen, Slovakia, 134 p. Reproduced by permission of TU Zvolen

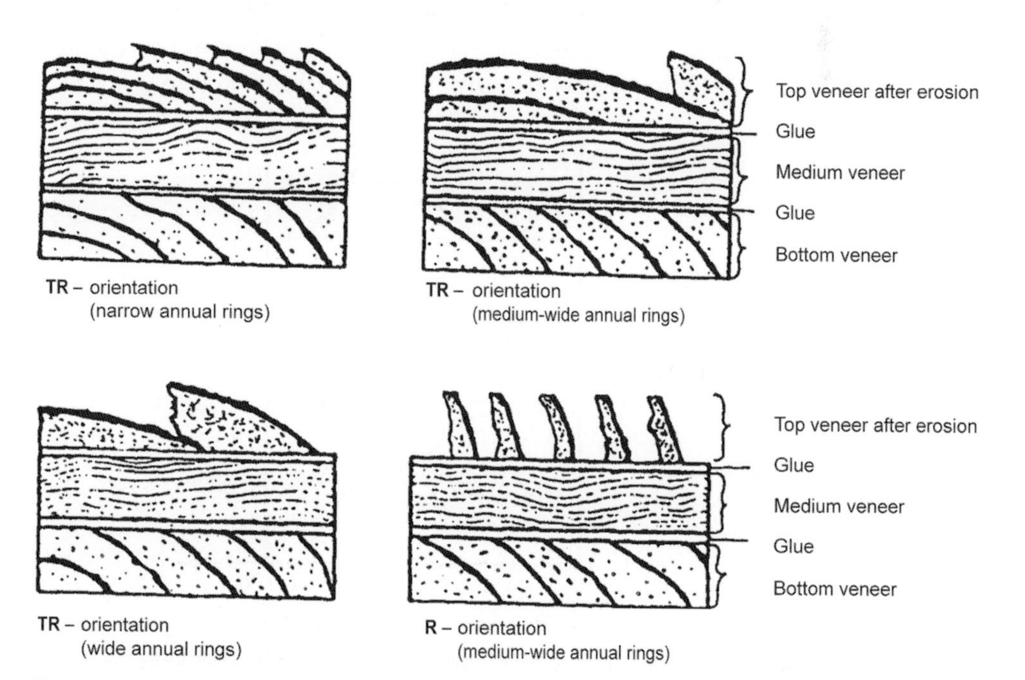

Figure 2.4 Plastic structure of the surface of weathered spruce plywood arising from the orientation of the annual rings in the top veneers (TR: tangential–radial; R: radial)

2.2 Wood damaged thermally and by fire

Wood easily ignites and burns. The cause is its chemical composition, since it is comprised of ~49–51% carbon, 43–44% oxygen and 6–7% hydrogen. The high energy content in the covalent bonds of wood polysaccharides and lignin was accumulated during photosynthesis (i.e. during production of glucose from carbon dioxide and water) and during subsequent endothermic reactions (i.e. during transformation of glucose to polysaccharides and lignin). This chemical energy can be retrospectively released from wood by supplying the necessary activation energy. Thus, the ignition and burning of wood is the opposite process to that of the origin of wood, and its components with high values of chemical energy are gradually split up into flammable gases and then further to carbon dioxide and water.

2.2.1 Thermal wood decomposition

The *thermal degradation of wood* is a set of chemical reactions (depolymerization, dehydration, etc.) initiated by its heating, and achieving the activation energy (Figure 2.5). At temperatures below 66 °C, the chemical reactions of wood in air do not in fact take place. At temperatures from 66 to 110 °C, some of them may take place depending upon the duration of wood heating, but they usually have only a negligible importance for the structure and properties of

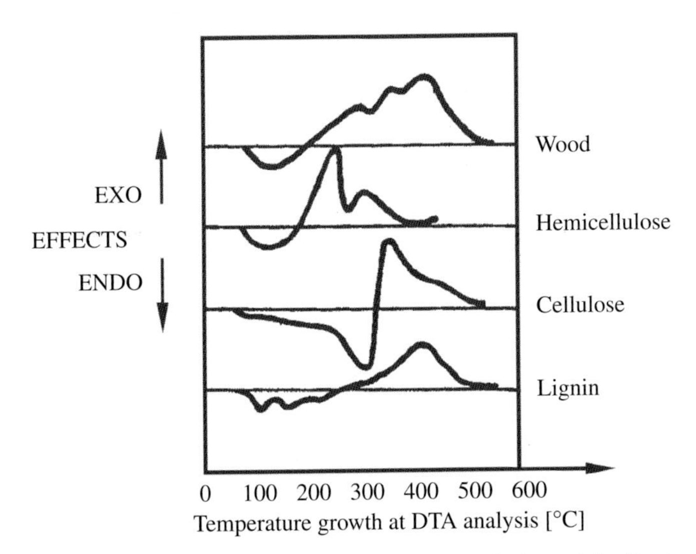

Figure 2.5 Energetic (exo: exothermic; endo: endothermic) effects upon thermal decomposition of wood and its components

Source: R., L. (2013) *Wood Protection*, Handbook, TU Zvolen, Slovakia, 134 p. Reproduced by permission of TU Zvolen

wood from a practical point of view. More apparent thermal disruption of the components of wood takes place only at temperatures exceeding 150 °C, when hemicelluloses are decomposed at first, then cellulose, while the most thermally stable is lignin.

The following general knowledge on the thermal decomposition of wood polymers was implied by the differential thermal analysis (DTA) (Figure 2.5), thermogravimetric and differential thermogravimetric, gas chromatography–mass spectrometry, ^{13}C nuclear magnetic resonance (NMR), Fourier transformed infrared spectroscopy (FTIR), and also other convenient thermal, physical, physical–chemical and chemical analyses (e.g. Beall & Eickner, 1970; Reinprecht & Mihálik, 1988; Marková & Klement, 2003; Kačíková et al., 2008; Brebu & Vasile, 2010; Poletto et al., 2012, 2015; Hrablay & Jelemenský 2014; Alangi & Malucelli, 2015):

- *Hemicelluloses* are already decomposed at temperatures above 150 °C, though with more significant exothermic degradation above 200 °C.
- *Cellulose* is significantly depolymerized at temperatures above 300 °C, from which levoglucosan (1,6-anhydro-β-D-glucopyranose) is generated and subsequently transformed to flammable gases (Figure 2.6).
- *Lignin* is significantly exothermically decomposed only at temperatures of 300–400 °C or higher. At these temperatures there is more significant cleavage of alkyl–alkyl bonds in the propane parts of phenyl-propane units, and of alkyl–aryl bonds, ether bonds and C–C bonds in the aromatic nucleus of the phenyl-propane units (Figure 2.7).

Oxygen plays a significant role in the thermal decomposition of wood. Oxygen easily reacts with thermally activated components of wood in the form

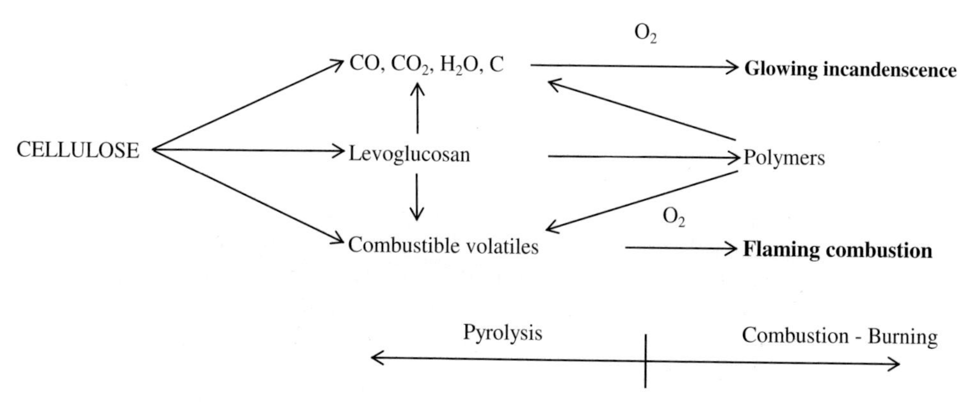

Figure 2.6 Thermal decomposition of cellulose in pyrolysis, and the burning phase (modified from Shafizadeh (1984))

Source: Reproduced by permission of Shafizadeh, F. (1984) The chemistry of pyrolysis and combustion. In: *The Chemistry of Solid Wood* (ed. R. M. Rowell), pp. 489–529. Adv Chem Ser 207, Am Chem Soc, Washington, USA. Copyright © 1984, American Chemical Society

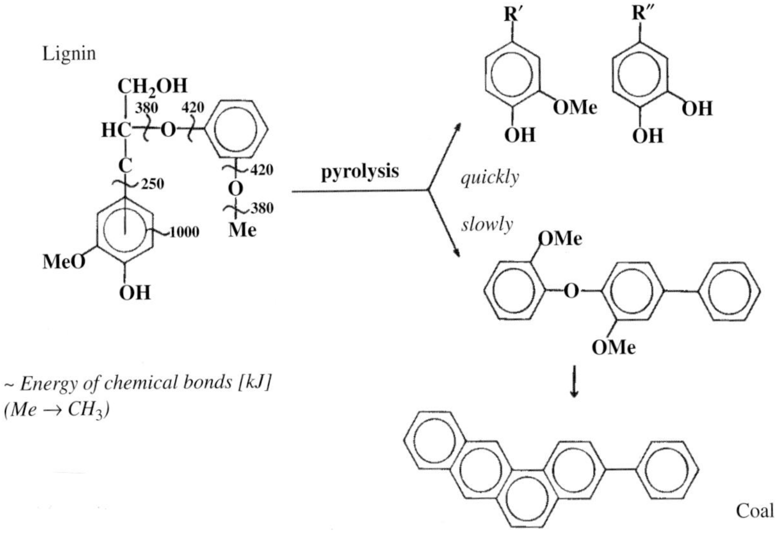

Figure 2.7 Thermal decomposition of lignin with production of charcoal (modified from Blažej and Košík (1985))

Source: Blažej, A. and Košík, M. (1985) *Fytomasa ako Chemická Surovina* (*Phytomass as a Chemical Source*). Veda, Bratislava, Czechoslovakia, 402 p. Reproduced by permission of VEDA, Vydavatel'stvo SAV

of exothermic thermo-oxidation reactions with the production of heat. Thus, wood overheats even more, new free radicals are created in it and gaseous flammable products are generated from it. Upon reaching a critical state, the flammable gases reactions with oxygen generate a quantity of heat that is sufficient for their self-ignition and they start burning; that is, a fire is started (Section 2.2.2).

2.2.2 Wood burning: fire

Fire is a good example how to recover the chemical energy from wood, the polysaccharides and lignin. Of course, fire is not only a good servant, but also a bad master. Fires in forests and in log yards often lead to the devaluation of an even greater material stock of wood, often with a negative impact on the biotope. Fires occasionally also destroy buildings, mines, bridges and other immovable and movable products made of the wood or wooden composites, entire panel structures comprised of wood, textiles and plastics, or even inflammable materials from minerals and metals. With fires also comes the loss of human life, in particular due to incineration, suffocation and falls. There are many examples fires of structures based on wood; for example, 180 houses burned down in 1590 in Bratislava (Slovakia), and 6000 buildings were lost in 1980 in Mandalaj – Burma (Malaysia). Fires are often accompanied by tragedies; for example, 187

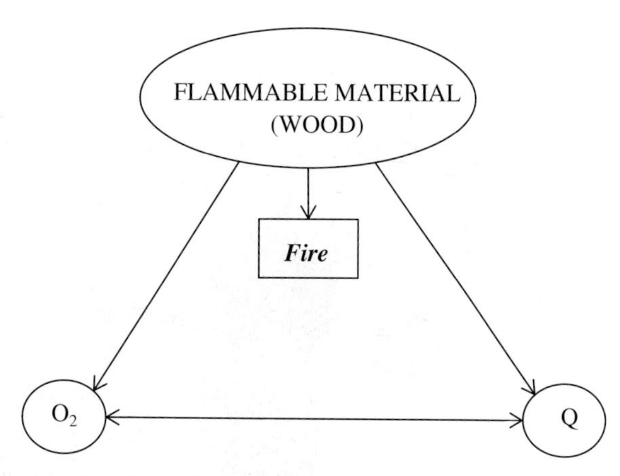

Figure 2.8 Combustion triangle of wood burning

people died in Sao Paolo in 1974 in a fire in a 21-storey building. There are thousands of fires in Europe every year, and 75% of them start in residential houses usually at night. Approximately 4000 people die annually due to fires (mainly children up to 5 years old and the elderly over 60 years of age). However, the dominant cause of death is not incineration, but suffocation with toxic gases (Nikolaos, 2003).

Combustion triangle – this interconnects the three factors that are needed for ignition and burning processes (Figure 2.8):

- flammable material, in which is accumulated sufficient chemical potential energy (e.g. wood, coal, oil);
- oxygen (O_2), which is a common component of air;
- activation energy Q, which is produced from a flame or radiating thermal source.

When the combustion triangle is disrupted (e.g. due to absence of oxygen or presence of fire retardants), the flame burning of wood is significantly slowed down and charcoal production is preferred. Subsequently, the charcoal created on the wood surface becomes a natural fire retardant barrier to further burning of the wood (Figures 2.9, 2.10 and 2.11).

2.2.2.1 Process of wood burning

The *burning of wood and wooden materials takes place in three stages*:

- initiation (i.e. combustion of wood by its ignition or inflammation);
- propagation – that is, flame spreading on wood surfaces, an intense thermal decomposition of wood (pyrolysis, as a mixture of exothermic reactions or also some endothermic reactions), and the actual burning of

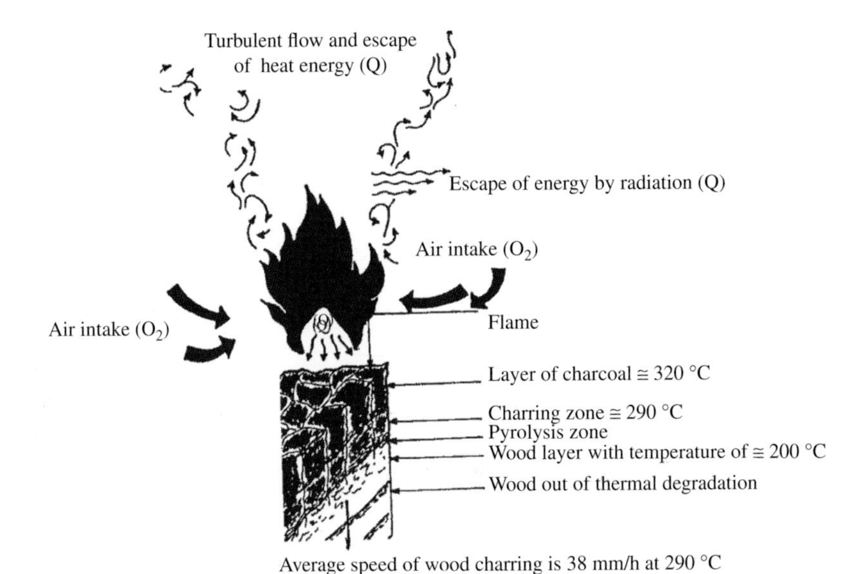

Figure 2.9 Basic mechanism of wood burning

Source: R., L. (2013) *Wood Protection*, Handbook, TU Zvolen, Slovakia, 134 p. Reproduced by permission of TU Zvolen

flammable gases originated from the thermally decomposed wood (burning, as exothermic thermo-oxidation reactions of flammable gases with oxygen, while releasing thermal and light energy);
- termination (i.e. incandescence of wood charcoal).

Combustion of wood occurs by its ignition or flashing, both in the presence of oxygen and initiation energy from thermal sources having sufficient input energy (IE) per area unit of wood surface A, usually $\sim 10^4$ W/m^2. The initiation energy input depends upon the type of energy source; that is, in the case of a radiant source it must be ~100–150% greater than for a flame source.

The temperature of wood combustion – that is, ignition (by a radiant source) or inflammation (by a flame source) – is usually in the range from 200 to 400 °C and is affected by several factors:

- heating time – its prolongation decreases the temperature of wood combustion (e.g. the ignition time is 12–40 min for radiant heating of wood to 200 °C, but it is only from 1.5 to 3.5 min for heating to 300 °C);
- moisture of wood – moist or wet wood is substantially harder to light up than dry wood is (causes: liquid water increases thermal conductivity of wood and thus also heat dissipation from its surface to its internal zones; water vapour is generated during heating, thus heat is consumed for phase transformations; flammable gases are diluted by water vapour);

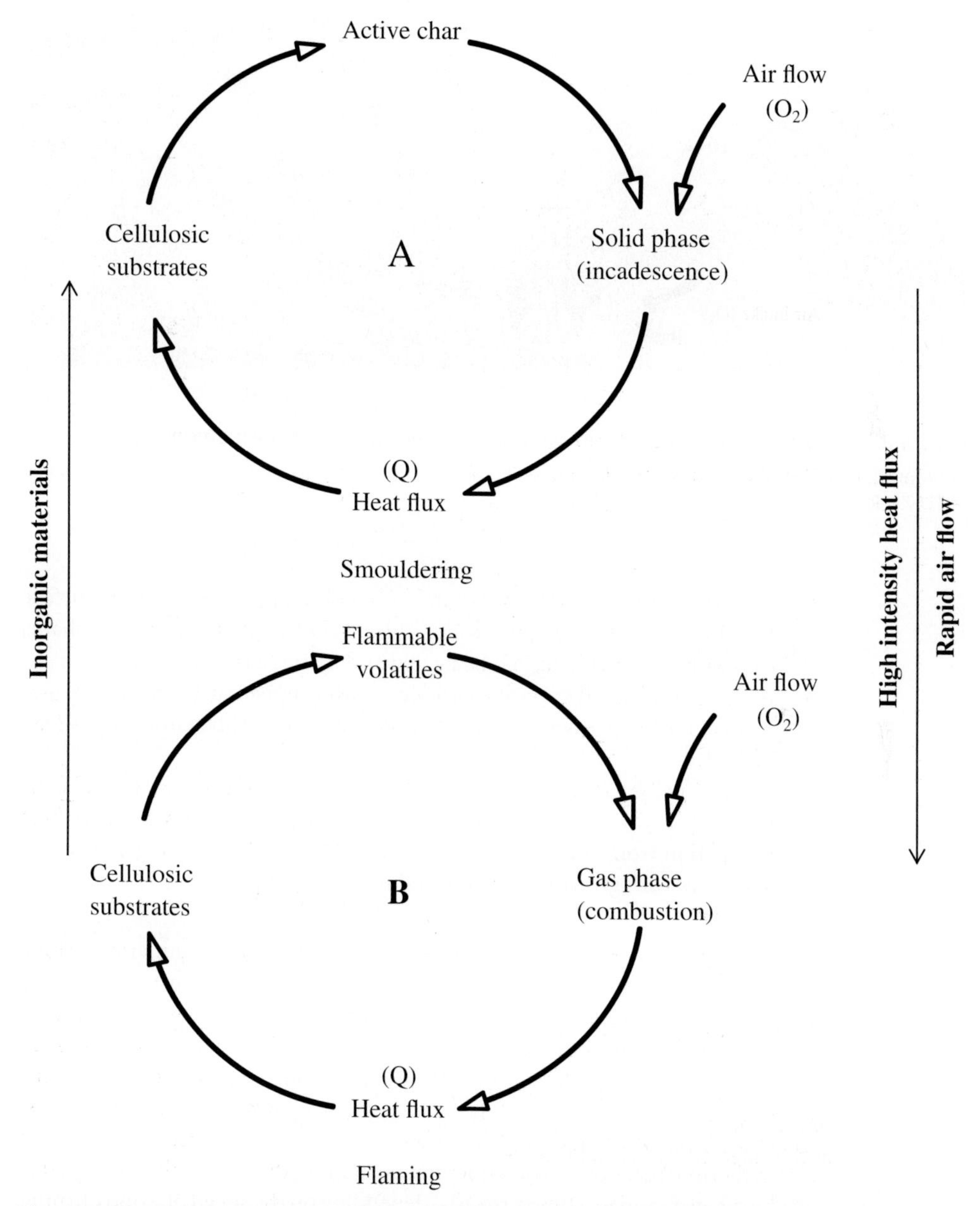

Figure 2.10 Thermal decomposition of wood and its burning depend upon the supply of heat, air flow and presence of inorganic additives in wood (modified from Shafizadeh (1984)): (A) preference for the production of charcoal; (B) preference for the production of flammable gases

Source: Reproduced by permission of Shafizadeh, F. (1984) The chemistry of pyrolysis and combustion. In: *The Chemistry of Solid Wood* (ed. R. M. Rowell), pp. 489–529. Adv Chem Ser 207, Am Chem Soc, Washington, USA. Copyright © 1984, American Chemical Society

(a) (b)

Figure 2.11 Wood trusses damaged by a fire – charcoal on wooden elements

Source: R., L. (2013) *Wood Protection*, Handbook, TU Zvolen, Slovakia, 134 p. Reproduced by permission of TU Zvolen

- ratio of wood surface to its volume – wood with greater ratios of surface to volume catches fire more easily (cause: assortments smaller as for shape, e.g., with less thickness, can be heated and ignited faster);
- roughness of wood surface – products with rougher surfaces with greater contact area (e.g. sawn) will ignite faster than smoothly worked ones (e.g. planed);
- wood density – denser wood is harder to set fire to (causes: thermal conductivity of wood grows with its density and so does the most intense heat dissipation from surface to internal zones; the amount of oxygen in wood decreases with growth of density and reduction of porosity).

Flame spreading and burning of wood follows after the fire initiation. Flame spreads most promptly on surfaces of a dry wood with lower density and higher roughness. The speed of wood burning depends upon the internal conditions of the wood (e.g. moisture, density), and also upon the external conditions in its vicinity (e.g. wind). The role of outside energy source needed for wood combustion is taken over by wood itself or by other flammable materials; that is, materials from which another heat necessary for fire development is generated by exothermic reactions. Wood does not burn immediately to ash, but it gradually burns away and in a time τ (min) – depending on the speed of wood burning β_0 (mm/min) – is changed to the carbonated layer of charcoal with a thickness of d_{char} (Figure 2.9):

$$d_{char}\ (\text{mm}) = \beta_0 \tau \tag{2.1}$$

The speed of wood burning away depends upon several factors. For example, on fire, spruce wood with a density of 350–440 kg/m^3 shows an average speed of burning away $\beta_0 = 0.65–0.8$ mm/min. However, beech wood, with a lower content of lignin and a higher density of 700 kg/m^3, reaches the same speed of burning away. For wooden composites, the speed of burning away depends also upon their composition and type of adhesive; for example, β_0 for glulam spruce wood bonded with phenol-formaldehyde adhesive is ~0.68 mm/min and for beech plywood 0.77 mm/min. This implies that the speed of burning away depends not only upon the density of the wood, but also on the thickness and type of wooden material (plywood, particleboard, etc.), and also in relation to its chemical composition, thermal conductivity, permeability for gases, and so on (White & Nordheim, 1992; EN 1995-1-2).

Incandescence of wood and creation of charcoal is the last stage of thermal degradation of wood. The carbonized layer 'charcoal' (Figure 2.9) acts as natural fire retardant because: (1) its density is just ~20% of the original wood density and therefore it also has approximately five times lower thermal conductivity, and (2) it is no longer decomposed to flammable organic gases during the incandescence. However, this carbonized layer forms flammable carbon monoxide gas in an exothermic oxidation reaction with oxygen and from this subsequently the nonflammable carbon dioxide:

$$C + \tfrac{1}{2}O_2 \rightarrow CO + 110.74 \text{ kJ mol}^{-1} \quad (2.2)$$

$$CO + \tfrac{1}{2}O_2 \rightarrow CO_2 + 284.75 \text{ kJ mol}^{-1} \quad (2.3)$$

When a layer of charcoal peels away, thermal decomposition and flame wood burning are restored. Wood burns in several cycles of 'initiation–propagation–termination'.

2.2.2.2 Parameters characterizing wood burning

Wood burning is characterized by the following parameters:

- ignition (combustion) temperature;
- speed of flame spreading;
- intensity of thermal decomposition (speed of wood burning away);
- smoking;
- production of a layer of charcoal.

The parameters of burning depend upon several factors in the proximity of wood, as well as directly in the wood:

- *External conditions in the proximity of wood:*
  - radiant or flame thermal source;
  - intensity of thermal source, distance of thermal source from wood, and time of its effect;

- chemical composition of air, in particular the ratio of oxygen and water molecules, as well as the intensity and forms (laminar, turbulent) of air flowing.

- *Internal conditions in the wood* (i.e. the structure and flammability of wood):
  - molecular structure (e.g. presence of terpenoids and other easily flammable accompanying substances);
  - morphological and anatomical structure (e.g. oxygen and flammable gases penetrate better in woods with a higher permeability);
  - geometric structure (e.g. wood with rougher surface will catch fire sooner);
  - moisture content (wood with higher humidity is more difficult to set fire to),
  - presence of additives (fire retardants, biocides, water repellents, adhesives and paints affect the parameters of burning specifically).

The parameters of internal conditions needed for burning of solid wood cannot be defined so easily for wooden composites, panels and other parts of constructions made that are based on wood, since there are also other factors that can be important; for example, material composition of the wooden composite in its individual layers, as well as the presence of metals and thermal insulation in panels. Therefore, it is more appropriate to use general term *material flammability* for products based on wood.

2.2.2.3 Flammability of wooden materials and constructions

Flammability is a material value that is evaluated by laboratory and large-dimensional tests (Osvald, 1994). Nowadays, in Europe, construction materials, including wood, are classified according to the European standard EN 13501-1 into seven classes: A1, A2, B, C, D, E and F (Table 2.2). They are based on several testing methods and their mutual combination (Vandevelde, 1999):

- flammability test by EN ISO 1182, means 'burns–does not burn' at 750 °C;
- flammability test by EN ISO 1716, based on caloric value, measuring heat of combustion in pure oxygen in a calorimetric bomb;
- single burning item test by EN 13823, which measures speed of fire development, lateral flame spreading, creation of smoke and burning drops;
- flammability test by EN ISO 11925-2, which measures flame spreading.

For materials in classes A1 to B there may not be total inflammation, and flame spreading also may not occur or occur only to a limited extent. For example, materials in class A2 may only partially spread flame, but for not more than 20 s, while their increase in temperature may not be greater than 50 °C and their decrease in weight may not exceed 50%. When classifying materials in classes C to E, the question of the speed of fire development defined by the unit (watts per second) is important, but very important also is flame spreading from a small

Table 2.2 Classes of response to fire of wood and other construction materials (in accordance with EN 13501-1, classes A1 to F)

Class of response to fire	Products from wood and other materials
A1	Concretes, natural stones, mortars, metals, glass Boards from inorganic material (e.g. asbestos-cements – compressed and non-compressed) Boards from glass and mineral fibres (some more fire resistant types)
A2	Plasterboard panels **Wood–cement particleboards** Polyvinyl chloride, nonsoftened and sturdy Boards from mineral fibres (from basalt felt, etc.) Boards from glass fibres (from glass mat, etc.)
B	**Hardwoods (broadleaves): beech, oak, etc.** **Wooden boards from veneers:** • plywood for general use • plywood, water-resistant for building industry • laminated veneer lumber (LVL) **Special wood-particleboards** Polystyrene, lightweight with a fire retarder Laminated paper with melamine decorative surface
C, D	**Softwoods (coniferous): pine, spruce, fir, etc.** **Classic wood-particleboards** (polished and rough) **Hard fibreboards** Boards made of vegetable matter (cork boards, cork parquet blocks)
E, F	**Sawdust boards and sawdust-particle boards** **Soft fibreboards** Polystyrene, sturdy and lightweight – standard Polyurethane, lightweight, soft and lightweight, hard, standard Polyvinyl chloride, lightweight

Wooden materials are highlighted in bold.

torch acting on the material for 30 or 15 s. Materials not meeting the criteria for E, or untested materials, are placed in class F. Wood is usually classified in classes B to E. Materials in class A2 can be prepared from wood only by adding a certain quantity of fire retardants or inorganic binding agents, an example of which is wood–cement particleboard (Table 2.2).

Special flammability tests are used for the evaluation of wooden floors. The auxiliary tests for the evaluation of flammability of construction materials are EN ISO 5660-1 (cone calorimeter), EN ISO 5659-2 (smoke chamber), EN ISO 5658-2 (flame spreading), EN ISO 4589-2 (critical and limit oxygen number), EN ISO 5657 (flammability test) and UL 94 (flame chamber).

Fire resistance of wooden constructions is determined by means of large-scale tests in the stage of fire developed. In addition to the material composition,

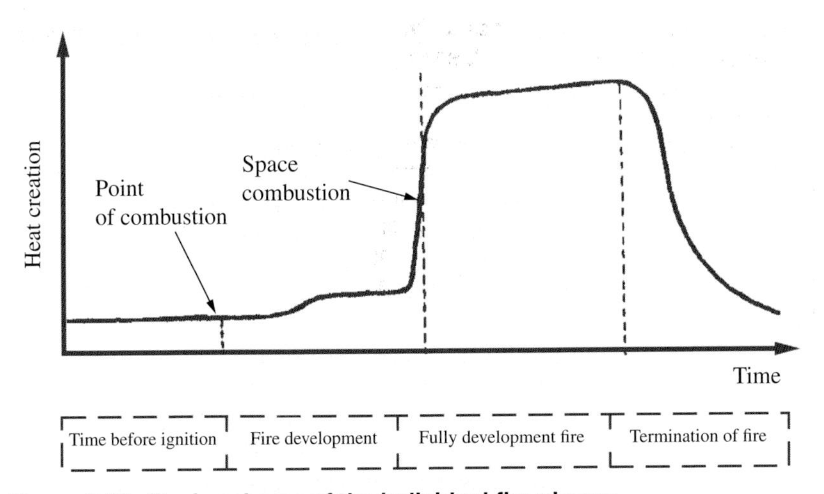

Figure 2.12 Basic scheme of the individual fire phases

Source: R., L. (2013) *Wood Protection*, Handbook, TU Zvolen, Slovakia, 134 p. Reproduced by permission of TU Zvolen

the construction design and efficiency of fire-safety equipment in the object are taken into account as well. Circumferential supporting constructions, stairways, windows, doors, ceilings or roofs are tested on 1 : 1 models. In model tests of fire resistance of a structural element (ISO 9705 – corner test), or also the entire object with residential and fire-safety equipment, the changes in the element or object are evaluated in the respective minute, while stability, deformations and burning through are assessed together with the generation of toxic gases and flammable gases.

The course of fire in a wooden construction is characterized similar to that of a burning process; that is, by three stages. However, the propagation stage is divided into two separate sub-stages – fire development and fully developed fire (Figure 2.12):

- initiation (ignition or inflammation);
- fire development (transfer of fire by flame spreading to the surrounding flammable materials);
- fully developed fire (burning of all flammable materials);
- interruption of fire (reduction of temperature after majority of flammable materials burnt away to charcoal or ash).

More practical findings, experimental results, and pyrolysis models show that wooden constructions have a fire resistance comparable to or greater than that of many noncombustible alternatives, such as steel or masonry (e.g. Odeen, 1985; Janssens, 2004).

Table 2.3 Effect of gaseous components created during burning of wood on human health

Gaseous component of fire	Concentration (vol.%)			
	5 min		30 min	
	Blackout	Death	Blackout	Death
CO	0.6–0.8	1.2–1.6	0.14–0.17	0.25–0.4
HCN	0.015–0.02	0.025–0.04	0.009-0.012	0.017–0.023
O_2	<10–13	<5	<12	<6–7
CO_2	7–8	>10	6–7	>9

Fire hazards with direct impacts on the lives and health of people are defined by several indicators, such as increased temperature, presence of wood dust, smoking, gas toxicity, oxygen insufficiency, and so on (Table 2.3).

2.3 Wood damaged by aggressive chemicals

2.3.1 Corrosion of wood by chemicals under aerobic conditions

Chemical corrosion of wood is caused by several types of aggressive chemicals. They are in particular oxidizers, alkalis, acids and their salts, which can come into contact with the wood in the form of gases, liquids, salts or pastes. Various accidental and deliberate effects of aggressive chemicals on wood depend upon their type, concentration, temperature and time of impact (Goldstein, 1984; Le Van, 1984; Reinprecht, 1988, 1991; Pandey, 1998) (Figures 2.13 and 2.14).

Sulphuric acid, hydrochloric acid, nitric acid and other aggressive acids will degrade wood at room temperature in rather a short time. Polysaccharides of wood are hydrolysed by acid salts substantially slowly, an example of which is the substantially longer corrosion of the bottom parts of wooden utility poles due to sodium chloride (Böttcher, 1989). The degree and range of wood degradation by chemicals depends also upon its humidity; for example, water-soluble salts can diffuse deeper into moister wood (Erler, 1990; Reinprecht & Makovíny, 1990). Also, aggressive gases, such as sulphur dioxide or ammonia, damage wood more intensively where there is a higher humidity.

The intensity of chemical corrosion of coniferous woods is usually minor compared with broadleaved woods (Figure 2.14). Of course, this depends on the particular species that has undergone the corrosion. However, for the majority of European species it is valid and can be explained by the following facts:

- wood of broadleaves (not only sap-wood) is usually more permeable to both liquids and gases (e.g. beech, poplar, hornbeam, lime tree, maple or birch wood), but this is not the case for heartwood of acacia, oak, or false heart of beech;

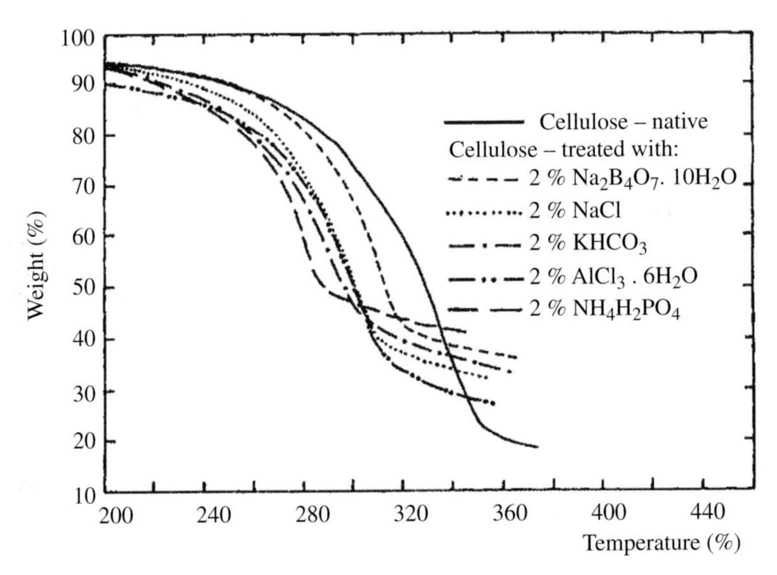

Figure 2.13 Impact of inorganic fire retardants on the thermal and chemical degradation of cellulose (modified from Le Van (1984)). Note: cellulose modified by fire retardants is dehydrated at temperatures up to 300 °C, when it carbonizes and shows greater decreases in weight than unmodified cellulose does

Source: Reproduced by permission of Le Van, S. L. (1984): Chemistry of fire retardancy. In: *The Chemistry of Solid Wood* (ed. R. M. Rowell), Adv Chem Ser 207, Am Chem Soc, pp. 531–574. Washington DC, USA. Copyright © 1984, American Chemical Society

- wood of broadleaves in comparison with conifers contains slightly more hemicelluloses (~25% compared with ~20%), which are the least stable components of wood at hydrolyses;
- wood of conifers in comparison with broadleaves contains rather more lignin (~30% compared with ~20%), which resists non-oxidizing chemicals rather well.

Generally, the intensity of chemical corrosion of wood increases with (1) higher temperatures and concentrations of aggressive chemicals, (2) prolongation of time of chemical effect, (3) increased permeability of wood and (4) a higher share of frontal areas of attacked wood.

Atmospheric-chemical corrosion of wood is actually the weather degradation of wood with the co-effect of aggressive emissions. The typical example is the contamination of wood by acid rain, which contains sulphur dioxide, nitrogen oxides and the acids produced from them (e.g. H_2SO_3, H_2SO_4 and HNO_3).

Chemical corrosion of wooden products in atmospheric conditions is often more moderate than that of products made from concrete, steel and some plastics. With advantages, wood is applied in industrial and agricultural objects, where it is exposed to aggressive gases, vapours, condensates and accidentally

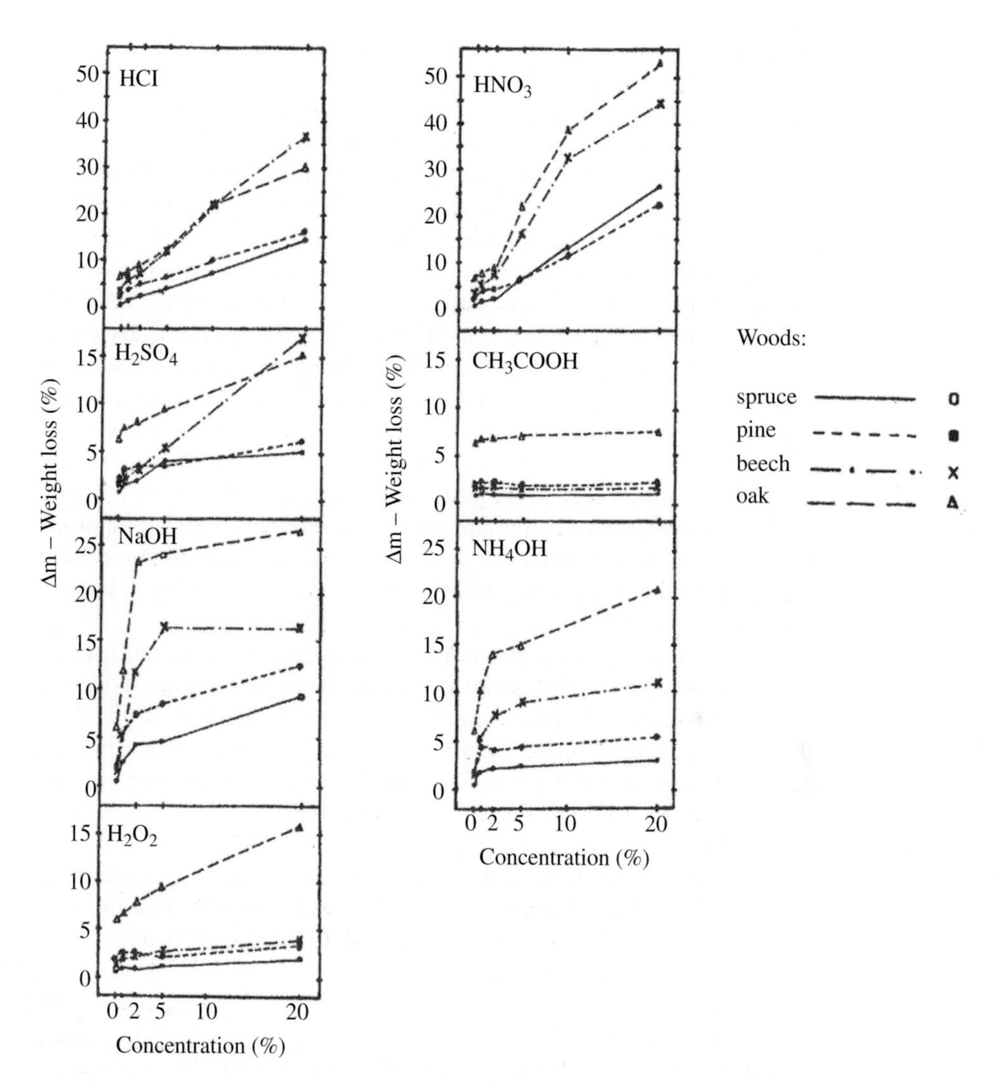

Figure 2.14 Impact of aggressive chemicals (HCl, HNO_3, H_2SO_4, CH_3COOH, NaOH, NH_4OH, H_2O_2) on rate of chemical corrosion of selected woods (spruce, pine, beech, oak) at 20 °C for 30 days (modified from Reinprecht (1988, 1991))

Sources: (a) R., L. (1991) Corrosion of wood due to NaOH, HCl and HNO_3 (in Slovak, English abstract). In: *Scientific Works of DF VŠLD in Zvolen 1991*, pp. 105–123. Alfa Bratislava, Czechoslovakia. (b) R., L. (1988) Selected properties of wood after its treatment with acids, alkalis or oxidizing agents (in Slovak, English abstract). *Drevársky Výskum – Wood Research* 119, 43–62. Reproduced by permission of TU Zvolen and Pulp and Paper Research Institute Bratislava

spilled liquids. It is used for wooden tiles, tanks, barrels, ceilings, roof frames, and so on. However, these wooden products do lose strength after prolonged exposure in the presence of aggressive chemicals, in particular their surface layers (Wegener & Fengel, 1986; Fengel & Wegener, 2003).

Unintentional chemical corrosion of wooden products is caused by some aggressive additives added to woods or wooden composites (Figure 2.13). These are usually substances able to disrupt hemicelluloses and cellulose. Several more aggressive types of substances were used for chemical protection of wood against biological pests, predominantly in past, such as inorganic fungicides based on copper, zinc and other salts of sulphuric acid (e.g. $CuSO_4 \cdot 5H_2O$) and hydrochloric acid (e.g. $ZnCl_2$). Similarly, inorganic fire retardants to reduce wood flammability (e.g. those based on the ammonium salts of sulphuric acid and phosphoric acid) can be aggressive on anatomical elements of wood and cause defibrillation of its surfaces (Le Van & Winandy, 1990, Kučerová et al., 2007, Kloiber et al., 2010). In the past, various corrosive additives were applied in adhesives and paints with the aim to improve and speed up the technologies of bonding and surface treatment of wood; for example, ammonium chloride for curing of amino resins or *p*-toluene-sulphuric acid for curing of phenolic resins at normal temperatures.

Inorganic additives hydrolyse in aqueous environment and produce acid or alkali solutions. Their effect on wood depends upon their concentration, dissociation constant and time of impact. There are examples of defibrillation (maceration) of wood surfaces down to depths of 1–12 mm in older objects due, for example, to the impact of inorganic fire retardants based on (1) acid salts, such as Glauber's salt ($Na_2SO_4 \cdot 10H_2O$), Epsom salt ($MgSO_4 \cdot 7H_2O$) or mixtures of ammonium salts ($(NH_4)_2SO_4$ and $(NH_4)_2HPO_4$), and (2) strongly alkaline substances, such as a mixture of sodium carbonate and potassium carbonate (Becker, 1986; Kučerová et al., 2007). Wood treated by these fire retardants gradually loses its strength and functionality (Le Van & Winandy, 1990; Le Van et al., 1996; Hodgin & Lee, 2002; Ayrilmis, 2007; Kartal et al., 2008; Kloiber et al., 2010).

Corrosive aesthetic defects in wood are also caused by several metals, particularly in wood with moisture over 20–30%. The presence of iron connecting and other elements, such as screws, nails or S-hooks, colours the wood to grey (beech, birch, hornbeam, pine, spruce), to blue and grey (larch, Douglas fir) or even to black (oak, walnut). The discoloration of wood results mainly from the characteristic reaction of its extractive substances with iron ions (Makovíny et al., 1992). On the other hand, extractives often corrode metals present in wood (Zelinka & Stone, 2011). Some reaction products of metals with wood also have a biocide effect (e.g. fungal rot is less intense in the proximity of iron element).

Degradation of wood by chemicals may be partially limited or its effect removed. First, the aggressive chemical must be washed out from the wood or neutralized in the wood. Second, the fiberized surface of the wood can then be reinforced with, for example, epoxy, acrylic or any other suitable painting system.

2.3.2 Corrosion of wood by chemicals under anaerobic conditions: wood fossilization

Anaerobic-chemical corrosion (fossilization) of wood is a chemical corrosion without the presence of oxygen that take place deeper in the ground or in water. In fossilization the wood is in contact with water and also with certain inorganic substances for a very long time. Remnants of wood with higher and lower degradation stages have been found during archaeological investigations and earth or survey works; for example, during exploitation of peat bogs, brown coal beds exposure, regulation of water streams, use of gravel pits, investigation of sea gulfs.

The level of wood degradation depends mainly upon its age and the conditions of the wet environment, in particular whether it found in shallower or deeper layers of soil, on the bottom of seas or rivers, and so on. There is a generally accepted opinion in scientific circles that the rather more significant degradations of wood take place after it is placed on the bottoms of seas, lakes and rivers, compared with its placement in peat bogs, under layers of wet clay soils or under permanent freezing conditions (Horský & Reinprecht, 1986; Solár et al., 1987; Kúdela & Reinprecht, 1990; Schniewind, 1990; Fengel, 1991). The cause is the reduced access of oxygen, inorganic salts and other substances to the wood wrapped by a solid phase.

The process of wood degradation in a permanently wet anaerobic environment, the 'fossilization process', takes place from its surface zones to the internal zones (Horský & Reinprecht, 1986; Hoffmann & Jones, 1990). The fossilization process is influenced by the presence of salts and other inorganic substances, range of temperatures, oxygen dissolved in water, anaerobic bacteria, marine organisms, and so on. Anaerobic bacteria attack wood usually only moderately, when they disrupt more or less just its bordered pits and thus increase its permeability, or slowly degrade its cell walls, as well (see Section 3.1). Oak wood is considered to be one of the most resistant species against the complex of degradation impacts acting in water under anaerobic conditions.

The *mechanisms of wood fossilization* are principally two (Figure 2.15):

- mineralization
- carbonization.

One or the other mechanism is preferred depending upon the different environmental conditions; however, both in a real wet anaerobic environment are mutually mingled.

Wood mineralization is connected with diffusion and deposition of mineral substances to the lumens of wood cells and subsequently also to the cell walls. Mineral substances get to the wood from the surrounding external environment. In continental exposures, wood is usually enriched by silicates (SiO_2), as well as other minerals such as calcites ($CaCO_3$) and dolomites ($CaCO_3{\cdot}MgCO_3$), ferrous salts ($FeCO_3$, Fe_2O_3) or sulphur compounds (FeS_2, ZnS). There are various phosphates, silicates and calcites in the sea

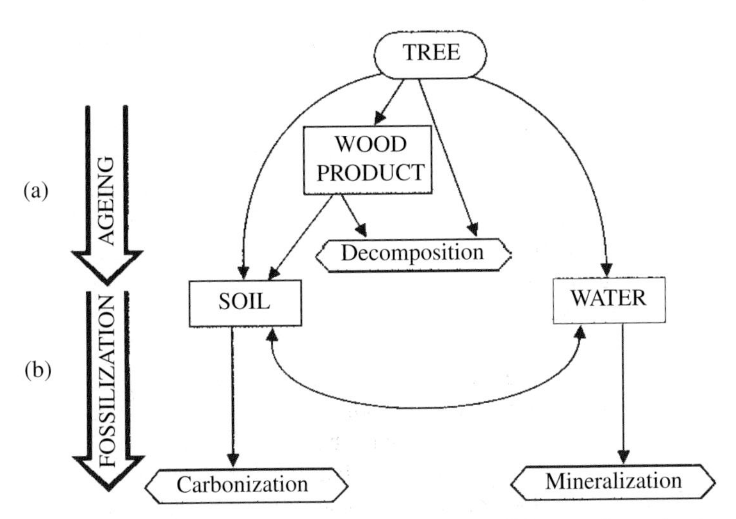

Figure 2.15 (a) Wood ageing under aerobic conditions – fungal rot and other decay; (b) scheme of wood fossilization under anaerobic conditions – mineralization and carbonization (modified from Fengel (1991))

Source: Fengel, D. (1991) Aging and fossilization of wood and its components. *Wood Science and Technology* 25, 153–177. Reproduced by permission of Springer

environment. Furuno et al. (1986) proposed a four-stage model of wood mineralization (Figure 2.16). Lumens start filling with minerals in the first stage. Deposition of minerals in lumens continues and the decomposition of cell wall components starts in the second stage. The components of wood are partially replaced with minerals in the third stage, and a significant to total replacement of wood polymers with minerals takes place in the fourth stage. The anatomical details remain copied in the petrified wood, although only when the mineral substance has not crystallized in the cell walls.

Wood carbonization is related to the decrease in the contents of hydrogen and oxygen in its components (polysaccharides and lignin), and with an aliquot increase in the content of carbon. Finally, carbon is concentrated in particular in the non-hydrolysable substances similar to lignin. This is a long-term process, in which hydrolytic reactions take place in polysaccharides and partially also in lignin at first, gradually followed by other reactions (e.g. of condensation type).

2.4 Properties of abiotically damaged wood

2.4.1 Properties of weathered wood

Wood degraded by weather changes in particular on its surface. The cell walls are weakened, and some cells get deformed, coloured and cracked, which is

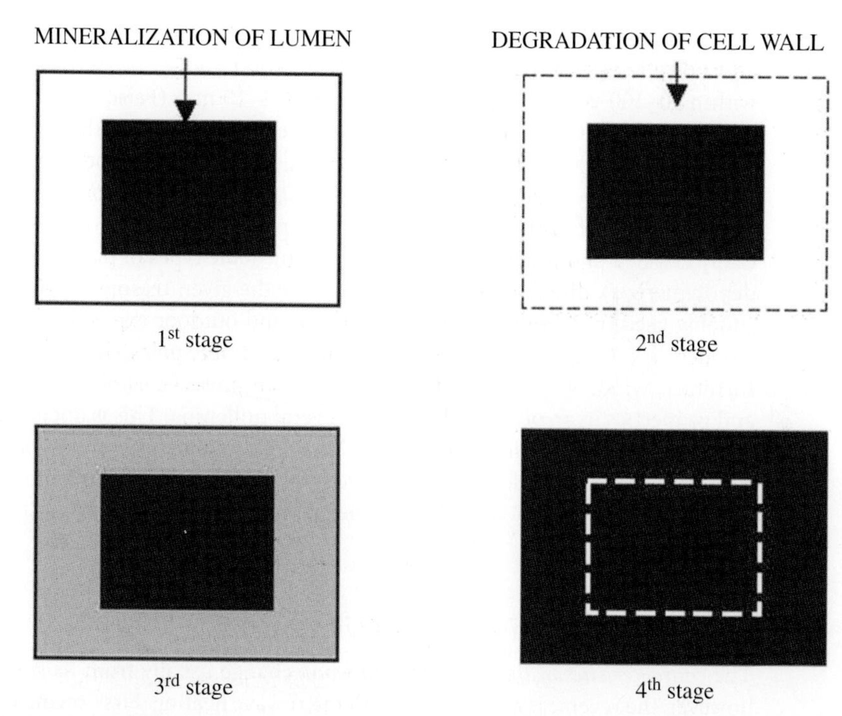

Figure 2.16 Model of mineralization of lumens and cell walls of wood (modified from Furuno et al. (1986))

Source: Furuno, T., Watanabe, T., Suziki, N., Goto, T. and Yokoyama, K. (1986) Microstructure and silica mineralization in the formation of silicified woods. II. Distribution of organic carbon and the formation of quartz in the structure of silicified woods. *Mokuzai Gakkaishi* 32, 575–583. Reproduced by permission of The Japan Wood Research Society Tokyo

reflected also in the change of other characteristics of weathered wood in its surface layers:

- *colour changes* – yellowing and browning caused by the consequence of photo-oxidation reactions in lignin and partially also in hemicelluloses (interior, sheltered exterior), or greying caused by the subsequent/simultaneous washout of decomposed substances from lignin and hemicelluloses by rainfall water with the possibility of adsorption of emissions, dusty particles and other contaminants on cellulose (unsheltered exterior);
- *roughening* – the consequence of washing out of photo-oxidized lignin and hemicelluloses by water and release of cellulose fibres;
- *plastic texture* – the consequence of more intense erosion of early wood having less density than late wood;
- *micro-cracks* – the consequence of weather and temperature stresses in cells and cell walls;
- *shape deformations* – the consequence of the effects of several weather factors.

Where due to just abiotic weather factors (i.e. without the effect of moulds and other biological pests), the structure of wood and its properties change within 10–100 years more or less to a depth of 2–10 mm (Feist, 1990). This is less than the destruction experienced by some inorganic or synthetic organic materials. Therefore, the weathering durability of wooden products can be higher than in products based on (1) unprotected iron, steel or other non-durable metals (if easily oxidized or rusts), (2) poor-quality types of concrete (if easily crushed and totally destroyed) and (3) unstable types of plastic (if easily depolymerized, discoloured and brushed). For the given reasons, wood is still suitably used in various products for interior and outdoor exposures, such as wooden structures, telecommunication and other utility poles, sleepers, garden furniture, windows, doors, and so on. Wood can preserve its native structure and properties in an optimal climate for several millennia. This is documented by the finding of historic woods exposed in an internal dry environment, such as with the 3300-year-old oak sarcophagus from the Tutankhamun tomb burial chamber and the 1800-year-old teak beam from a Buddhistic monastery in India.

2.4.2 Impact of increased temperature and fire on wood properties

The *characteristics of thermally loaded wood* change usually from its surfaces; however, the reverse is also the case with microwave heating. First, owing to relatively lower temperatures of 130–150 °C the cell walls of wood may get thicker (Yildiz et al., 2004), but then at higher temperatures they are weakened by cracks and thinning, and finally totally disappear and only ash remains from them (Zicherman & Williamson, 1982). At the same time, the physical–mechanical, technical and aesthetic properties of wood are changed (Le Van & Winandy, 1990; Reinprecht et al., 1992). The intensity of changes is greatest in the surface zones of wood exposed to radiant or flame sources of heat. The internal zones of wood are degraded faster only when heated by microwave radiation. Withal wood gets dark, from yellow through brown up to black colour, or when transformed to ashes it becomes grey. At the same time, its density and hygroscopicity decrease, while its strength properties apparently worsen (Militz, 2002; Reinprecht & Vidholdová, 2008).

The *biological resistance of thermally loaded wood* increases against wood-decaying fungi, moulds or wood-damaging insects. This significant characteristic is utilized in the preparation of thermally modified woods (see Section 6.2). By an intentional thermal treatment of wood at temperatures of 160–260 °C, in an environment with air or limited access of air (e.g. in vegetable oils or vacuum), materials of the 'ThermoWood', 'PlatoWood' and 'OHT-Wood' types are prepared. They are characterized by better dimensional stability, lower absorption of moisture and soaking in water, higher biological resistance, connected usually only with moderate 10–20% decrease in strength. They are used for various wooden products, such as saunas, windows, claddings, and so on, for which an increased resistance against dimensional changes and biological pests is required.

The *decrease in strength of thermally loaded wood* depends in particular upon the damage grade of its structural polymers, in a direct connection with the temperature and time of thermal effects (Militz, 2002). For example, bending strength of the wood decreases by 20%: (1) at 155 °C after approximately 3 days, (2) at 135 °C after ~2 weeks and (3) at 115 °C after ~15 weeks (Millett & Gerhards, 1972). Bending strength of wood exposed to 220 °C for 5 h, with stages of preheating and cooling down within the total time of 4 days, is decreased up to 50% (Bengtsoon et al., 2002). In contrast, the modulus of elasticity (MOE) of wood can even increase during its heating at lower temperatures of 100–160 °C, but it decreases due to the impact of higher temperatures over 160 °C, although this decrease is substantially milder than for the wood strength. Kubojima et al. (2000) documented changes in the modulus of elasticity of wood heated at a temperature of 160 °C; there was a moderate decrease the modulus of elasticity when heat treated in air and no changes when heat treated in pure nitrogen. In general, the reduction in strength of thermally treated wood is greater than the reduction of its mass or density, and reduction of its rigidity.

There were more significant reductions in rigidity and strength of wood in the case of its thermal and chemical loading. An example is the experiment with maple wood that was first impregnated with various concentrations of sulphuric acid before exposing it to radiant heat of 160 °C/3 h or 190 °C/3 h, when oxidation and hydrolysis processes in the wood occurred. Degradation of wood took place to several degrees and such wood was subsequently subjected to the rigidity tests. Decreases of the static modulus of elasticity in bending (MOE_S) and the dynamic modulus of elasticity (MOE_D) were approximately 2.5 times greater than the decreases in its mass (Figure 2.17).

Fires cause a significant, but not always exactly definable reduction in strength of wood and the violation of statics of a wooden structure as a whole. Roof frames – trusses, ceilings and other wooden elements damaged by fire must be thoroughly reinforced or replaced in full (see Chapter 7).

2.4.3 Impact of water and other chemicals on wood properties

The *characteristics of wood attacked by aggressive chemicals* will change in a specific way according to the type and concentration of chemical acting. Most frequently, this is the reduction of degree of polymerization of polysaccharides and the break-up of the three-dimensional network of lignin. Damage of these polymers is reflected in microscopic and macroscopic defects of wood and in the change of its properties. The mechanical properties of chemically degraded wood get worse in the prevailing majority of cases. Its physical properties are changed individually; for example, alkali causes cell collapse and a growth in wood density.

Alkali treatments – for example, with sodium hydroxide (NaOH) or ammonium hydroxide (NH_4OH) – cause the wood to lose mainly hemicelluloses. The weakened cell walls collapse (analogous to the collapse of walls of fossil wood), and this results in increased wood shrinkage and then also in a higher density of

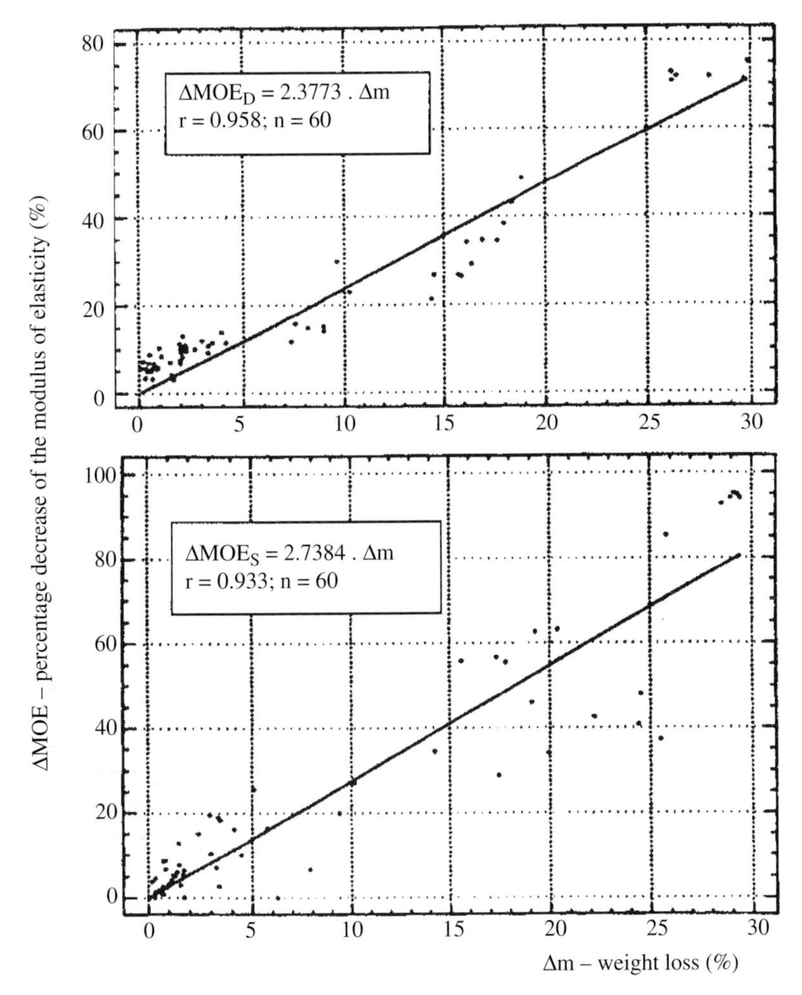

Figure 2.17 Impact of the weight loss of thermally–chemically loaded maple (*Acer pseudoplatanus*) wood on the reduction in its rigidity: MOE_D: dynamic modulus of elasticity; MOE_S: static modulus of elasticity in bending (Reinprecht et al., 1996)

Source: Reinprecht, L., Rajčan, E. and Čudková, M. (1996) Comparison of the dynamic and static moduli of elasticity of the chemically and thermally degraded maple. In: Acoustics '96, 3rd International Colloqium, pp. 25–31. TU Zvolen, Slovakia. Reproduced by permission of TU Zvolen

air-dried wood. Such wood with only slightly degraded cellulose, but with significantly increased density, is usually stronger in the dry condition than native wood is. In particular, in dry conditions, there is an increase in its hardness, compression strength and shear strength. In contrast, its strength is really significantly reduced in wet conditions, since it swells more than native wood does and even the reinforcing hydrogen bonds and van der Waals interactions are not

Table 2.4 Properties of chemically corroded beech wood (*Fagus sylvatica*) (modified from Reinprecht (1988, 1991))

Chemical substance	Wood property			
			Shear strength across fibres (MPa)	
	Weight loss (%)	Volume shrinkage (%)	Dry (w = 10%)	Wet (w > FSP)
None (reference)	0	16.8	51.8	45.4
HCl	12.5	22.7	38.9	19.6
HNO_3	15.5	27.6	41.9	24.8
H_2SO_4	5.3	19.8	45.7	23.7
CH_3COOH	1.6	17.9	49.0	40.9
NaOH	16.5	35.1	60.4	26.8
NH_4OH	9.1	27.3	54.6	40.3
H_2O_2	2.7	16.2	49.4	40.1

Source: R., L. (2013) *Wood Protection*, Handbook, TU Zvolen, Slovakia, 134 p. Reproduced by permission of TU Zvolen.
Corrosion by 5% aqueous solutions of selected chemical substances: 30 days, 20 °C, volume hydromodule 2 : 1.
w: absolute moisture content of wood; FSP: fibre saturation point of wood, $w \approx 30\%$.

applied in it in the presence of water molecules (Table 2.4). In summary, wood attacked by alkalis is very similar in structure and properties to fossil or subfossil wood, in which there is hemicellulose degradation and only minimal cellulose and lignin damage (Table 2.5).

Acid treatments – for example, sulphuric acid (H_2SO_4), hydrochloric acid (HCl), nitric acid (HNO_3) or acetic acid (CH_3COOH) – cause the wood to lose various portions of polysaccharides and/or lignin. In spite of changes in the molecular structure of wood, its density and shrinkage will not usually significantly increase or decrease. The mechanical properties of wood damaged by acids are reduced not only in its wet state, but partly also in its dry state (Table 2.4). Strength decrease in the dry state of wood damaged by acids can be explained by a significant depolymerization of hemicelluloses (due to the impact of HNO_3) or also of cellulose (due to the impact of H_2SO_4 and HCl), as well as degradation of lignin (due to the impact of HNO_3), connected only with a minimal change (increase) in its density. With organic acids, such as acetic acid or gallic acid, wood at common temperatures corrodes only moderately and its properties change minimally. However, at higher temperatures of 60–100 °C organic acids are more aggressive and they more evidently change the wood structure and properties. Characteristics of wood degraded by acids (except for HNO_3 and others with an oxidizing effect) are similar to the wood attacked by brown-rot fungi (Table 2.5).

Table 2.5 Changes in molecular structure and in properties of wood degraded by aggressive chemicals – analogy of chemical degradations with other types of wood degradation (modified from Reinprecht (1999))

Characteristics	Aggressive chemical				
	$NaOH$ NH_4OH	H_2SO_4 HCl	HNO_3	H_2O_2	CH_3COOH
Significant degradation					
Hemicelluloses	Yes	Yes	Yes	No	No
Cellulose	No	Yes	No	No	No
Lignin	(?)	No	Yes	Yes	No
Wood property					
Density (at $w = 10\%$)	+				
Volume shrinkage	+	+	+	(+)	(+/–)
Shear strength across fibres					
$w = 10\%$	+	–	–	(–)	(+/–)
$w \geq$ FSP(30%)	–	–	–	–	(–)
Degradation analogy	Fossilization	Brown rot	White rot	White rot	Hydrothermal modification

Source: R., L. (2013) *Wood Protection*, Handbook, TU Zvolen, Slovakia, 134 p. Reproduced by permission of TU Zvolen.
+ significant growth, – significant reduction, (–) moderate reduction, (+/–) ambiguous changes.

On exposure to *oxidizers* – for example, hydrogen peroxide (H_2O_2) or HNO_3 – the wood loses mainly lignin. This is similar to wood damaged by white-rot fungi (Table 2.5).

Wood in anaerobic conditions in a long-term contact with water – for example, underground, under ice, in oceans – is subject to fossilization processes. The typical feature of wood mineralization is the filling of lumens with substances of inorganic character (Figure 2.16). Cell walls gradually acquire a porous consistency, in particular in the higher fossilization stages, when just lignin remains in them (Fengel, 1991). Wet archaeological and other fossil wood with weakened cell walls deforms when dried and its dimensions are significantly diminished (Horský & Reinprecht, 1986; Schniewind, 1990). After drying, it may have even higher strength than recent wood, resembling a wood degraded by alkalis (Table 2.5). Its higher strength in the dry condition may be contributed to by its increased density, with the support effect of hydrogen bonds and van der Waals hardening interactions (Figure 2.18). Wet wooden artefacts must be conserved in a suitable way before exposing them to a dry environment, usually using polyethylene glycols, saccharose, and so on, whereby their deformations and collapses can be prevented (see Section 7.3).

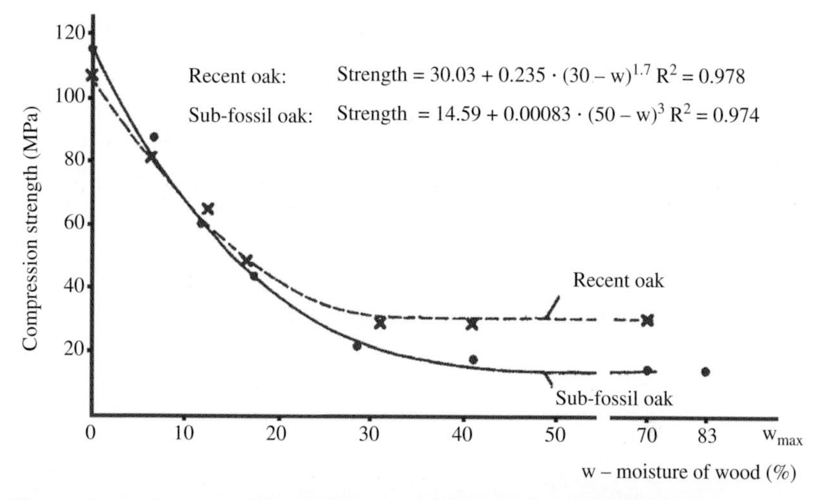

Figure 2.18 Impact of humidity on compression strength in parallel with grains of 8150-year-old subfossil oak found 16 m underground in Gabčíkovo near the Danube river and recent oak (Horský and Reinprecht, 1986)

Source: Horský, D. and Reinprecht, L. (1986) *Study of Subfossil Oak Wood* (in Slovak, English abstract). Monograph, VŠLD Zvolen, Slovakia, 70 p. Reproduced by permission of TU Zvolen

References

Alangi, J. & Malucelli, G. (2015) Thermal degradation of cellulose and cellulosic substances. In *Reactions and Mechanisms in Thermal Analysis of Advanced Materials*, Tiwari, A. & Raj, B. (eds). John Wiley & Sons, Inc., Hoboken, NJ, chapter 14, pp. 301–332.

Ayrilmis, N. (2007) Effect of fire retardants on internal bond strength and bond durability of structural fiberboard. *Building Environment* 42(3), 1200–1206.

Beall, F. C. & Eickner, H. W. (1970) Thermal degradation of wood components: a review of the literature. Forest Service Research Paper FPL 130. Forest Products Laboratory, Madison, WI.

Becker, H. (1986) Umwelteinflüsse auf holz. *Praktic Schädlingsbekämpfer* 5, 63–66.

Bengtsoon, C., Jermer, J. & Brem, F. (2002) Bending strength of heat-treated spruce and pine timber. In *IRG/WP 02*, Cardiff, Wales, UK. IRG/WP 02-40242.

Blažej, A. & Košík, M. (1985) *Phytomass as a Chemical Source.* Veda, Bratislava (in Slovak).

Böttcher, P. (1989) Unterschuchungen zur dauerhaftigkeit von gründungspfählen. *Holz als Roh- und Werkstoff* 47, 179–184.

Brebu, M. & Vasile, C. (2010) Thermal degradation of lignin – a review. *Cellulose Chemistry and Technology* 44(9), 353–363.

Chang, S. T., Hon, D. N. S. & Feist, W. C. (1982) Photodegradation and photoprotection of wood surfaces. *Wood and Fiber Science* 14, 104–117.

Derbyshire, H. & Miller, E. R. (1981) The photodegradation of wood during solar irradiation. *Holz als Roh- und Werkstoff* 39, 341–350.

Erler, K. (1990) Korrosion un anpassungsfaktoren für chemisch-aggressive medien bei holzkonstruktionen. *Holztechnologie* 30, 228–233.

Evans, P. D. (1989) Effect of angle of exposure on the weathering of wood surfaces. *Polymer Degradation and Stability* 24(1), 81–87.

Feist, W. C. (1990) Outdoor wood weathering and protection. In *Archaeological Wood – Properties, Chemistry, and Preservation*, Rowell, R. M. & Barbour, J. R. (eds), Advances in Chemistry Series, vol. 225. American Chemical Society, Washington, DC, chapter 11, pp. 263–298.

Fengel, D. (1991) Aging and fossilization of wood and its components. *Wood Science and Technology* 25, 153–177.

Fengel, D. & Wegener, G. (2003) *Wood – Chemistry, Ultrastructure, Reactions*. Verlag Kessel, Remagen.

Furuno, T., Watanabe, T., Suziki, N., et al. (1986) Microstructure and silica mineralization in the formation of silicified woods. II. Distribution of organic carbon and the formation of quartz in the structure of silicified woods. *Mokuzai Gakkaishi* 32(8), 575–583.

Goldstein, I. G. (1984) Degradation of wood by chemicals. In *The Chemistry of Solid Wood*, Rowell, R. (ed.). Advances in Chemistry Series, vol. 207. American Chemical Society, Washington, DC, chapter 15, pp. 577–588.

Hodgin, D. A. & Lee, A. W. C. (2002) Comparison of strength properties and failure characteristics between fire-retardant-treated and untreated roofing lumber after long-term exposure: a South Carolina case study. *Forest Products Journal* 52(6), 91–94.

Hoffmann, P. & Jones, M. A. (1990) Structure and degradation process for waterlogged archaeological wood. In *Archaeological Wood – Properties, Chemistry, and Preservation*, Rowell, R. M. & Barbour, J. R. (eds). Advances in Chemistry Series, vol. 225. American Chemical Society, Washington, DC, pp. 35–65.

Horský, D. & Reinprecht, L. (1986) *Study of Subfossil Oak Wood*. Monograph. VŠLD Zvolen, Slovakia (in Slovak, English abstract).

Hrablay, I. & Jelemenský, Ľ. (2014) Kinetics of thermal degradation of wood biomass. *Chemical Papers* 68(12), 1725–1738.

Janssens, M. L. (2004) Modeling of the thermal degradation of structural wood members exposed to fire. *Fire and Materials* 28, 199–207.

Kačíková, D., Kačík, F., Bubeníková, T. & Košíková, B. (2008) Influence of fire on spruce wood lignin changes. *Wood Research* 53(4), 95–104.

Kartal, S. N., Hwang, W. J. & Imamura, Y. (2008) Combined effect of boron compounds and heat treatments on wood properties: chemical and strength properties of wood. *Journal of Material Processing and Technology* 198, 234–240.

Kloiber, M., Frankl, F., Drdácký, M., et al. (2010) Change of mechanical properties of Norway spruce wood due to degradation caused by fire retardants. *Wood Research* 55(4), 23–38.

Kubojima, Y., Okano, T. & Ohta, M. (2000) Bending strength and toughness of heat-treated wood. *Journal of Wood Science* 46(1), 8–15.

Kučerová, I., Michalcová, A., Novotná, M. & Ohlídalová, H. (2007) Examination of damaged wood by ammonium phosphate and sulphate-based fire retardants: the results of the Prague castle roof timber examination. In *COST Action IE0601 – Wood Science for Conservation of Cultural Heritage*, Florence, Italy.

Kúdela, J. & Reinprecht, L. (1990) Einflus der Holzfeuchte auf die Druckfestigkeit von rezentem und subfossilem Eichenholz – *Quercus robur* L. *Holzforschung* 44(3), 211–215.

Le Van, S. L. (1984) Chemistry of fire retardancy. In *The Chemistry of Solid Wood*, Rowell, R. (ed.). Advances in Chemistry Series, vol. 207. American Chemical Society, Washington, DC, chapter 14, pp. 531–574.

Le Van, S. L. & Winandy, J. E. (1990) Effect of fire retardant treatments on wood strength – a review. *Wood and Fiber Science* 22, 112–131.

Le Van, S. L., Kim, J. M., Nagel, R. J. & Evans, J. W. (1996) Mechanical properties of fire-retardant-treated plywood after cyclic temperature exposure. *Forest Products Journal* 46(5), 64–71.

Lindergaard, B. & Morsing, N. (2003) Natural durability of European wood species for exterior use above ground. In *IRG/WP 03*, Brisbane, Australia. IRG/WP 03-10499.

Lionetto, F., Del Sole, L., Cannoletta, D., et al. (2012) Monitoring wood degradation during weathering by cellulose crystallinity. *Materials* 5, 1910–1922.

Makovíny, I., Solár, R. & Reinprecht, L. (1992) Corrosion of steels with water extracts of wood. *Drevársky Výskum (Wood Research)* 134, 39–52 (in Slovak, English abstract).

Marková, I. & Klement, I. (2003) Thermal analysis (TG, DTG and DSC) of hornbeam wood after drying. *Wood Research* 48(1–2), 53–61.

Matsuo, M., Umemura, K. & Kawai, S. (2012) Kinetic analysis of color changes in cellulose during heat treatment. *Journal of Wood Science* 58(2), 113–119.

Militz, H. (2002) Thermal treatment of wood – European processes and their background. In *IRG/WP 02*, Cardiff, Wales, UK. IRG/WP 02-40241.

Millett, M. A. & Gerhards, C. C. (1972) Accelerated aging. Residual weight and flexural properties of wood heated in air at 115° to 175°C. *Wood Science* 4(4), 193–201.

Nikolaos, T. (2003) *Modern Fire Retardant Coatings for Wood*. Kokkola, Greece.

Odeen, K. (1985) Fire resistance of wood structures. *Fire Technology* 21(1), 34–40.

Osvald, A. (1994) Burning of wood and wooden materials. *In VP-DF-VŠLD of Zvolen 1993/94*. Alfa Bratislava, Slovakia, pp. 197–216 (in Slovak, English abstract).

Pandey, K. K. (1998). Reaction of wood with inorganic salts. *Holz als Roh- und Werkstoff* 56, 412–415.

Pandey, K. K. & Vuorinen, T. (2008) Comparative study of photodegradation of wood by a UV laser and a xenon light source. *Polymer Degradation and Stability* 93, 2138–2146.

Pánek, M. & Reinprecht, L. (2016) Effect of vegetable oils on the colour stability of four tropical woods during natural and artificial weathering. *Journal of Wood Science* 62(1), 74–84.

Poletto, M., Zattera, A. J. & Santana, R. M. C. (2012) Thermal decomposition of wood: kinetics and degradation mechanisms. *Bioresource Technology* 126, 7–12.

Poletto, M., Júnior, H. L. O. & Zattera, A. J. (2015) Thermal decomposition of natural fibres: kinetics and degradation mechanisms. In *Reactions and Mechanisms in Thermal Analysis of Advanced Materials*, Tiwari, A. & Raj, B. (eds). John Wiley & Sons, Inc., Hoboken, NJ, chapter 21, pp. 515–546.

Reinprecht, L. (1988) Selected properties of wood after its treatment with acids, alkalis or oxidation agents. *Drevársky Výskum* (*Wood Research*) 119, 43–62 (in Slovak, English abstract).

Reinprecht, L. (1991) Corrosion of wood due to NaOH, HCl and HNO_3. In *Scientific Works of DF VŠLD in Zvolen 1991*. Alfa Bratislava, Czechoslovakia, pp. 105–123 (in Slovak, English abstract).

Reinprecht, L. (1999) Comparative study on some physical and mechanical properties of old waterlogged oaks and those deteriorated in artificial conditions. In *Drewno Archeologiczne: Badania i Konserwacja: Sympozjum: Biskupin-Wenecja, 22–24 Czerwca*

1999 (Archaeological Wood: Investigation and Conservation: Symposium, Biskupin-Wenecja, 22–24 June 1999), Babiński, L. (ed.). Muzeum w Biskupinie, pp. 263–276.

Reinprecht, L. (2013) *Wood Protection*, Handbook. TU Zvolen, Slovakia.

Reinprecht, L. & Makovíny, I. (1990) Diffusion of inorganic salt $CuSO_4 \cdot 5H_2O$ in wood structure. In *Latest Achievements in Research of Wood Structure and Physics – Proceedings*, Pozgaj, A. (ed.). Faculty of Wood Technology, University of Forestry and Wood Technology, Zvolen, Czechoslovakia, pp. 203–214 (in Russian, English abstract).

Reinprecht, L. & Mihálik, A. (1988) On problem of phosphorous and lignin–saccharidic composition function in the course of thermic wood analyses. In *Horenie Dreva – Wood Burning, 1st Conference*, High Tatras–TU Zvolen, Czechoslovakia, pp. 185–197 (in Slovak, English abstract).

Reinprecht, L. & Vidholdová, Z. (2008) Mould resistance, water resistance and mechanical properties of OHT-thermowoods. In *Sustainability through New Technologies for Enhanced Wood Durability: Socio-economic Perspectives of Treated Wood for the Common European Market*, Final Conference Proceedings of COST E37. Ghent University, Belgium, pp. 159–165.

Reinprecht, L., Chovanec, D. & Horský, D. (1992) Thermal degradation of chemically pre-treated wood. Part 1. Course of thermal degradation of wood valued on the basis of loss of mass, impact strength in bending and cutting strength. Part 2: Course of thermal degradation of wood valued on the basis of anatomical, coloured and shape changes. In *Wood Burning '92, International Conference*, High Tatras–TU Zvolen, Czechoslovakia, pp. 57–67, 69–77.

Reinprecht, L., Rajčan, E. & Čudková, M. (1996) Comparison of the dynamic and static moduli of elasticity of the chemically and thermally degraded maple. In *Acoustics '96, 3rd International Colloqium*. TU Zvolen, Slovakia, pp. 25–31.

Schniewind, A. P. (1990) Physical and mechanical properties of archaeological wood. In *Archaeological Wood – Properties, Chemistry, and Preservation*, Rowell, R. M. & Barbour, J. R. (eds). Advances in Chemistry Series, vol. 225. American Chemical Society, Washington, DC, chapter 4, pp. 87–109.

Sell, J. & Feist, W. C. (1986) Role of density in the erosion of wood during weathering. *Forest Products Journal* 36, 57–60.

Shafizadeh, F. (1984) The chemistry of pyrolysis and combustion. In *The Chemistry of Solid Wood*, Rowell, R. M. (ed.). Advances in Chemistry Series, vol. 207. American Chemical Society, Washington, DC, chapter 13, pp. 489–529.

Solár, R., Reinprecht, L., Kačík, F., et al. (1987) Comparison of some physico-chemical properties of carbohydrate and lignin part of contemporary and subfossile oak wood. *Cellulose Chemistry and Technology* 21, 513–524.

Tolvaj, L., Persze, L. & Albert, L. (2011) Thermal degradation of wood during photodegradation. *Journal of Photochemistry and Photobiology – B* 105(1), 90–93.

Vandevelde, P. (1999) Classification of fire performance. In *Fire Safe Products in Construction*. EGOLF, Luxembourg, pp. 1–12.

Wegener, G. & Fengel, D. (1986) Untersuchungen zur beständigkeit von holzbauteilen in aggressiven atmosphären. *Holz als Roh- und Werkstoff* 44, 201–206.

White, R. H. & Nordheim, E. V. (1992) Charring rate of wood for ASTM 119 exposure. *Fire Technology* 28(1), 5–30.

Williams, R. S. (2005) Weathering of wood. In *Handbook of Wood Chemistry and Wood Composites*. CRC Press, Boca Raton, FL, pp. 139–185.

Williams, R. S., Knaebe, M. T., Evans, J. & Feist, W. C. (2001) Erosion rates of wood during natural weathering: part III. Effect of exposure angle on erosion rate. *Wood Fiber and Science* 33(1), 50–57.

Yildiz, Ü. C., Gercek, Z., Serdar, B., et al. (2004) The effects of heat treatment on anatomical changes of beech wood. In *IRG/WP 04*, Ljubljana, Slovenia. IRG/WP 04-40284.
Zelinka, S. L. & Stone, D. S. (2011) The effect of tannins and pH on the corrosion of steel in wood extracts. *Materials and Corrosion* 62(8), 739–744.
Zicherman, J. B. & Williamson, R. B. (1982) Microstructure of wood char. Part 1: whole wood. *Wood Science and Technology* 16, 237–249.

Standards

EN 1995-1-2 (2004) Eurocode 5: Design of timber structures – Part 1-2: General – Structural fire design.
EN 13501-1+A1 (2009) Fire classification of construction products and building elements – Part 1: Classification using data from reaction to fire tests.
EN 13823 (2010) Reaction to fire tests for building products – Building products excluding floorings exposed to the thermal attack by a single burning item.
EN ISO 1182 (2010) Reaction to fire tests for products – Non-combustibility test.
EN ISO 1716 (2010) Reaction to fire tests for products – Determination of the gross heat of combustion (calorific value).
EN ISO 4589-2 (1999) Plastics – Determination of burning behaviour by oxygen index – Ambient-temperature test.
EN ISO 5657 (1997) Reaction to fire tests – Ignitability of building products using a radiant heat source.
EN ISO 5658-2 (2014) Reaction to fire tests – Spread of flame – Part 2: Lateral spread on building and transport products in vertical configuration.
EN ISO 5659-2 (2012) Plastics – Smoke generation – Part 2: Determination of optical density by a single-chamber test.
EN ISO 5660-1 (2002) Reaction-to-fire tests – Heat release, smoke production and mass loss rate – Part 1: Heat release rate (cone calorimeter method).
EN ISO 11925-2 (2010) Reaction to fire tests – Ignitability of products subjected to direct impingement of flame – Part 2: Single-flame source test.
ISO 9705 (1993) Fire tests – Full-scale room test for surface products.
UL 94 Standard for Tests for Flammability of Plastic Materials for Parts in Devices and Appliances.

3 Biological Degradation of Wood

Wood in suitable environmental conditions can be attacked by bacteria, fungi, algae, insects, marine borers and some other biological pests (Table 3.1). Wood-damaging pests already degrade ~15–20% of wood in forests and in stocks of logs and sawn-timbers in industrial plants. However, they have a very important role in deterioration of wooden products as well. In Europe, but also in other regions of world, to the most dangerous pests acting on wood products belong the decay-causing fungi, boring insects and marine borers, while rather less dangerous are the bacteria, wood-staining fungi, moulds, birds and mammals.

3.1 Wood damaged by bacteria

A lot of various bacteria colonize living, harvested and processed wood in different ways and with different intensities (Liese, 1992; Schmidt & Liese, 1994; Milling et al., 2005; Schmidt, 2006; Daniel, 2014). Common bacteria are subtle, measuring 0.4–5 μm in size. Bacteria are rod, sphere, helix, or comma shaped. In living trees they attack the phloem cells, but may also be associated with the development of false frost cracks as a consequence of action in the ray parenchyma cells, or the discoloration of sawn wood during its air-drying.

Bacteria penetrate into a dead wood via xylem parenchyma cells of ray tissues and resin canals. Subsequently, they enter libriform fibres, vessels, tracheids and other wood cells, mainly using pits in cell walls for their penetration. Bacteria live on pectin, monosaccharides and other easily accessible nutrients and may prepare the wood for fungi (Rayner & Body, 1988). The strength of wood is reduced only by some species of bacteria that are able to attack not only the non-lignified pits in tracheids or other tissues but also the cell walls of wood cells.

Wet wood having a certain amount of oxygen (i.e. logs protected with water in basins or spraying), wooden products above the ground or in contact with the

Wood Deterioration, Protection and Maintenance, First Edition. Ladislav Reinprecht.

Table 3.1 Biological damage of wood by organisms – pests

Type of organism	Character of wood damage
Bacteria	No or less-intense decomposition of wood cells
Fungi	Rot of wood cells caused by wood-decaying (wood-destroying) fungi
	Moderate impairment of wood by wood-staining fungi and moulds
Insects	Damage of wood by feeding marks produced by larvae (or imagoes)
Marine borers	Damage of wood by boreholes and tunnels
Parasitic spermatophytes	Growth of plant roots in the wood of a living tree
Synanthropic vertebrate	Damage of wood by biting it out, hiding places in panels, excrement
Human	Unsuitable approach to harvesting, storage and processing of wood, as well as bad protection and maintenance of wooden products; following creation of suitable conditions for fungal-rot, insect damage, atmospheric, thermal and other forms of wood damage

ground (e.g. sleepers, utility poles and palisades) can be attacked by *aerobic* bacteria – for example, *Bacillus brevis*, *B. firmus*, *B. subtilis*, *Cellulomans flavigena*. Wet wood totally soaked with water (i.e. wood placed long-term in water or in wet ground – cooling towers, sea harbour structures or archaeological wood) can be attacked by *anaerobic* bacteria – for example, *Acetivibrio cellulolyticus*, *Clostridium thermocellum*, *Clostridium papyrosolvents*.

Wood-colonizing bacteria have already in the past been classified into four groups (Greaves, 1971):

- Bacteria living off the accompanying substances in ray tissues of wood, able to impair the non-lignified pits in cell walls and thus increase the permeability of wood, though without the impact on its strength. These bacteria were later studied in more detail by, for example, Efransjah et al. (1989), Despot (1993), and Pánek and Reinprecht (2011).
- Bacteria destroying cell walls and reducing the strength of wood – erosion, cavitation and tunnelling bacteria. Experiments in this topic were later made by, for example, Singh et al. (1992), Daniel and Nilsson (1998), Wakeling (2006), Daniel (2014), and others.
- Bacteria acting only in wood primarily or concurrently attacked by another biological pest, or in the wood primarily delignified by abiotic factors; for example, in archaeological wood. They were later studied intensively by, for example, Björdal et al. (2005).
- Antagonistic bacteria suppressing the development and activity of other biological pests in wood. Among these today are the well-known *Bacillus* spp., *Pseudomonans* spp. and *Streptomyces* spp., used for biological modification of wood (see Section 6.4).

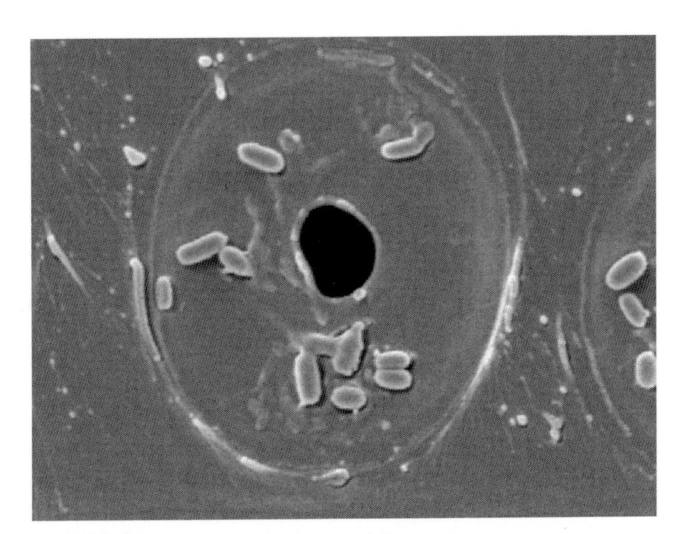

Figure 3.1 ***Bacillus subtilis*** **bacteria intentionally degraded margo and released torus in a bordered pit of a tracheid in the sap-zone of Norway spruce wood (*Picea abies*)**

Source: Pánek, M. and Reinprecht, L. (2008) Bio-treatment of spruce wood for improving of its permeability and soaking – Part 1: Direct treatment with the bacterium *Bacillus subtilis*. *Wood Research* 53(2), 1–12. Reproduced by permission of Pulp and Paper Research Institute Bratislava

Bacteria living from accompanying substances of wood destroy the pit membranes (margo) of tracheids (Figure 3.1) and ray parenchyma. Conifer wood is attacked more than the wood of broadleaves, and sapwood more than heartwood. When protecting round wood by soaking in water basins or spraying with water, these bacteria can significantly impair membranes in pits of wood cell walls after several days to weeks. Owing to their impact, round wood quickly soaks up water and sometimes sinks under the water level. Uncontrolled and uneven opening of pits leads to technological problems in wood processing. For example, at an artificial drying of sawn timber damaged by bacteria are created cracks, or at a surface painting of wood since failures in coatings can be created. In contrast, the destruction of pits can be intentionally used to improve the permeability of difficult to permeate refractory woods (e.g. spruce and fir poles) using convenient species of bacteria; for example, *Bacillus subtilis* (Blanchette et al., 1990; Pánek & Reinprecht, 2008, 2011).

Bacteria can destroy the cell walls of wood by causing erosion, cavities or tunnels in the cell walls (Figure 3.2). Erosion decomposition takes place from the S_3 layer towards the S_2 and other interior layers of cell walls. However, in most cases, the middle lamella remains intact. Tunnel and cavity decomposition is localized in particular to the S_2 layer of secondary walls – each tunnel being headed by a single bacterium along the long axis of the cell; cavities extend at a right angles to the long axis of the wood cell (Daniel & Nilsson 1986, 1998).

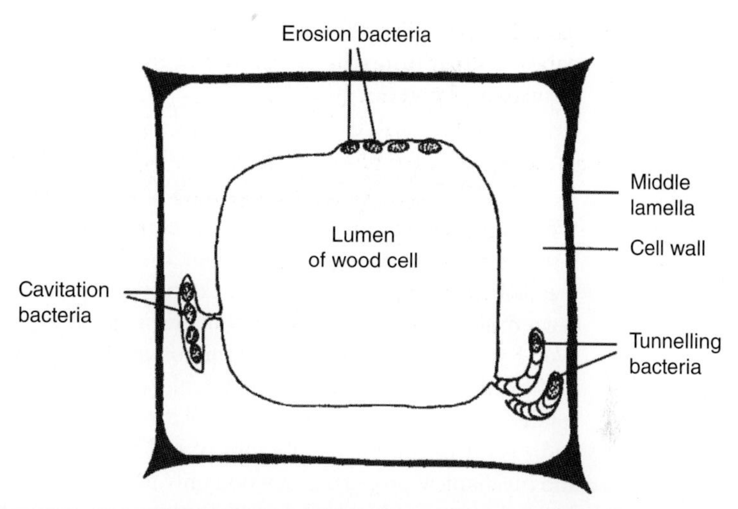

Figure 3.2 Overview of bacterial decomposition of wood cell wall – erosion bacteria, tunnelling bacteria and cavitation bacteria

Source: Modified from Eaton, R. A. and Hale, M. D. C. (1993) *Wood – Decay, Pests and Protection*. Chapman & Hall, London, UK, 546 p

Bacteria acting in wood damaged by other pests are living in symbiosis with fungi or insects. For example, during degradation of mine timbers or cooling tower slats, the wood polymers are decomposed first by enzymes of fungi to low molecule weight organic substances with a high content of carbon (glucose, xylose, etc.), and these subsequently become the food for bacteria. Fungi record this sensitively and adapt their metabolism to the current situation, that is, they will increase the production of 1,4-β-glucanases needed for depolymerization of cellulose.

Antagonistic bacteria, such as *B. subtilis*, *Bacillus asterosporus* and other species, produce antibiotics and toxic or repellent substances to suppress the growth of fungi and other pests in wood (see Section 6.4).

3.2 Wood damaged by fungi

Fungi are carbon-heterotrophic organisms without pigments for photosynthesis and without the ability to transform carbon dioxide (CO_2) from air to organic substances. They for life must obtain carbon from other organic substrates; for example, wood-damaging fungi from wood.

Wood-damaging fungi attack live or non-live wood (Gáper & Pišút 2003):

- parasitic fungi act on live trees from which they draw nutrients and energy, and sometimes they even cause various disease in them;
- saprophytic fungi damage dead wood (e.g. wooden products);

- parasitic-saprophytic fungi primarily attack live trees, and saprophytic-parasitic fungi primarily attack non-live wood from which can be transformed on live trees.

Wood-damaging fungi cause failures in wood in various ways:

- Wood-decaying fungi disrupt the structural components of wood (cellulose, hemicelluloses, lignin) and thus they also worsen its physical and mechanical properties. They cause extensive rot damages on live trees, logs, sawn timber and various products made of wood.
- Wood-staining fungi and moulds cannot impair the structural components of the cell walls of wood in more notable way and they feed only on starch, soluble sugars, fats and other protoplasmic remnants in the cell lumens, and pectin in the pit membranes of cell walls. They impair the density and mechanical properties of wood only to a negligent extent; on the other hand, they more notably change the colour and permeability of wood.

3.2.1 Reproduction, classification and physiology of wood-damaging fungi

3.2.1.1 Reproduction of wood-damaging fungi

Staining fungi and soft-rot fungi, and also some species causing white rot, belong to the Ascomycota phylum. They mainly reproduce via gamogenesis with ascospores created in the asci of fruiting bodies (Figure 3.3), or via asexual reproduction with conidia created in conidiophores (Table 3.2).

White-rot and brown-rot decay-causing fungi belong to the Basidiomycota phylum. They are mainly reproduced by reproductive basidiospores (Figure 3.4). Four, or sometimes just two, (+) and (−) reproductive spores are created in the basidia located in the hymenium; that is, in the top ~10– 80 μm thick layer of the hymenophore of fruiting bodies. Basidiomycota sometimes also reproduce via asexual spores (conidia, oidia and chlamydospores), or via fragments of vegetative-hyphae formations (Table 3.2).

Moulds, and also some species of staining fungi and soft-rot fungi, belong to the Deuteromycota phylum (Fungi imperfecti – asexual forms of Ascomycota and Basidiomycota fungi). They only reproduce asexually, mainly via conidia formed in conidiophores (Figure 3.5). They more rarely reproduce using fragments of mycelia or via oidia, chlamydospores and arthrospores.

3.2.1.2 Classification of wood-damaging fungi

After insects, the fungi are the second most frequent group of organisms in terms of species, with approximately 1.5 million to 3 million taxa, from which the great majority have not yet been identified (Robson, 1999; Schmidt, 2006). They are currently classified into three kingdoms: Protozoa, Chromista and Fungi (Gáper & Pišút, 2003). Wood-decaying fungi are included in the kingdom Fungi. This has four basic phyla – protists (Chytridiomycota), true

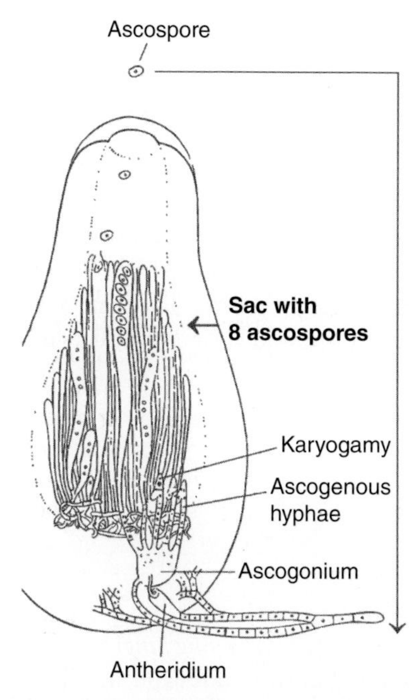

Figure 3.3 A scheme of the creation of asci (sacs) with ascospores in fungus from the Ascomycota phylum

- ***The germination of ascospores*** **creates single-nucleus hyphae and mycelia**
- ***Plasmogamy*** **takes place in the new fruiting body of the fungus, where the male formation (antheridium) contacts the female formation (ascogonium), subsequently creating multinuclei and later diploid nuclei (dikaryotic '*n* + *n*') ascogenous hyphae**
- ***Karyogamy*** **takes place in the top cells of ascogenous hyphae in the form of merging two sexually different nuclei, creating a diploid nucleus '2*n*'**
- ***Meiosis and mitosis*** **are subsequent processes where diploid nuclei present in the ascus of a fungus fruiting body divide, usually creating eight ascospores**
- ***After maturity and rupturing of the sac*** **the ascospores are discharged from the fruiting body and germinate on a suitable substrate**

Source: Modified from Urban, Z. and Kalina, T. (1980) *System and Evolution of Lower Development Plants* (in Czech). SPN Praha, Czechoslovakia, 415 p

moulds (Zygomycota), ascomycota (Ascomycota) and basidiomycota (Basidiomycota) – as well as one specific phylum that includes mitosporic fungi with an asexual mode of reproduction; that is, moulds (Deuteromycota). The cell wall of these fungi is mainly formed of chitin and β-glucans.

Table 3.2 Spores and formations for asexual reproduction of Basidiomycota and other fungi

Asexual reproduction	Characteristics
Conidia	Created exogenously – by detaching from the top cell of a specific fungal strand called a conidiophore
Oidia	Created exogenously – via decomposition of the top cells of fungal strands into individual cells
Chlamydospores	Created endogenously – inside special cells of fungal strands in the form of thickened cells. They are very durable cells, 'the resting stage of fungi', which even after several years are able to create new hyphae and mycelia
Vegetative formations	Parts of hyphae or mycelia – able to grow and act on a new wooden substrate; for example, even pieces of mycelia taken from rotten wood and transferred to healthy wood are able to grow in the humid environment and induce its decay

The criteria for classifying fungi in kingdoms and other systematic groups have not been and are still not explicit. The classification of fungi into taxonomic groups (species → genus → family → order → class → phylum → ...) is based on knowledge from several scientific disciplines:

- a genetic viewpoint, based on molecular biology; that is, on the principle of a different sequence of adenine (A), thymine (T), guanine (G) and cytosine (C) in macromolecules of the helixes of deoxyribonucleic acid (DNA) in individual species of fungi;
- a structural and ultrastructural viewpoint, based on the morphological structure and cytology of fungi hyphae and fruiting bodies;
- a development viewpoint, based on the growth and reproduction of fungi.

The gene apparatus of fungi has been investigated by scientists since the 1980s. The molecular systematics of fungi and the identification of individual species of fungi and their sequenced isolates using protein, DNA and other molecular methods are based on knowledge of their genes (Schmidt, 2006; Kirker, 2014; and see Section 7.3).

The morphological structure and cytology of fungi address the composition of their cells and tissues (i.e. spores, hyphae, mycelia and fruiting bodies), which are typical for each individual species of fungus. Fungi spores have varying shapes (spherical, cylindrical, oval) and size (usually 1–10 μm). Germinated spores form fibrous cells called hyphae with a diameter of 0.1–0.4 μm (e.g. microhyphae of the *Phellinus pini*) to 60 μm (vessel hyphae in the mycelial strand 'rhizomorphs' of the true dry-rot fungus *Serpula lacrymans*). The growth of individual hyphae is apical – their tip (apex) comes into contact with the cells of wood first. The mycelia and fruiting bodies of fungi are created from single

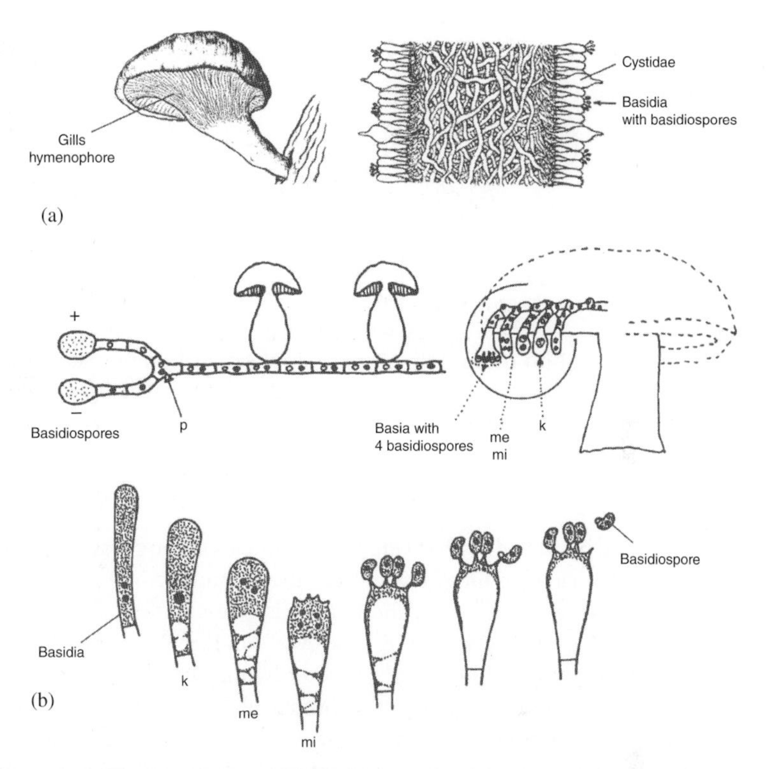

Figure 3.4 Fruiting body of Basidiomycota with a hymenophore (a) and a scheme of the creation of basidiospores on basidia in a hymenphore (b). (a) The capped fruiting body with a leaf-shaped hymenophore and a cross-section with a double-sided layer of hymenia containing basidia with basidiospores. (b) The creation of basidiospores in a hymenophore:

- ***The germination of (+) and (−) type basidiospores*** **forms single-nucleus fungal strands (primary mycelium)**
- ***Plasmogamy (p)*** **lies in the initial contact between two sexually different single-nucleus fungal strands, subsequent merging of their cytoplasms together with the transfer of two separate nuclei into a new cell, and finally in the creation of dual-nuclei fungal strands (secondary mycelium)**
- ***Karyogamy (k)*** **takes place in each basidium in the form of merging two sexually different nuclei, '$n + n$', creating a diploid nucleus '$2n$'**
- ***Meiosis (me) and mitosis (mi)*** **are subsequent processes where diploid nuclei present in the basidia are gradually dividing, creating in each basidia four haploid nuclei**
- ***Basidiospores*** **are created in such a way that four protuberances, called sterigmata, grow from the ends of the basidia, into which haploid nuclei are transferred in the cytoplasm. Four independent cell formations called 'basidiospores' are created exogenously on the top of each basidium and, under the influence of wind, rain or other factors, they separate from the fruiting body and the circle of sexual reproduction of the fungi continues. Basidiospores have a size and shape typical for each individual species of fungi, which is used for their identification**

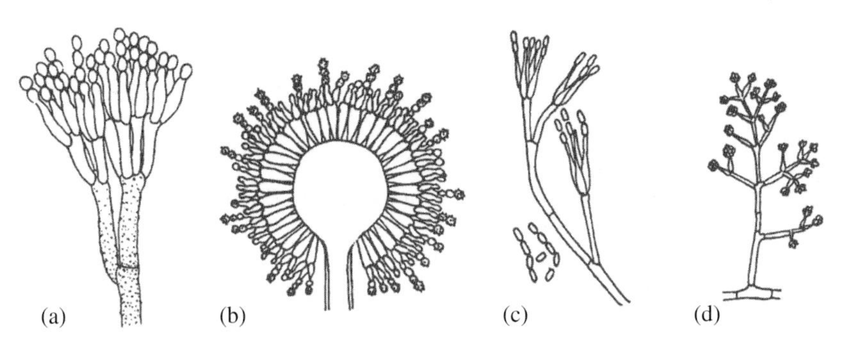

Figure 3.5 The creation of conidia on conidiophores of some moulds: (a) *Penicillium cyclopium*; (b) *Aspergillus niger*; (c) *Paecilomyces variotii*; (d) *Trichoderma viride*

Source: Modified from Fassatiová, O. (1979) *Moulds and Microscopic Fungi in Technical Microbiology* (in Czech). SNTL Praha, Czechoslovakia, 211 p

and branched hyphae (Figures 3.3 and 3.4). The substrate mycelia of decay- and stain-causing fungi grow inside the wood (Sections 3.5 and 3.6). The strength of cell walls of spores and hyphae is given by their composition – they consist mainly of chitin (poly-β-(1,4)-*N*-acetoamido-2-deoxy-D-glucopyranose), and never of cellulose. The surface of hyphae can be encrusted and covered with resins, oils and crystals of calcium oxalate. The inside of hyphae cells contains a protoplast consisting of a nucleus, vacuole, mitochondrion, endoplasmic reticulum, lipid drops and other specific organs. All metabolic processes of fungi are implemented within the protoplast, including the production of extracellular and intracellular enzymes and digestion of soluble sugars, phenols and other products primarily created from wood and transported into the cells of hyphae.

In terms of a *development viewpoint of fungi*, the wood-damaging fungi are classified in the phyla Ascomycota, Basidiomycota and Deuteromycota (Table 3.3), and very few of them in the Zygomycota phylum (e.g. *Mucor* sp. Div). Fungi from these phyla differ in the modes of reproduction; that is, the creation of spores followed by the creation of hyphae, mycelia and fruiting bodies (Figures 3.3, 3.4 and 3.5).

3.2.1.3 Physiology of wood-damaging fungi

The physiology (i.e. life and activity conditions) of wood-damaging fungi is influenced by trophic, abiotic, biological and anthropogenic factors present in an environment of their activity. Optimum physiology conditions needed for biodegradation activity of wood-damaging fungi are not always identical with optimum conditions of their reproduction and development (Rypáček, 1957; Rayner & Boddy, 1988; Schmidt, 2006).

Trophic factors of wood-damaging fungi are related to the chemical composition of wood. Fungi obtain nutrients and energy for growth from wood components. For life they need mainly carbon, nitrogen, hydrogen, oxygen, minerals and other specific substances. Fungi consist of about 90% water and 10% dry

Table 3.3 Classification of wood-damaging fungi from the Fungi kingdom – shortened form (phylum → ...mycota, class → ...mycetes, order → ...ales, family → ...aceae)

Ascomycota
• Saprophytes and parasites on substrates rich in sugars (e.g. on fruit, in sap or in wounds on trees), or also mycorrhizas contacting the roots of trees with which they live in symbiosis. Overall, ~32,270 species are classified into 46 orders, but from the viewpoint of foresters and woodmen, only a few are important; for example, species from the Xylariales order and dangerous soft-rot fungi
Basidiomycota
• Saprophytes, parasites and often also mycorrhizal symbiotic species. From a total of ~22,300 species described to date, several also attack living trees (from the classes Teliomycetes, Ustomycetes and mainly Basidiomycetes), and processed wood (from the class Basidiomycetes)
Deuteromycota
• Saprophytes in soils, organic substrates and fresh and salt water which are usually part of the ontogenesis of fungi from Ascomycota phylum; that is, they are their anamorphs

Source: Data modified from Gáper and Pišút (2003).

mass of chitin and other substances. The hydrogen and oxygen are currently present in the environment, usually in air and water.

Carbon is the basic element of dry wood, which creates ~50% of its mass. Carbon is needed for life of all wood-damaging fungi. It is present in polysaccharides, lignin and accompanying components of wood. The carbon-to-nitrogen ratio in components of wood is high (350 :1 to 1250 : 1). Lignin and polysaccharides are not soluble in water, and thus the wood-damaging fungi have to impair them – decay-causing fungi do so; however, moulds and staining fungi prefer if these components of wood are previously depolymerized by steaming, γ-radiation and chemical or atmospheric corrosion.

Nitrogen is also a prerequisite for life in fungi. Chitin, the structural component of cell walls of fungi, can be synthesized only in the presence of nitrogen. Nitrogen is needed for biosynthesis of proteins in their protoplast and enzymes necessary for wood decomposition. An increased amount of nitrogen in wood above 0.3% positively affects the growth and activity of the majority of fungi. For example, wood-staining fungi primarily colonize parenchyma cells of wood with the highest content of nitrogen, or parasitic fungi attack just the wood of live trees with a permanently increased content of nitrogen.

Mineral substances (inorganic extractives) are inorganic nutrients necessary for fungi; in particular, they are substances that contain phosphorus, sulphur, calcium, magnesium, potassium, iron, zinc and copper. Some brown-rot fungi, such as the true dry-rot fungus *S. lacrymans* and cellar fungus *Coniophora puteana*, need mainly calcium (Palfreyman et al., 1996). Iron, magnesium and copper are required by several brown-rot fungi in order to generate a nonenzymatic system for decomposition of cellulose; for example, the iron in Fenton's $H_2O_2/Fe^{2+}/(COOH)_2$ reagent (Henry, 2003; Messner et al., 2003; Arantes et al., 2012).

Accompanying organic substances (organic extractives) have an inhibitive or supportive effect on fungi. We know more than 10,000 types of extractive substances in various wood species and other plants (Duchesne et al., 1992). Inhibitors of the growth of fungi are, for example, phenols, tannins, flavonoids, pinosylvin, stilbenes and quinones, located mainly in heartwood. Wood with a high proportion of these substances, such as very resistant species (e.g. azobe, black locust, greenheart, ipe, jarrah, teak) or resistant species (e.g. black walnut, cedar, chestnut, iroko, kapur, oak, wallaba) have rather high resistance against wood-decaying fungi and other biological pests. On the other hand, some organic substances act as stimulators of the growth and vitality of wood-damaging fungi. For example, the parasitic brown-rot fungus birch polypore (*Piptoporus betulinus*) attacks just birch wood, or some other wood-damaging fungi need vitamin B1 (thiamine) or vitamin H (biotin).

The acidity of a wood substrate depends upon its chemical composition. The pH value affects the germination of spores, growth of mycelia, enzymatic and degradation ability of fungus and fruit-body formation. Wood-decaying fungi require a more acid environment and they are the most active at a wood of pH 3.5–6.5, with an optimum of pH 4–6. Several of them can also attack wood with critical values of pH from 2.5 to 9 (Thörnqvist et al., 1987). Mould *Aspergillus niger* is also active on a very acid wood with pH 1.5. In contrast, soft-rot fungi also attack alkaline wood, with pH 11. Fungi self-regulate the acidity of wood to the range of their optimum, when they produce acids (acetic, oxalic, citric) or alkalis (ammonia).

Abiotic factors include mainly the amount of water and air in the wood, the temperature of the wood, and the climatic conditions in the proximity of the wood. Fungi are aerobic organisms that decompose organic substrates to CO_2 and water when respiring. Consequently, they produce heat, which also increases the temperature of wood. This, together with the presence of solar radiation, affects their life and the entire process of wood damage.

Air (presence of oxygen O_2) must be not only in the proximity of wood when it rots, but also directly in the wood. Wood-decaying fungi need at least 5–20 vol.% of air in the wood (i.e. ~1–4% oxygen). Growth of fungi is stopped if the volume of air in the wood drops down under the 'critical minimal limit of air volume', which is at least 5–20% of the pore volume. It corresponds to the 'critical maximum moisture of wood' $w_{crit.max}$ over which the activity of fungi is stopped (Bech-Andersen, 1995). This limit can be achieved in the wet storage of logs using technologies such as their ponding in water pools, spraying with water, and so on (Reinprecht, 2008). The value of $w_{crit.max}$ functionally depends upon the porosity of wood or its density (Figure 3.6). It is higher for more porous woods (e.g. $w_{crit.max}$ for spruce is 180–200%) than for denser woods (e.g. $w_{crit.max}$ for beech is 80–90%).

CO_2 is the product of respiration and metabolic processes of fungi. For example, in a healthy stem of maple, the concentration of CO_2 varies around 4.8%,

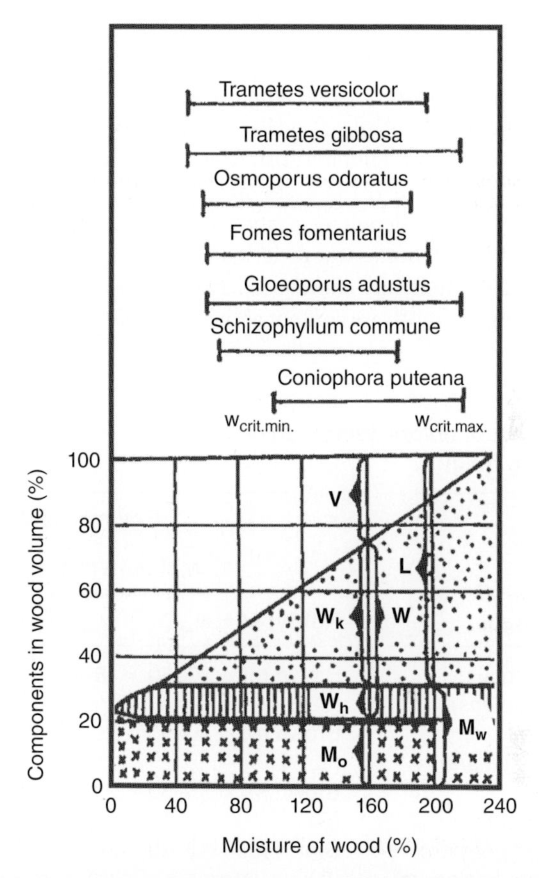

Figure 3.6 Limit moisture conditions in spruce wood with density $\rho_o = 350\ kg/m^3$ for activity of decay-causing fungi, with the specification of $w_{crit.min}$ and $w_{crit.max}$. Note (all values in volume%): M_o, amount of wood in the oven-dry state; M_w, amount of wood containing bounded water; V, air; W_h, bounded water in cell walls; W_k, free water in lumens of cells; W, total water; L, lumens filled with air and/or free water

Source: Schánĕl, L. (1975) Influence of environmental conditions on wood decay by fungi (in Czech, English abstract). *Drevársky Výskum – Wood Research* 20(1), 59–79. Reproduced by permission of Pulp and Paper Research Institute Bratislava

but in the stems decayed by fungi it grows to 17% (Schánĕl, 2005). The increased concentration of CO_2 in wood often affects the speed and mode of fungus growth and enzymatic activity. CO_2 in concentrations up to 10% speeds up the growth of all fungi, in concentrations up to 20% it stimulates the growth of some fungi, while in concentrations of 30% it suppresses the growth of the majority of fungi. This knowledge is used also in the new technologies for protection of logs in plastic bags in a CO_2 atmosphere.

Water (H_2O) is necessary in wood for its attack by fungi since it (1) comprises 90% of the fungus body (hyphae, mycelia, fruit body), (2) is the medium for diffusion of extracellular enzymes from fungus hyphae to components of the wood in its cell walls, and also for the reverse diffusion of glucose and other water-soluble substances created during rotting of the wood by fungus hyphae, and (3) is a chemical substance directly included in the enzymatic hydrolytic depolymerization of cellulose and hemicelluloses in the cell walls of wood. Some wood-decaying fungi are able to attack wood even at its low (20%) moisture content (such as *S. lacrymans*); however, the majority of them require moisture content above 25–30% (Figure 3.6, Table 3.4). For the majority

Table 3.4 Effect of abiotic factors in wood on the activity of selected species of wood-decaying fungi

Species of wood-decaying fungus	Rot type[a]	Environmental conditions in wood				
		Moisture (%)		Temperature (°C)		
		w_{opt}	w_{min}	t_{opt}	$t_{min-max.}$	pH_{opt}
True dry rot fungus (*Serpula lacrymans*)	Brown rot (C, B)	30–60	18–20	18–22	3–26	5–7
Cellar fungus (*Coniophora puteana*)	Brown rot (C, B)	40–90	26	23	3–35	5.7–6.3
Mine polypore (*Antrodia vaillantii*)	Brown rot (C)	35–50	28	27	3–37	7
Yellow–red gill polypore (*Gloeophyllum sepiarium*)	Brown rot (C)	40–60	30	35	5–45	3.8–6
Fir gill polypore (*Gloeophyllum abietinum*)	Brown rot (C)	30–50	25	26–29	5–36	
Scaly lentinus (*Lentinus lepideus*)	Brown rot (C)	35–60	30	27–29	8–37	6
Stalkless paxillus (*Paxillus panuoides*)	Brown rot (C)	50–70	40	23–26	5–29	
Many-zoned polypore (*Trametes versicolor*)	White rot (B)	80	>30	26–29	5–38	6
Oak maze-gill (*Daedalea quercina*)	Brown rot (B)	40–50	>30	23–29	5–38	
Hairy parchment (*Stereum hirsutum*)	White rot (B)		>30	25	3–35	
Common split-gill (*Schizophyllum commune*)	White rot (B, C)	50		30–35	10–44	

Source: R., L. (2008) *Ochrana Dreva* (*Wood Protection*), Handbook, TU Zvolen, Slovakia, 453 p. Reproduced by permission of TU Zvolen.
Selected data from Bech-Andersen (1995), Cartwright and Findlay (1958), Mirič and Willeitner (1984) and Reinprecht (2008).
[a] B: broadleaves; C: coniferous.

of rotting fungi the optimum wood moisture is between 35 and 70%. Moulds grow on wood if the relative humidity of air is more than 80%; that is, when at normal temperatures of ~20 °C the moisture content of wood surfaces is at least 20%.

Temperature affects the enzymatic activity and life of fungi. Wood-decaying fungi are most active at 18–35 °C, staining fungi at 18–29 °C and moulds at 27–37 °C. The temperature optimum for fungus growth may or may not be close to the temperature optimum of its degradation activity. Wood-damaging fungi belong to eurythermic organisms with rather high temperature tolerance, and thus the majority of them are able to live in wider temperature intervals. However, the very dangerous fungus *S. lacrymans* is not amongst these, as it stops growing at a temperature above 26 °C and its substrate mycelium dies at a temperature of 37 °C within 30 min (Bech-Andersen, 1995). Temperature also significantly affects the speed and mechanism of decomposition of polysaccharides by this fungus (Mička & Reinprecht, 1999). Generally, the most resistant against high or low temperatures are spores of fungi, then surface mycelia (rhizomorphs, sclerotia); in contrast, the least resistant is substrate mycelium. A variation in temperature is often necessary for the fructification processes; that is, the transitional cooling or heating of the wood.

The majority of wood-damaging fungi have the following limit temperatures:

- The optimum growth temperature is within the range 18–35 °C (Table 3.4);
- The critical minimum growth temperature, at which the fungus temporarily stops growing, is usually within the range of −2 to +3 °C. In addition to water, there are also other substances in fungus hyphae (e.g. glycerol) that prevent its freezing below 0 °C (Jennings & Lysek, 1999); this enables some wood-staining fungi and moulds to grow even at −8 °C.
- The lethal minimum temperature, under which the fungus dies, is notably lower. For mycelia it is within the range −5 to −10 °C/24 h, and for spores with better thermal stability it can be even lower; for example, for spores of fungus *S. lacrymans* it is −70 °C/24 h.
- The critical maximum growth temperature, at which mycelium stops its growth, is specific for the particular fungus species and it usually varies within the interval 35–45 °C. Exceptions include the mycelium of decay-causing fungus *S. lacrymans*, which stops growing at 26 °C, whereas decay-causing fungus *Gloeophyllum sepiarium* and mould *A. niger* stop growth at temperatures above 45 °C only.
- The lethal maximum temperature, above which fungi die, is specific to the species. For example, it is 50–55 °C lasting 1 h for *S. lacrymans* mycelium, but 70–80 °C lasting 1 h for *G. sepiarium* mycelium. However, the spores of *S. lacrymans* with greater thermal stability are eradicated only after a prolonged time of 32 h at 60 °C or at substantially higher temperature of 100 °C for 1 h (Hegarty et al., 1986).

Solar radiation affects the growth and activity of fungi in a specific way. The UV components of this radiation are tolerated better by those fungi that contain darker pigments; for example, by black mould *A. niger*. In contrast, indirect UV radiation (i.e. the radiation with changed wavelengths acting in a shadow) and the visible components of solar radiation usually support the degradation activity of fungi and have a positive impact also on the production of their fruiting bodies. The fructification process does not usually occur in absolute darkness, or there only sterile fruiting bodies are created, which is typical for brown-rot fungi *G. sepiarium* and *Lentinus lepideus* on mine timber. The position and orientation of fruiting bodies often depends upon the direction of incoming light (Jennings & Lysek, 1999), while it is always affected also by gravitation (Kern & Hock, 1996).

Biological factors include the presence and activity of other organisms in wood and in the surroundings of the wood with antagonistic or synergetic impacts. Organisms having antagonistic impacts slow down or stop the growth of wood-damaging fungi. Antagonistic impacts are characteristic for several microscopic fungi (e.g. *Trichoderma viride*), lichens (e.g. *Usnea hirta*) and bacteria (e.g. *B. asterosporus*). These organisms produce antibiotics, mycotoxins or other antifungal substances (Tichý, 1975; Rayner & Boddy, 1988; Score et al., 1998) that are effective as biological modifying agents for wood protection (see Section 6.4.1). In contrast, synergic impacts are bilaterally favourable for the life of both the wood-damaging fungus and the other organism. Some bacteria fix atmospheric nitrogen necessary for the life of fungi. Synergic impacts also exist among various moulds acting on wooden sculptures (Šimonovičová et al., 2005). Similarly, the primary rot of wood caused by one species of decay-causing fungus can accelerate its rotting with other fungus species (Reinprecht & Tiralová, 2001).

Anthropogenic factors are interventions by humans to wood (e.g. its chemical protection) and to the surrounding environment (e.g. air conditioning). By using the correct regulations for storage and exposure conditions of wood (temperature, humidity, pH value, etc.) one is able to prevent its rotting and moulding. If environmental regulations using structural principles of wood protection are limited or impossible (Chapter 4), then it will be necessary to also use a chemical or modifying approach to protect the wood with fungicides or some modification process (Chapters 5 and 6).

3.2.2 Wood-decaying fungi

Wood-decaying fungi transform the wood mass (polysaccharides, lignin, extractives) to water and CO_2 and simultaneously synthetize various organic substances needed for their life and growth: (1) enzymes – catalysts of protein character necessary for decomposition of the lignin–polysaccharide matrix of wood; (2) chitin – polysaccharide creating the cell walls of mycelia and fruiting bodies of fungi; (3) mycotoxins – for example, aflatoxin and ochratoxin to suppress the growth of other organisms, and which can cause allergic reactions or even intoxication of humans; and (4) compounds of cytoplasm (e.g. alkaloids and fats).

Wood-decaying fungi degrade cell walls of wood and cause it to rot:

- brown rot (by brown-rot Basidiomycota fungi);
- white rot (by white-rot Basidiomycota fungi, and less often also by white-rot Ascomycota fungi);
- soft rot (by soil Ascomycota and some Deuteromycota fungi).

Mechanisms of brown, white and soft rot of wood derive mainly from possibility of fungi to produce enzymes and low-molecular weight catalytic systems needed for degradation of wood. During brown rot the fungi attack hemicelluloses and cellulose, but lignin only at a minimum rate (Box 3.1, Figure 3.7).

Box 3.1 Basic data on the brown rot of wood

Types of fungi

- Brown-rot Basidiomycota.

Degradation of wooden components

- Intensively – cellulose and hemicelluloses.
- Minimally – lignin.

Principle of brown rot

- The brown-rot fungi produce hydrolase enzymes – hydrolases (i.e. *endo*-1,4-β-glucanases or more seldom the synergic system of *endo-exo*-1,4-β-glucanases, furthermore 1,4-β-glucosidase, *endo*-1,4-β-xylanases, 1,4-β-D-mannanases and others) – by means of which they depolymerize by hydrolysis mainly the amorphous polysaccharides of wood.
- Crystalline cellulose is degraded in particular thanks to the activity of the low molecular weight Fenton oxidation system consisting of hydrogen peroxide, iron ions and oxalic acid [$H_2O_2/Fe^{2+}/(COOH)_2$]. The Fenton system – and also some others containing Cu^{2+}, Ni^{2+}, quinone-type chelators, and so on – rather easily penetrates from the fungus hyphae to the inside parts of elementary fibrils of cellulose and induces radical splitting of its pyrane circles.
- A significant reduction of polymerization degree of polysaccharides as well as strength of rotten wood is related to both enzymatic hydrolytic and nonenzymatic oxidation degradation processes upcoming even in the initial phases of rot.

Structure of rotten wood

- Wood with brown-rot is typically changed from the molecular up to the geometric structural levels. Substrate hyphae grow in the lumens of cells even via cellular walls. The S_2 layer with a high proportion of cellulose and hemicelluloses is degraded in the most intense way, while it totally disappears at the advanced stages of rot. Cell walls become fragile; they shrink and crack. Despite this they remain coherent up to the high stages of rot, since their strongly lignified middle lamella and primary wall do not decompose. Wood gradually becomes yellow and brown in colour. It becomes more fragile and less strong, cracks into cubes and has also lower adsorbing capacity for bounded water and other polar liquids.

Source: R., L. (2013) *Wood Protection*, Handbook, TU Zvolen, Slovakia, 134 p. Reproduced by permission of TU Zvolen.

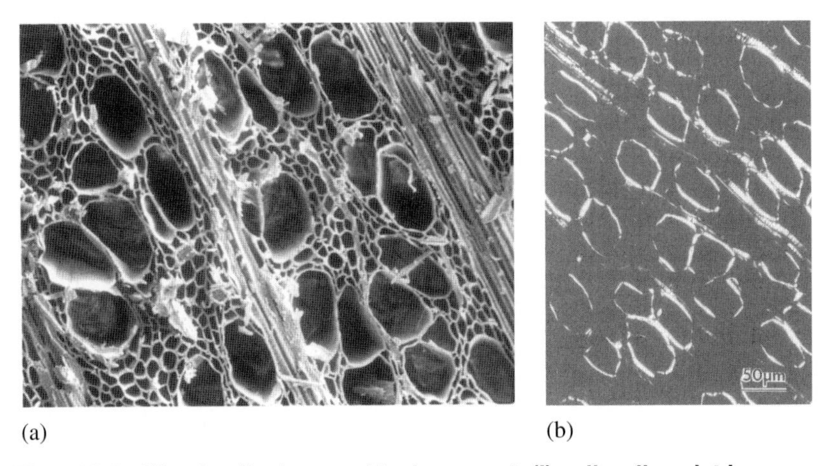

(a) (b)

Figure 3.7 Wood cells damaged by brown rot: (i) cell walls exist in a very high degree of decay, and they are significantly thinned (a, b); (ii) crystalline cellulose is totally broken down in libriform elements, while it is retained in vessels and parenchyma cells (b). (Fig. 3.7a) Scanning electron micrograph showing 63.8% weight loss in beech wood *Fagus sylvatica* attacked with fungus *Serpula lacrymans*. (Fig. 3.7b) Polarized-light microscopy image showing 68% weight loss of *Liquidambar styraciflua* wood attacked with fungus *Poria placenta*

Source: (a) Reinprecht, L. and Lehárová, J. (1997) Microscopic analyses of woods – beech (*Fagus sylvatica* L.), fir (*Abies alba* Mill.) and spruce (*Picea abies* L. Karst.) in various stages of rot caused by fungi *Serpula lacrymans, Coriolus versicolor* and *Schizophyllum commune* (in Slovak, English abstract). In: *Drevoznehodnocujúce huby '97 Wood-damaging fungi 97*, pp. 91–113. TU Zvolen, Slovakia. Reproduced by permission of TU Zvolen

Source: (b) Wilcox, W. W. (1968) Changes in wood microstructure through progressive stages of decay. US For Ser Res Pap FPL-70, USA, 46 p

In white rot the fungi usually completely destroy all components of wooden cell walls (i.e. lignin, hemicelluloses and cellulose – Box 3.2, Figure 3.8). Soft rot is typically caused by ascomycetes in a wet environment with their ability to destroy all structural polymers of wood, but usually only in the S_2 layer (Box 3.3, Figure 3.9). In addition to ascomycetes, soft rot can also be caused by several asexual forms from Deuteromycota (Fungi imperfecti), such as *Allescheria* sp., *Graphium* sp., *Humicola* sp., *Chaetomium* sp., *Paecilomyces* sp. and around another 120–300 species. Mechanisms of wood rot on its molecular level are disassmbled more detaily in Section 3.5.

Box 3.2 Basic data on the white rot of wood

Types of fungi

- White-rot Basidiomycota and Ascomycota.

Degradation of wooden components

- Intensively – lignin, cellulose and hemicelluloses.

Principle of white rot

- The white-rot fungi degrade lignin and polysaccharides by:
  - erosion form (i.e. concurrently);
  - delignification form (i.e. gradually, first lignin and then polysaccharides).
- Polysaccharides are decomposed enzymatically via hydrolase enzymes, usually also with the support of some oxidation and oxidation–reduction enzymes.
- Crystalline cellulose is decomposed by means of synergic system of *endo-exo*-glucanases, as these fungi, in contrast to brown-rot fungi, do not have the nonenzymatic Fenton system of $[H_2O_2/Fe^{2+}/(COOH)_2]$.
- The activity of oxidation and oxidation–reduction enzymes in the decomposition of polysaccharides is conditioned by the presence of specific substances created upon the biodegradation of lignin. For example, the activity of cellobiose:quinone oxidoreductase enzyme, by means of which cellobiose is converted to cellobiono-lactone, depends upon the presence of quinones created from lignin.
- The polymerization level of cellulose due to the impact of white-rot fungi decreases substantially less than in the case of brown-rot fungi, since the crystalline cellulose is depolymerized more slowly in the case of only purely enzymatic forms of its decomposition.
- White-rot fungi decompose lignin by means of an oxidizing and ring-splitting enzymatic system, while Fenton and similar reagents can also be involved in lignin degradation. The enzymatic system for lignin decomposition comprises lignin-peroxidases, Mn^{2+}

peroxidases, laccases, arylalcohol-oxidases, methanol-oxidases, dioxygenases and some other enzymes. During lignin biodecomposition, phenyl-propane, quinone and other units are created, methoxyl groups are removed from phenyl units, aromatic rings of phenyl units are cleaved, various intermediate products are oxidized, and so on.

Structure of rotten wood

- Wood with white rot is characterized with a typical changes from the molecular up to geometric structural levels. Substrate hyphae promptly colonize the wood. They primarily penetrate parenchyma cells and subsequently or concurrently all the types of wood cells, while disrupting cell walls in the form of erosion or delignification (Figure 3.8):
  - Erosion form of white rot is characterized by simultaneous concurrent degradation of all structural components of the cell walls of wood in the direction from the S_3 layer to the middle lamella. First, hollows are created in cell walls; subsequently, by disappearance of individual cells. The decomposition of wood is often irregular; for example, one type of cell can remain undamaged while others disappear totally from wood (e.g. by *Trametes versicolor*).
  - The delignification form of white rot is characterized by sequential degradation of wood (i.e. primarily of lignin together with small amounts of polysaccharides), while the main portion of cellulose and the remnants of hemicelluloses are degraded only during the second stage of rot. Delignification starts in particular in middle lamella and only then does it take place in the other layers of the cell wall, which leads to the mutual isolation of wood cells (e.g. by *Hymenochaete rubiginosum*). Delignification sometimes concentrates only to the local zones of wood, whereby 'honeycomb rot' take pace (e.g. by *Trichaptum abietinum*).

Source: R., L. (2013) *Wood Protection*, Handbook, TU Zvolen, Slovakia, 134 p. Reproduced by permission of TU Zvolen.

3.2.2.1 Wood-decaying fungi in exteriors

Stocks of logs and wooden products subject to outdoor exposure can be attacked by several wood-decaying fungi, usually by saprophytic species. Moist mine wood, sleepers, utility poles, bridges, fences, pergolas, cladding, garden furniture, log cabins, windows, external doors and other products in the outdoor environment are often damaged by them (Figures 3.10 and 3.11). Some of these fungi also attack wood in interiors (e.g. in cellars, ceilings and trusses) if favourable environmental conditions exist for them to live.

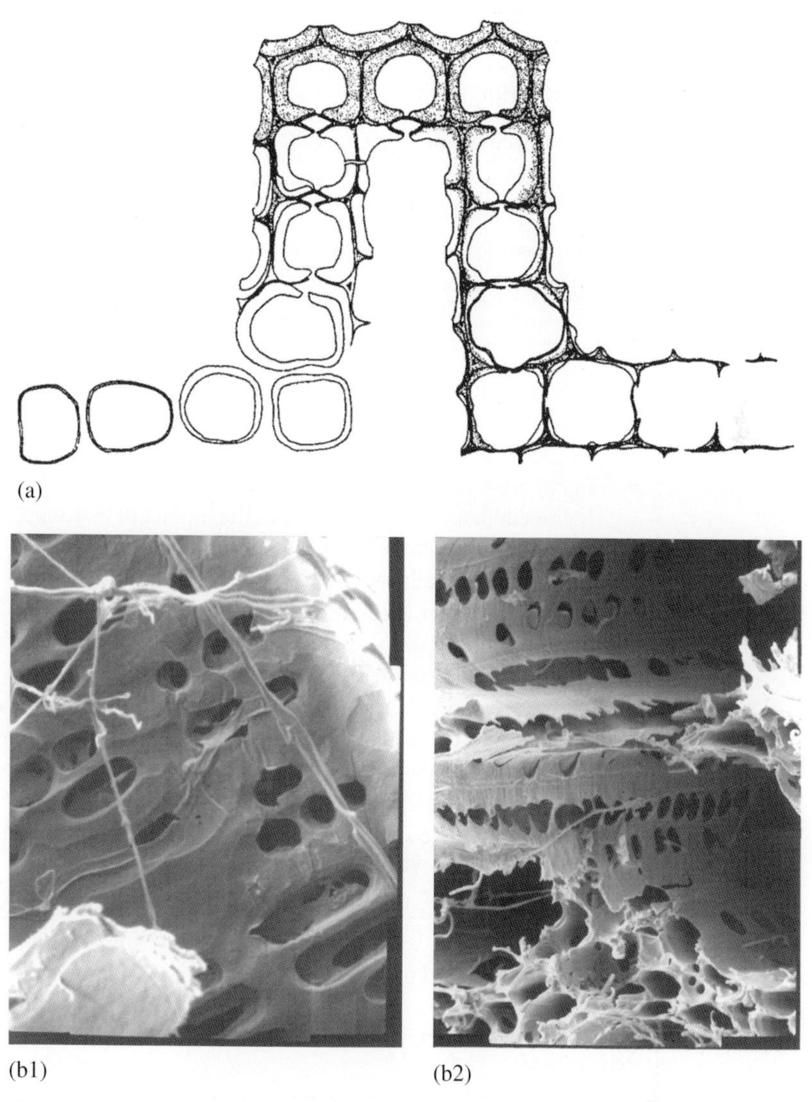

Figure 3.8 Wood cells damaged by white rot. (a) Scheme of cell wall delignification (left) and erosion (right). (b) Erosion of cell walls of beech wood *F. sylvatica* due to fungus *Trametes versicolor,* shown for its weight losses 20.6% (b-1) and 41.8% (b-2)

Source: (a) Eriksson, K-E., Blanchette, R. A. and Ander, P. (1990) Reproduced by permission of Springer

Source: (b) Reinprecht, L. and Lehárová, J. (1997) Reproduced by permission of TU Zvolen

Box 3.3 Basic data on the soft rot of wood

Types of fungi

- Ascomycota, and their asexual forms from Deuteromycota.

Degradation of wooden components

- Polysaccharides (in particular cellulose) and lignin, usually in the S_2 layer.

Principle of soft rot

- Soft-rot fungi attack wood of coniferous and broadleaves species, particularly on or below the soil surface and in wet conditions, such as sleepers, poles, fences, cooling towers, and so on.
- In the case of soft rot, all polymers of wood are degraded (cellulose, hemicelluloses and lignin) due to the impact of hydrolase enzymes and oxidation–reduction ligninolytic enzymes.
- Soft rot is typically localized in the S_2 layer of wood cell walls, where micro-hyphae of diameter of <5 μm create L-bends or T-branching, and decompose cellulose in particular around them to create longitudinally aligned cavities.
- Wood treated with fungicides resists soft rot not always enough, because the soft-rot fungi are still able to spread in the wood by micro-hyphae growing within cell walls and thus avoid contact with creosote oil or other fungicides in the lumens of wood cells.

Structure of rotten wood

- Soft rot is connected with damage of the S_2 layer. However, in woods of broadleaves, sometimes even entire secondary walls of cells are degraded. Wood gradually becomes brown, is fragile and less strong.

Source: R., L. (2013) *Wood Protection*, Handbook, TU Zvolen, Slovakia, 134 p. Reproduced by permission of TU Zvolen.

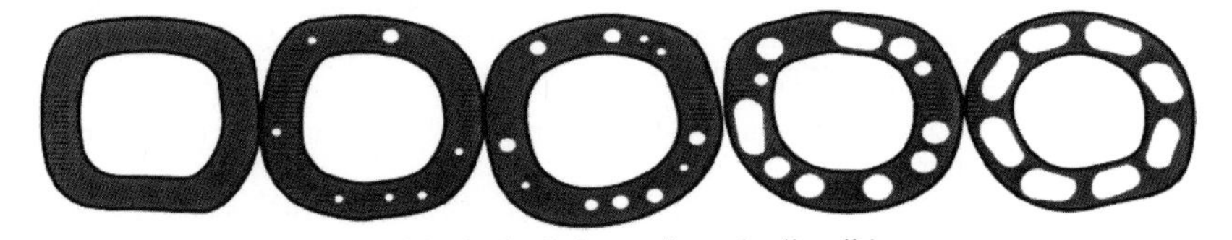

(Gradual formation of cavities in the S_2 layer of wood cell walls)

→

Figure 3.9 Wood cells damaged by soft rot

Source: R., L. (2013) *Wood Protection*, Handbook, TU Zvolen, Slovakia, 134 p. Reproduced by permission of TU Zvolen

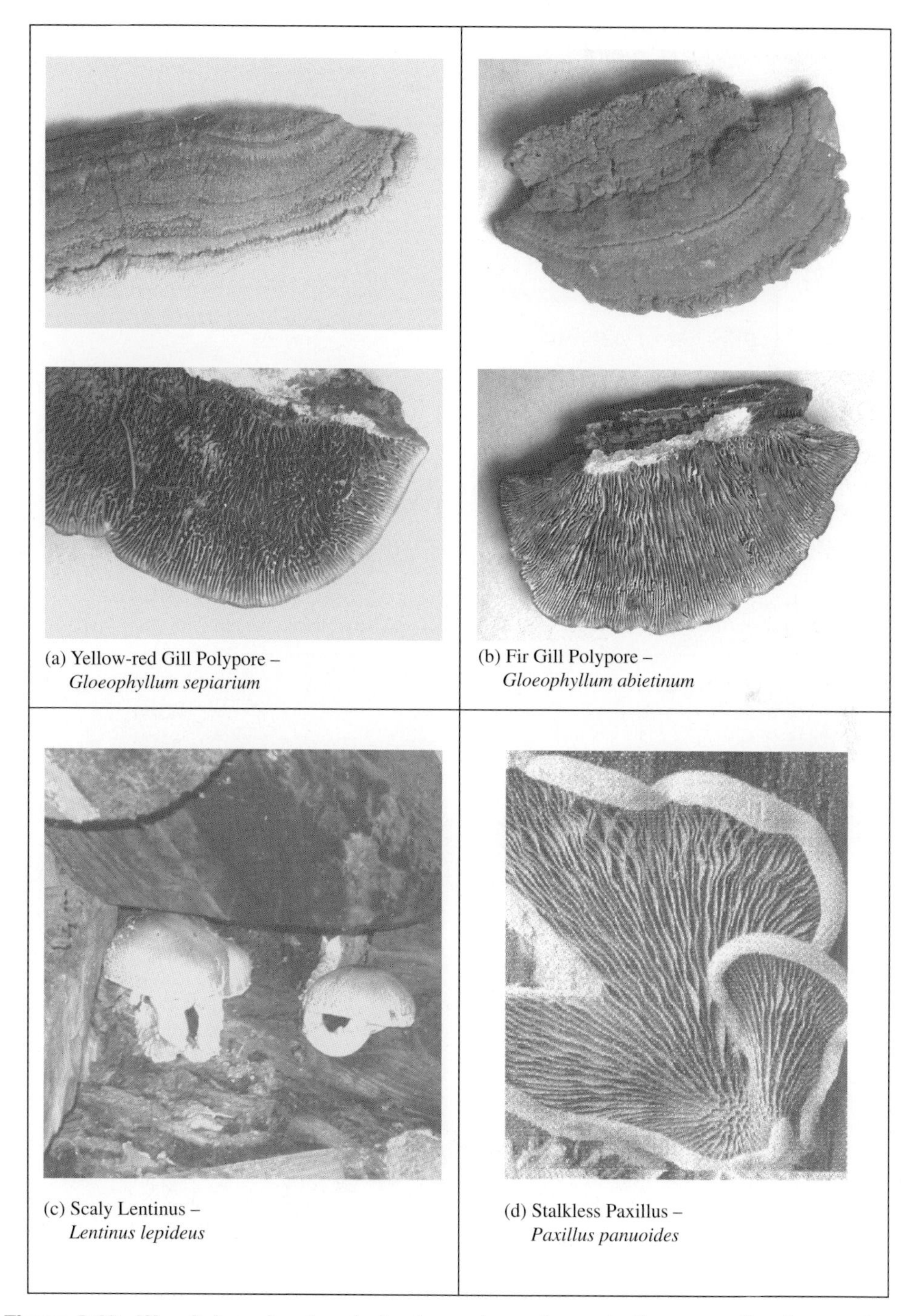

(a) Yellow-red Gill Polypore – *Gloeophyllum sepiarium*

(b) Fir Gill Polypore – *Gloeophyllum abietinum*

(c) Scaly Lentinus – *Lentinus lepideus*

(d) Stalkless Paxillus – *Paxillus panuoides*

Figure 3.10 Wood-decaying fungi of primary importance in Europe acting in exteriors on conifers – softwoods

Source: R., L. (2013) *Wood Protection*, Handbook, TU Zvolen, Slovakia, 134 p. Reproduced by permission of TU Zvolen

Figure 3.11 Wood-decaying fungi of primary importance in Europe acting in exteriors on broadleaves – hardwoods

Source: R., L. (2013) *Wood Protection*, Handbook, TU Zvolen, Slovakia, 134 p. Reproduced by permission of TU Zvolen

3.2.2.2 Wood-decaying fungi in interiors

Wood-decaying fungi acting in buildings are saprophytes. They cause significant technical and economic damage on built-in structural wood, but also on wallpapers and other lignin-polysaccharide materials. For example, in Great Britain they caused damage valued to ~£3 million every week in 1977 (Rayner & Boddy, 1988). Mainly brown-rot species act in interiors. They act in particular in badly designed or older constructions with the presence of overflow rainwater, water condensate or sources of capillary water. These are buildings with poor insulation and ventilation, mostly cellars, floors, ceilings or trusses, but also wooden windows and doors. Some of these fungi are also capable of acting in an external environment.

In Europe, ~50–85% of rot in wood exposed in buildings is caused by three house wood-decaying fungi: *Serpula lacrymans*, *Coniophora puteana* and *Antrodia vaillantii* (Figure 3.12). *S. lacrymans* is active mainly in the northern and central parts of Europe, but also in Russia, Japan and North America. For example, in Poland, during a survey of 3050 buildings, wood rot was caused in 53.8% cases by *S. lacrymans*, in 22.4% by *C. puteana*, and in 11.3% by *A. vaillantii* (Wažny & Czajnik, 1963). In Belgium, when assessing 749 rotted wooden interior elements, *S. lacrymans* was identified in 59.4% of cases, *C. puteana* or *Coniophora marmorata* in 10.1%, and *A. vaillantii* with other mine fungi in 2.3% of cases (Schmidt, 2006). In the northern parts of Germany, 36 species of Basidiomycota fungi were identified on wooden elements in 63 buildings, of which *S. lacrymans* was found in 57 buildings, *C. puteana* in seven buildings, and *A. vaillantii* in six buildings (Huckfeldt & Schmidt, 2005).

- True dry rot fungus (*Serpula lacrymans* (Wulfen) J. Schröt., Syn: *Merulius lacrymans*).
  - *Activity:* It is brown-rot fungus that breaks up wood into relatively large cubes with length of ~50 mm. Rotten wood is grey–yellow to grey–orange at first. Then, both longitudinal and crosswise cracks are created in its structure, and it becomes dark brown in colour in the later stages of rot. Finally, it changes to a dark brown dust consisting just of lignin. *S. lacrymans* intensively attacks the softwoods and less durable hardwoods. Hardwoods containing tannins (e.g. oak, acacia or walnut) are rather resistant to it. In addition to the wooden elements in constructions (beams in ceilings and trusses, floor boards, parquets, wall plates, window frames, furniture, mining wood, etc.), it also decomposes other materials containing polysaccharides (paper – often books or wallpapers, cardboards, particleboards, plywood, textile, etc.). It is also able to grow in buildings through inorganic materials by means of special strands of grey mycelia 'rhizomorphs' up to 6 mm in diameter (e.g. through brick and stone masonry or plaster), even to distances of several metres from its place of origin, spreading in the direction of healthy wood. Rhizomorphs of this fungus survive under dry conditions even for several years, and upon repeated increase in wood humidity they

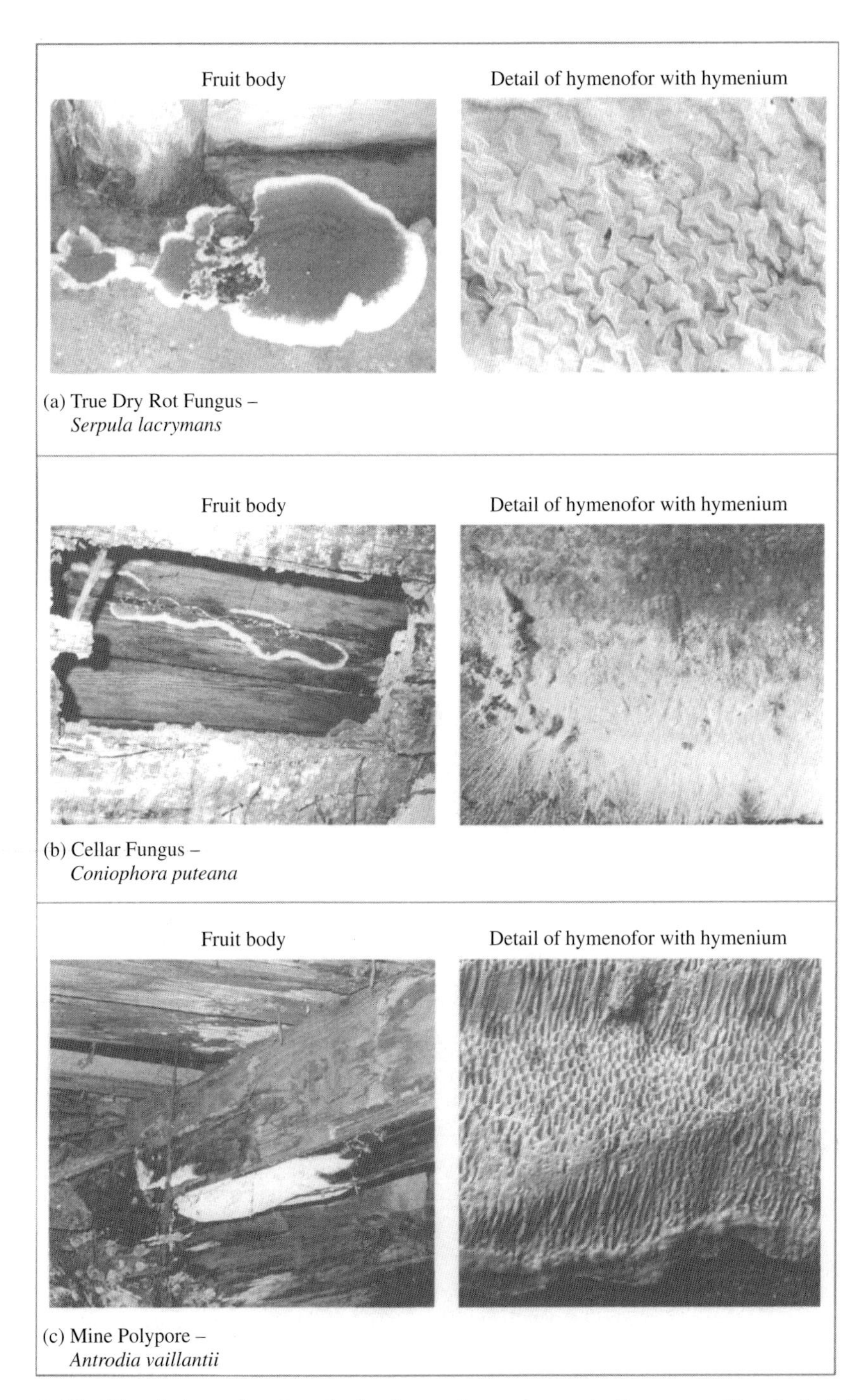

Figure 3.12 Wood-decaying fungi of primary importance in Europe acting in buildings

Source: R., L. (2013) *Wood Protection*, Handbook, TU Zvolen, Slovakia, 134 p. Reproduced by permission of TU Zvolen

will activate again. *S. lacrymans* is considered to be the most dangerous house wood-decaying fungus worldwide (Bech-Andersen, 1995). It is able to start development and activity at rather low wood moisture ($w_{crit.min}$ 18–20%). It more evidently attacks mainly wood near to sources of calcium; for example, the beams next to brickwork and plaster. In the first stages of rot it decomposes polysaccharides of wood to molecules of water, whereby it wets the wooden substrate up to the range of its moisture optimum (w_{opt} 30–60%), or sometimes even 140% (Huckfeldt & Schmidt, 2005). It is able to continue degradation activity even after the removal of the original source of humidity (e.g. after the repair of a damaged water pipe, after the repair or replacement of roofing – Section 7.2), particularly in the case of badly ventilated objects. This is thanks to its ability to transport water to the wood using rhizomorphs even from greater distances and also thanks to the intense generation of drops of water by which it wets healthy wood. *S. lacrymans* flourishes best at rather lower temperatures, within the interval of 3 to 26 °C, and optimally at 18–22 °C. Thermal sterilization of its substrate mycelia in wood can be achieved by heating to 60 °C for 1 h (Terebesyová et al., 2010); however, its spores are more thermally stable.

- *Fruiting bodies:* These are oval and flat or crusty, being 50–300 mm in diameter and of thickness up to 12–20 mm, often assembled into groups with length of more than 1 m (Figure 3.12a). They grow on wood as well as other materials, such as masonry. At first, fruiting bodies appear to be soft, cotton-like white–yellow pillows, and then a papular hymenophore is gradually created on their surface. Then, they are yellow–orange, orange–brown, rusty-brown and finally dark brown, usually covered with water drops in the form of tears. The edges of fruiting bodies having a width of 5–20 mm are constantly yellow–white. They grow in particular in dark, unlit spaces, usually in summer and autumn. The typical accompanying feature of this fungus is the characteristic smell even before the creation of fruiting bodies, which can be registered by trained dogs known as *rothounds* (Hutton, 1994).
- *Spores:* Basidia in the fungus hymenium have dimensions of 7–9 μm × 30–40 μm. Spores created in basidia (four in one basidium) are elliptical, bean-shaped, smooth, yellow–rusty to yellow–brown, non-amyloid, with size 9–12 μm × 4.5–6 μm.
- *Mycelium: S. lacrymans* has a rich surface mycelium in the form of white cotton-like coatings. It grows to greater distances with the help of rhizomorphs, consisting of fine fibres, thick-walled fibres and so-called vessels. Its vessels, of diameter of 5–60 μm, provide for the transport of water from places of immediate activity of the fungus to places of its future activity; for example, to healthy wood. Older mycelium is grey–white to grey–brown and fragile after drying.
- *Solution for rotten wood:* Wood attacked by mycelia of *S. lacrymans*, but also outwardly healthy wood at a distance of at least 1–1.5 m from visually rotten wood, must be removed from the object and

incinerated. This requirement is implied by the great stability and vitality of the spores and rhizomorphs of this fungus. Spores can survive under dry conditions even for 10 years and mycelia 1 year at 20 °C or 8 years at 7.5 °C (Chapter 7).

3.2.3 Wood-staining fungi and moulds

3.2.3.1 Wood-staining fungi

Wood staining fungi attack freshly felled timber and wet wood with a moisture content of 30–130%. Their spores have just a short germination period. Their mycelia grow at a temperature between −3 °C and 28–40 °C, and optimally within the range 18–29 °C. These fungi penetrate into wood in particular via the system of parenchyma cells. They attack mostly just sapwood and live from reserve substances stored in wood cells, such as sugars, proteins, fats, and so on. Hyphae of staining fungi live in the lumens of wood cells and they penetrate from one cell to another cell of wood through pits in cell walls. Micro-hyphae (having a diameter ~10 times smaller than the diameter of hyphae growing in lumens) are able to break through the cell walls of tracheids or other morphological elements of wood without their enzymatic decomposition (Figure 3.13) by means only of high pressure created at their top (apex). These fungi do not change wood strength significantly; neither do they impact strength in bending – a very sensitive indicator of damage in wood structure, which can be decreased about 10–15% at the most. However, in addition to colour changes of attacked wood, these fungi significantly increase the kinetics of water absorptivity and permeability (Fojutowski, 2005). The typical coloration of wood in

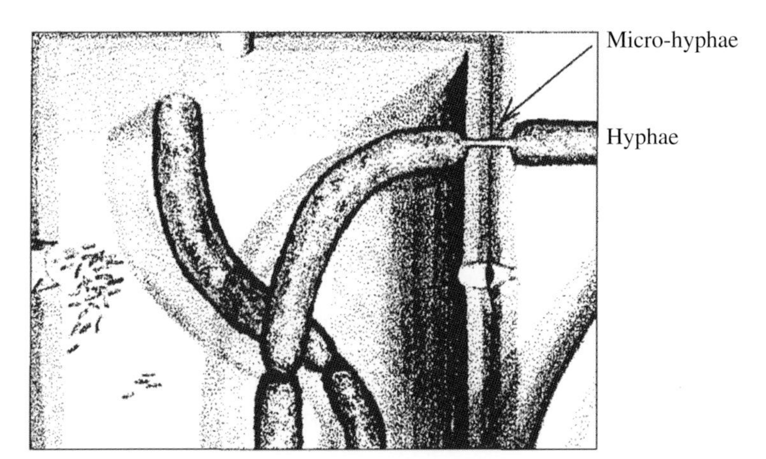

Figure 3.13 Micro-hypha of the wood-staining fungus can impair even the cell wall at the apex by mechanical pressure, and then it is able to grow into the adjacent wood cell

Source: Shigo, A. L. (1979) *Tree Decay – An Expanded Concept*. Agricultural Information 419, 73 p

Table 3.5 Significant species of wood-staining fungi on fresh logs and wet wooden products, with their typical colorations

Coloration	Fungus species	Comment
Blue	*Ceratostomella pilifera* Fries.	The name used for several fungus species causing a typical blueing of softwood
	Ophiostoma coeruleum (Münch.) Syd.	This intensively colours the sapwood of pine in a blue colour and less intensively also spruce wood
	Ophiostoma exiguum (Hedg.)	This colours pine and some other species, such as oak or ash, in a blue–black colour
	Phoma petersii B. et C.	This colours the sapwood of pine in a blue–grey colour
Grey	*Ophiostoma castanea* (Van. et Sol.)	This colours oak and chestnut wood in a blue–grey colour
	Ophiostoma quercus Georgiev	This colours wood of oaks in a dark grey colour
Brown	*Graphium album* (Corda) Sacc.	This colours hardwood, in particular beech, in a brown colour
Red	*Fusarium sambucinum* Fuck.	This colours pine, as well as other species, in a red colour
	Fusarium reticulatum Mont.	This creates dark red stripes in the wood of younger living hardwood species (e.g. maple, ash)
	Penicillium roseum Link.	This colours softwood in a red colour.
Yellow	*Eidamia catenulata* H. Will.	This colours oak wood in a yellow colour.
Violet	*Fusarium javanicum* Koord.	This colours the wood of several poplars in rose to violet shades
Green	*Hormonema dematioides* Lag. et Mel.	This colours softwood in a green to grey–green colour
Black	*Aureobasidium pullulans* (de Bary) Arnaud	This creates blue–black patches on coating and in wet wood to a depth of 1–2.5 mm (via rays)
	Daldinia concentrica De Not.	This creates black–purple lines and stripes in hardwood (birch, alder, ash)
	Stysanus stemonites (Pers.) Corda	This creates a surface black spottiness on processed beech

Source: R., L. (2008) *Ochrana Dreva* (*Wood Protection*), Handbook, TU Zvolen, Slovakia, 453 p. Reproduced by permission of TU Zvolen.

central Europe is blue (e.g. by *Ceratostomella pilifera*); however, other types of coloration are frequent as well (Table 3.5).

3.2.3.2 Moulds

Moulds are microscopic fibrous fungi that reproduce just asexually, in particular by conidia spores on conidiophores (Fassatiová, 1979). They grow on damp surfaces of various materials (wood, wooden composites, paper, textile, skin, plaster, masonry, etc.); whereas they deteriorate their aesthetic appearance, they usually do not reduce their strength. Moulds negatively affect the environment and also some technological operations of wood processing; for example, they prolong the time of natural timber drying. In general, they worsen so called 'breathing of wood – leakage of water from wood', and wood with higher moisture and covered with moulds is also more easily attacked by other biological pests. However, an exception is the cases of antagonistic relations between a specific mould species (e.g. *Trichoderma viride*) and other more dangerous pests (e.g. wood-decaying fungi), when the substances produced by mould (e.g. viridin) prevent the activity of other pests in wood (Ananthapadmanabha et al.; 1992; Phillips-Laing et al.; 2003).

Moulds grow on moist surfaces, optimally at high relative humidity of air (85–99%), and at rather higher temperatures of 27–37 °C (Viitanen & Ritschkoff, 1991). However, some moulds are also able to live at the temperature below 0 °C (e.g. *Alternaria alternata*). They easily grow on wood with increased contents of starch, and also on steamed or boiled logs and steamed furniture-making assortments having increased amounts of oligo- and monosaccharides created from hemicelluloses. Several mould species act on surfaces of wooden products, in particular those belonging to the genera *Alternaria*, *Aspergillus*, *Cladosporium*, *Fusarium*, *Gliocladium*, *Mucor*, *Paecilomyces*, *Penicillium* and *Trichoderma* (Table 3.6).

Table 3.6 Significant species of moulds active on the surfaces of wood and wooden composites

Alternaria alternata (Fr.) Keissl.	Brown or darker spots.
Aspergillus amstelodami (L. Mangin.) Thom. et Church	Yellow–green lower growths
Aspergillus niger Tiegh.	White cotton-like mycelium. Subsequently, upon the ripening of conidiae on conidiophores, the entire surface of mycelium becomes dark brown to black.
Paecilomyces variotii Bainier	Yellow–brown to dark olive–brown coatings, having filaceous and powder consistency
Penicillium brevicompactum Dierckx	Yellow–green velvet growths
Penicillium cyclopium Westling	Blue–green grainy coatings with white edge
Trichoderma viride Pers.	Green-mat growths with coconut smell. It produces the antibiotic gliotoxin (viridin), which is efficient against the growth of other wood-damaging fungi

Source: R., L. (2008) *Ochrana Dreva* (*Wood Protection*), Handbook, TU Zvolen, Slovakia, 453 p. Reproduced by permission of TU Zvolen.

The ecological risk of moulds is in their ability to produce ~100 various mycotoxins (Betina, 1990). The most dangerous one is the carcinogenic aflatoxin produced by *Aspergillus niger* and *Aspergillus flavus*. Other mycotoxins (e.g. zearalenon, gliotoxin, T-2 toxin, patulin) are dangerous as well, causing for example, skin allergies, damage to the respiratory ways or reduction of overall immunity of human and other warm-blooded organisms. Benko (1992) identified 24 mould species on wood with pathogenic effects on humans.

3.3 Wood damaged by insects

Wood-damaging insects attack live and dead trees, round wood, sawn timber and various products made of wood. They are usually classed into three groups based on their physiology:

- *primary physiological pests* – these act on living trees, where they nibble at leaves or needles; they attack bark, phloem and cambium, or even roots; a tree gets weaker and produces less wood;
- *secondary physiological pests* – these generate feeding marks in the wood of weakened or diseased trees, whereby they immediately reduce the quality of wood raw material;
- *technical pests* – these generate galleries (feeding marks) in logs, sawn timber and wooden products.

3.3.1 Reproduction, classification and physiology of wood-damaging insects

3.3.1.1 Reproduction of wood-damaging insects

Insects usually reproduce via eggs, which are laid either fertilized or unfertilized. Also known are species in which the females give birth to living individuals (vivipary).

The ontogenesis of insects (Figure 3.14) has two development stages (Hůrka & Čepická, 1978):

- Embryonic in an egg; that is, the creation of an embryo and the creation of the organs of a future larva.
- Post-embryonic from a larva to an adult (imago). In a complete metamorphosis, 'holometabolism', the larva transforms into an adult through the pupal inter-stage; in an incomplete metamorphosis, 'hemimetabolism', the larva (nymph) transforms into an adult with wings and genitals without the pupal inter-stage.

Eggs are laid by females in varying numbers depending upon the species and physiology of the insect. Some females only lay several dozens or hundreds of eggs (e.g. *Ips typographus* and *Hylotrupes bajulus*), whilst others, over several

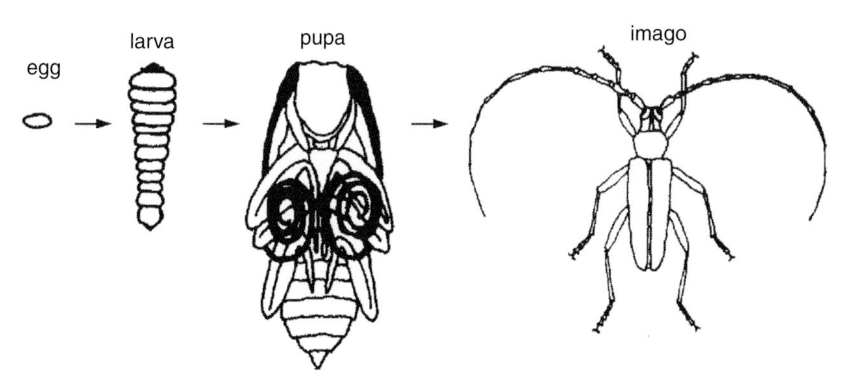

Figure 3.14 Ontogenesis of insects with a complete metamorphosis (holometabolism) (egg, larva, pupa, imago)

years, lay as many as thousands or millions of eggs (termites). The development of an embryo in an egg takes several days to weeks in optimum conditions.

Larvae, after hatching, receive and metabolize food (e.g. wood) and gradually grow. Their outer, firm body cover, the 'cuticula', is shed several times during growth. A new cuticula is created after shedding the old in such a way that the soft cover under the old cuticula hardens in several hours to days. The development stage of a larva between two sheddings is called 'instar'. Depending upon the appearance and type of metamorphosis, larvae are classified into three groups:

- Primary larvae, with a body shape similar to adults but smaller and with undeveloped genitals, and the organs of an adult (e.g. wings are created gradually during shedding).
- Secondary larvae, with specific organs that gradually disappear during development.
- Tertiary larvae, with a body shape completely different from an adult. Typical for insects having a complete metamorphosis and, depending upon the number of legs, they are classified into three groups: (1) polypods have three pairs of thoracic legs and, simultaneously, two to eight pairs of abdominal prolegs (e.g. the larvae of some hymenoptera); (2) oligopods, which have three pairs of thoracic legs and is typical for the majority of beetles and some hymenoptera; and (3) apodous larvae, which do not have legs and is typical for the *Scolytus* family in the beetle order.

A *pupa* occurs between the development stages in the complete metamorphosis of an insect (Figure 3.14). A pupa is usually located in a cocoon created by the larva (e.g. from particles of wood in the gallery). Development in a pupa mainly depends upon temperature and, in wood-boring beetles, it takes 1–3 weeks, but it stops completely in winter. Pupae of wood-boring beetles and hymenoptera have a so-called free shape in which we can clearly see the developing antennae, wings and legs of the future adult.

The *adult (imago)* at holometabolism differs completely from the larva in its body shape (Figure 3.14). Adults hatched from pupae emerge from the wood either using their mother's tunnels and the openings gnawed by their parents when establishing the new generation (*Trypodendron lineatum* – striped ambrosia beetle), or they gnaw their own tunnels and openings (*Ips typographus* – European spruce bark beetle; *Cerambycidae* – longhorn beetles). Adults do not grow further, but some species become sexually mature during maturation feeding (bark beetles). The basic task of an adult is to establish a new generation. Their live usually several hours to weeks, but the termite queen lives for a long as 10 years. Adults of a given species usually differ in appearance, exhibiting (1) sexual dimorphism (i.e. the differences between a male and female), (2) seasonal dimorphism (i.e. colour of a certain species depending upon the season, environment and other factors – e.g. *Tetropium castaneum* is black and also brown) and (3) polymorphism (e.g. in colonies of ants and termites who have a queen, workers, soldiers with a large head and strong mandibles, and other development forms).

An insect generation is an assembly of individuals of a certain species who live within the time period from eggs (or in the case of vivipary from born insects of higher development stages) to adults. The time period of a generation depends upon the species of insect and upon factors influencing its physiology (e.g. type of wood attacked, humidity, temperature). Some insect species have a 1-year generation but with the opportunity to multiply the number of generations within a year under optimum temperature and humidity conditions (e.g. *Ips typographus* sometimes has two to three generations in 1 year). Other species of insect normally have two generations per annum (bivoltine species) or more generations (polyvoltine species). On the other hand, other species of wood-damaging insect have one generation in a maximum of two, three or more years (semivoltine), a typical example of which is the house longhorn beetle (*H. bajulus*) operating in wooden ceilings, trusses and other constructions usually for 3–7 years.

3.3.1.2 Classification of wood-damaging insects

In the animal kingdom, insects are classified in a separate class: the Insecta. The most notable insects damaging wood substance by galleries belong to the following three orders (Gyarmati et al., 1975; Dominik & Starzyk, 1983; Langendorf, 1988; Gogola, 1993):

- beetles (Coleoptera) – with families Anobiidae, Bostrichidae, Cerambycidae, Lyctidae, Lymexylonidae, Scolytidae, and others;
- hymenoptera (Hymenoptera) – with families Formicidae and Siricidae,
- termites (Isoptera) – divided on the basis of their living conditions on subterranean termites (e.g. families Mastotermitidae, Rhinotermitidae and

Termitidae) and nonsubterranean termites (family of the damp-wood termites Hodotermitidae and of the dry-wood termites Kalotermitidae).

3.3.1.3 Physiology of wood-damaging insects

Physiological conditions of wood-damaging insects are influenced by trophic, abiotic, biological and anthropogenic factors. These factors, similar to those for wood-damaging fungi, can be different for the boring activity of insects and for their reproduction and development.

Trophic factors of wood-damaging insets are similar to any other heterotrophic organisms. It means, insects the organic and inorganic substances necessary for their development (sugars, starch, fats, proteins, nitrogen substances, vitamins, salts, etc.) found in consumed wood or leaves. Various wood-damaging and other insects are classified according to the type of consumed food:

- *biophagous* insects digest organic material in its original form – (1) *phytophagous*, primary consumers (i.e. of plant food) and (2) *zoophagous*, secondary consumers (i.e. of animal food);
- *necrophagous* insects digest dead organic material in the form of humus (Křístek & Urban, 2004).

To the phytophagous species of wood-damaging insects belong the phylophagous insects (which procure nutrients from leaves) and the xylophagous insects (which attack wood of live and dead trees, particularly feeding on its polysaccharide portion). For phytophagous species, monophagia (e.g. food is obtained just from one wood species) is less frequent than oligophagia and polyphagia (when food is obtained from several species and genera of wood). Some species of insects can find more suitable food source in a softer early wood (*Trypodendron lineatum* – striped ambrosia beetle), just sapwood suits others (*Callidium violacem* – violet tanbark beetle), and others prefer just wood primarily depolymerized by wood-decaying fungi (*Hadrobregmus pertinax* – house borer beetle). Bacteria that help in the enzymatic degradation of components of the gnawed wood are also present in the digestive tracts of wood-damaging insects (Schloss et al., 2006).

Wood-damaging insects can be attracted to living trees or to stored and processed wood by attractants (Harborne, 1988):

- primary attractants are volatile chemical substances released from bark, leaves, needles or wood (e.g. monoterpenes);
- secondary attractants are chemical substances biosynthesized by insects by which they mutually communicate upon the invasion of trees and harvested wood (e.g. aggregation pheromones, sexual pheromones).

Abiotic factors include the temperature and moisture content of wood as decisive conditions for activity of wood-damaging insects (Table 3.7). However, the

Table 3.7 Effect of abiotic factors on the activity of selected species of wood-damaging insects

Wood-damaging insect species	Galleries[a]	Environmental conditions in wood			
		Moisture (%)		Temperature (°C)	
		w_{opt}	$w_{min-max}$	t_{opt}	$t_{min-max}$
House longhorn beetle (*Hylotrupes bajulus*)	Interior (C)	30–35	9–65	28–30	12–38
Giant horned beetle (*Ergates faber*)	Exterior (C)	60		30	
Common furniture beetle (*Anobium punctatum*)	Interior (C, B)	28–30	10–50	21–24	12–29
House borer beetle (*Hadrobregmus pertinax*)	Interior (C – rotten)	30	19–55	25–26	
Brown powder post beetle (*Lyctus brunneus*)	Interior (B)	14–16	7–23	26–27	18–30

Source: R., L. (2008) *Ochrana Dreva* (*Wood Protection*), Handbook, TU Zvolen, Slovakia, 453 p. Reproduced by permission of TU Zvolen.
Selected data from Dominik and Starzyk (1983), Gogola (1993) and Langendorf (1988).
[a] B: broadleaved; C: coniferous.

impacts of surrounding climatic factors (i.e. temperature, relative humidity and flow of air, sunlight or even electromagnetic field) are also important.

Temperature of wood and air is of special importance for insects. Temperature affects all the forms of movement and physiological activity of insects, the time of their development, the fertility of females and the number of generations per year, and thus also their abundance. Every insect species has its thermal zones of vitality (wider thermal range) beyond which it dies, and also the thermal zones of activity (narrower thermal range) within which it moves, propagates, receives and metabolizes food. Between the vitality and activity zones there are the lower and upper zones of torpor. Maximal temperatures of vitality are almost the same for all the insect species and within the range 40–50 °C. The temperature dependence of the activity of insect larvae is well characterized, with an optimum of usually 20–30 °C (Table 3.7).

The humidity of the wood and environment is also very important for wood-damaging insects (Table 3.7, Figure 3.15). At least ~10% moisture in wood is necessary for technical insects. The optimum moisture ranges from 15% (e.g. *Lyctidae* – powder post beetles), 20–35% (e.g. *Anobiidae* – furniture beetles; *H. bajulus* – house longhorn beetle) up to 60% or even more (e.g. *Ergates faber* – giant horned beetle). The water content in wood affects the time of insect larvae development. For example, *H bajulus* develops in just 2 years at optimum humidity and temperature, but can take 5–12 years in a drier wood. This is attributed to the active role of symbiotic fungi acting on insect development,

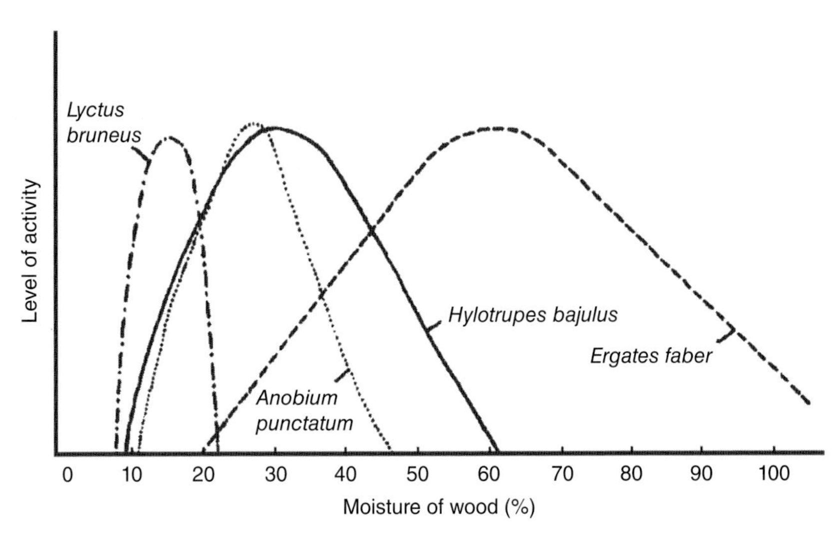

Figure 3.15 Limit and optimum moisture conditions in wood for development of wood-borer larvae (modified from Dominik and Starzyk (1983), based on Becker (1950))

Source: Dominik, J. and Starzyk, R. (1983) *Insects Damaging Wood* (*Owady Niszczace Drewno* – in Polish). Panstwowe Wydawnictwo Rolniczne i Lesne – PWRL, Warszawa, Poland, 440 p

since they assist in the decomposition of wood components at higher humidity, or contribute to the nourishment of insects otherwise.

Light contributes to easier orientation of insects. Light, at the same time, also influences its physiological processes, in particular the diapause, when the development of the insect is interrupted for various reasons (i.e. during wintering period). The larvae of technical insect pests avoid direct light.

Biological factors include mainly the intra- and interspecies relations among insects. These can be positive and negative. Positive intra-species relations of insects include the propagation and care of offspring. In contrast, competition for food and space affects the development of insects in a negative way. An example is the attack of a lone Norway spruce stem by a considerable number of larvae of the eight-toothed spruce bark beetle (*Ips typographus*). They start to be mutually aggressive due to the lack of food, their mortality increases and the females of a new generation have reduced fertility. Interspecies relations of insects may be mutually or unilaterally beneficial, detrimental and indifferent. The interspecies competition for food is usually bilaterally detrimental.

Pheromones are specific substances also produced by wood-destroying insects. They can be sensed by the particular insect species even in a minimum concentration of approximately one to four molecules in 1 cm^3 of air. Pheromones are classified according to their information function as follows:

- sexual – that is, females inform males that they are still virgins, or, in contrast, males attract females by aphrodisiacs;

- aggregation (or, on the contrary, anti-aggregation pheromones) – in the case of bark beetles, informing of the presence of food or even its lack at a certain location;
- alarm – warning of danger from predators;
- defensive – acting as repellents against other animals;
- dispersion – by which insects intentionally ensure an optimum abundance of the population within a certain space;
- identification – giving information of species affiliation of individual insects.

Anthropogenic factors are related to conditions on stocks of logs, created by humans with the aim of protection against feeding marks of insects; for example, by wetting them or rapid debarking and drying. According to need, humans also use other preventive protection measures, including insecticides for wooden products (Section 5.2.3) and sterilization of attacked wood (Section 7.4).

3.3.2 Wood-damaging insects

Wood-damaging insects of the technical type deteriorate wood by feeding marks – galleries. Wood is their food or a place to hide. Galleries in wood are made in particular by several beetles, hymenoptera and termites. Certain beetles may complete development and emerge several years after wood is dry, often raising a question as to the origin of the infestation (Clausen, 2010).

3.3.2.1 Galleries of beetles

For example, ~6500 species of beetles live in central Europe, mainly in a forest environment. Their number in Asia or South America is several times larger. Beetles have complete metamorphosis (i.e. holometabolism – egg, larva, pupa, adult). A pair of chitinized wing-cases is on the second thoracic element of adults. The other membranous pair of wings is on their third thoracic element, folded under the chitinized wing-case.

Larvae of wood-boring beetles attack living trees, logs, sawn timber yards and various products made of wood. Less important galleries from 0.5 to 3 mm in a wood structure are created by bark beetles (family Scolytidae). They attack trees and unbarked logs. They live under bark in monogamy or polygamy but some of them are also ambrosia beetles living in symbiosis with sap-stain fungi. Bark beetles living under bark and in phloem belong to the physiological pests (e.g. eight-toothed spruce bark beetle – *Ips typographus*). On the other hand, species that also develop in the wood parts of trees belong to the technical pests (e.g. striped ambrosia beetle – *Trypodendron lineatum*; European shot-hole borer – *Xyleborus dispar*). To avoid their activity, round wood should be debarked rapidly, sprayed with approved insecticides, stored in water basins or under water spray, or cut during the dormant season (Clausen, 2010). In

Europe, the most significant technical wood-damaging beetles belong to three families: Cerambycidae, Anobiidae and Lyctidae.

- Cerambycidae – long-horned beetles.
  - Approximately 240 beetles from the family Cerambycidae live in Europe. They are phytophagous, and usually xylophagous. Females lay ~60–240 eggs, most frequently in cracks or under bark. The galleries created by white, oval, fleshy larvae are of elliptic shape, perfectly filled with bore dust. Larvae pupate at the ends of feeding-mark corridors. The newly hatched imago (usually with longer antennae) nibbles an oval hole in the surface of the wood, through which it flies out. The most significant species are displayed in Figure 3.16.
- House longhorn beetle (*Hylotrupes bajulus* L.).
  - *Activity:* This is the most frequently occurring and the most dangerous insect pest of processed softwoods, particularly of beams in trusses and ceilings, but also in log cabins, half-timbered structures, or sometimes even in exterior structures such as balconies and bridges (Figure 3.16a). Its harmfulness in Europe can be compared to the harmfulness of termites in Africa and America. In contrast to the majority of other longhorn beetles, it attacks just processed debarked softwood (e.g. sapwood and mature wood of spruce, but not heartwood of larch). It attacks predominantly newer wooden structures, while 50–70-year-old or older structures only with a lower intensity due to the lower amount of proteins and terpenes (Kačík et al., 2012). Females lay ~140–180 eggs, of length 2 mm, in cracks and slots in wood to a depth of 20–30 mm. White larvae hatch after a week. Larvae develop optimally at 29 °C (t_{limits} 12–38 °C) and at a wood moisture content of 30–35% (w_{limits} 9–65% – however, fresh wood does not suit them). The generation of *H. bajulus* is rather long, usually 3–7 years, when larva development takes 3–5 years or more and the pupation period takes 2–4 weeks.
  - *Galleries:* Larval corridors are of oval shape, firstly in the longitudinal direction in the spring wood layer that is richer in nutrients with the contents of proteins. The galleries consist of feeding-mark corridors which are gradually enlarged and deviate from the longitudinal direction, and are often tortuous. In heartwood species (pine, larch) they are located just in sapwood, but in mature-wood species (spruce, fir) they form across entire wooden elements and seriously impair static function. Larvae gradually grow to a length of 15–22 mm. Galleries, perfectly filled with bore dust (small chips and wood fibres may be present), reach a cross-section of 12 mm × 7 mm. Larvae do not eat the very fine surface layer of wood, so it is intact (there are just flight holes from adults on the surface), and thus the damage to the wood often escapes our attention. It is possible to see a weak falling out of bore dust through flight holes and cracks in the later stage of wood damage only (i.e. usually caused by two or more generations) that accumulates on

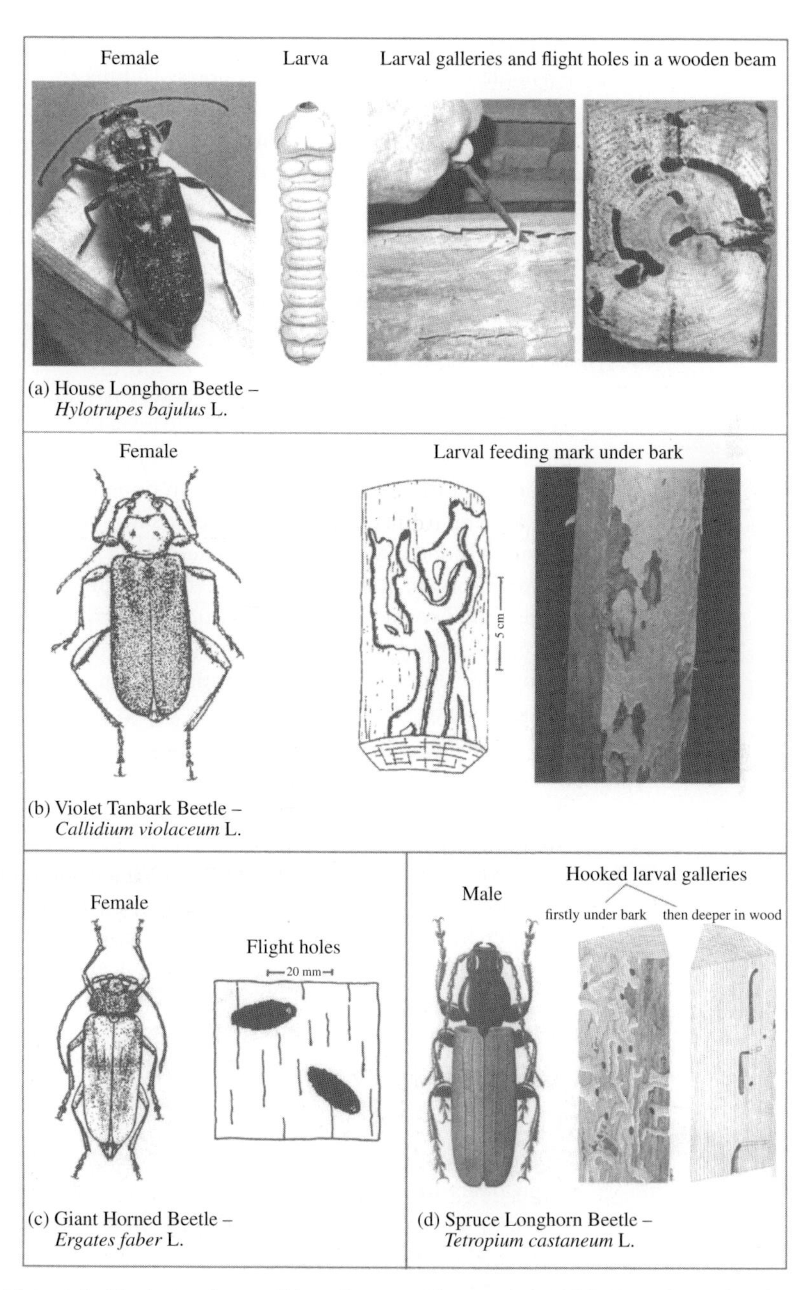

Figure 3.16 Long-horned beetles attack wood: (a, b) in interiors; (b–d) logs, sawn timber and wooden products exposed in exteriors (modified from Novák et al. (1974), Dominik and Starzyk (1983) and Reinprecht (2008))

Sources: (a,b) R., L. (2008) *Ochrana Dreva* (*Wood Protection*), Handbook, TU Zvolen, Slovakia, 453 p. (b,c) Dominik, J. and Starzyk, R. (1983) – PWRL, Warszawa, Poland, 440 p. (d) Novák, V., Hrozinka, F. and Starý, B. (1974) *Atlas of Insect's Pests of Forest Woods* (in Slovak). Príroda Bratislava, Czechoslovakia, 128 p

the floor. Larval feeding can be spotted acoustically as a scraping noise, particularly on warm days. Flight holes of adults are oval of size 7 mm × 4 to 10 mm × 5 mm. Several adults hatched from pupas can even fly out through a single hole, which means that the number of holes on the surface of wood do not indicate the actual scope of internal insect damage.
  - *Adult:* Having a length of 7–25 mm, coloured to brown–black in colour, with fine pale grey pubes, forming two cross spots on the wing-cases. Its antennae, in comparison with other Cerambicydae, are untypically very short.

- Anobiidae – woodworm beetles.
  - Approximately 60 species from the family Anobiidae live in Europe. They are xylophagous; some of them are mycetophagous, acting mostly in processed wood. Their larvae are white or coloured in yellowish shades with three pairs of legs. Prior to pupation, their length is from 6 to 10 mm. Adults are small, of size 2–8 mm, with a prolonged cylindrical body and rather short antennae. Adults have a hood-shaped protothorax (on the first element of the thorax), covering almost the entire head. Woodworm beetles make a ticking sound (i.e. the characteristic sound caused by the head of an individual beetle hitting the walls of corridors that serves for its communication with other individuals). Woodworm beetles usually have 1- or 2-year generations (Langendorf, 1988).

    Beetles from family Anobiidae damage wood by galleries of a circular shape with a diameter of 1–4 mm. Galleries are usually perfectly filled with bore dust comprising egg-shaped or lens-shaped excrements. Individual species act either as wide-spectrum beetles on products made from various hardwoods and softwoods (e.g. *Anobium punctatum*), or they attack just particular wood species, or just rotten wood (e.g. *Hadrobregmus pertinax*). They can cause a significant damage in particular on furniture, parquets and floorboards, windows, but also on beams in ceilings, trusses, log cabins or altars and benches in churches (Figure 3.17).

- Common furniture beetle (*Anobium punctatum* De Geer).
  - *Activity:* This is the most frequent beetle from the family Anobiidae in the whole of Europe (Figure 3.17a). It causes very serious damage in products from softwood and hardwood species (furniture, musical instruments, tools, windows, doors, floors, panels, stairs, ceilings, trusses, log cabins). However, this beetle seldom attacks heartwood zones of wood with a content of tannins, terpenoids and other substances toxic for its life; for example, such as the heartwood of eucalyptus (Creffield, 1996) or heartwood of oaks. The optimum temperature for development of its larvae is rather low at 21–24 °C (t_{limit} 12–29 °C), and optimum wood moisture is near to the saturation point of fibres at ~30% (w_{limit} 10–50%). Larvae are sensitive to frost, when ~80–100% of them died at −16°C at a depth of 1.5 cm from the wood surface over several days. Females adhere 30–60 oval eggs of size 0.5 mm ×

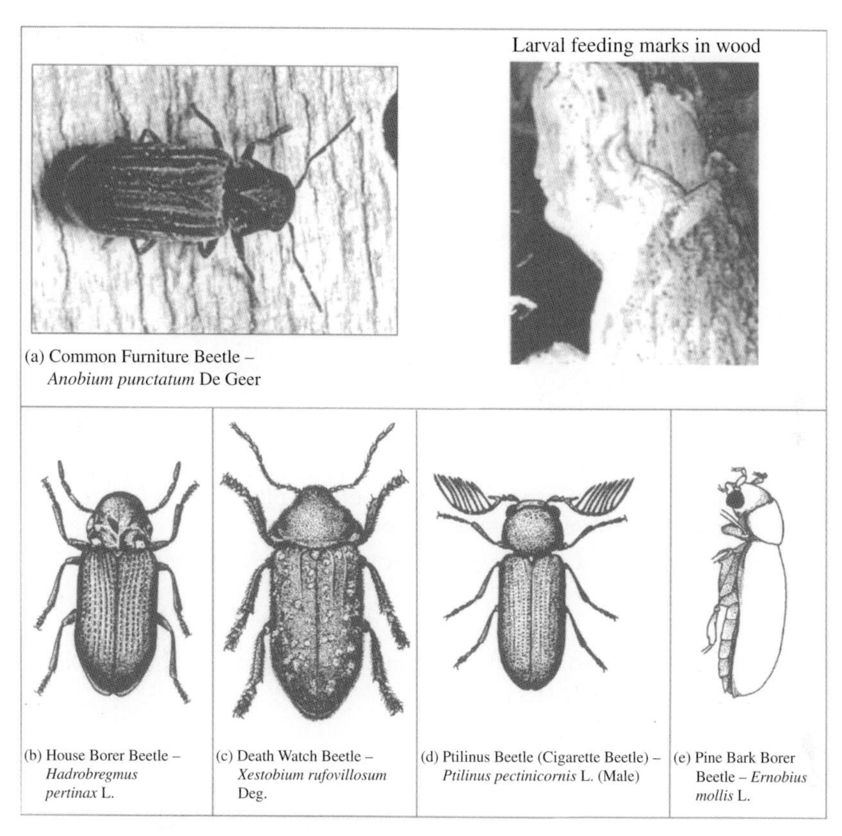

Figure 3.17 Anobiid beetles, damaging built-in wood (a–d) or attacking just non-barked wood (e)

Source: Dominik, J. and Starzyk, R. (1983) – PWRL, Warszawa, Poland, 440 p

0.2 mm to a wooden substrate – into cracks, slots, frontal areas, and so on – after night mating. After 2–3 weeks the larvae hatch from the eggs and grow to a length of 4–6 mm. Larvae develop well in the sap zone of softwoods, which is rich in proteins and starch. They digest mainly hemicelluloses from the cell walls; however, they also metabolize cellulose using yeast fungi present in their digestive system (Serdjukova & Toskina, 1995). Their generation lasts 2 or 3 years, but it may be even shorter in a favourable climate.

- *Galleries:* The larvae damage hardwoods and softwoods by a quantity of galleries having a diameter of 0.4–2 mm. Characteristic egg-shaped excrements 'pellets' are found in bore dust. The larvae feed close to the surface of wood at the beginning of spring before pupation, where they make their pupal chamber in the direction alongside wood fibres. The pupa stage takes ~2 weeks at 20 °C or just 10 days at 28 °C; however, there is substantially higher mortality of individuals at such a higher temperature. The flight holes of adults are circular, usually with a diameter of 1.5–2 mm.

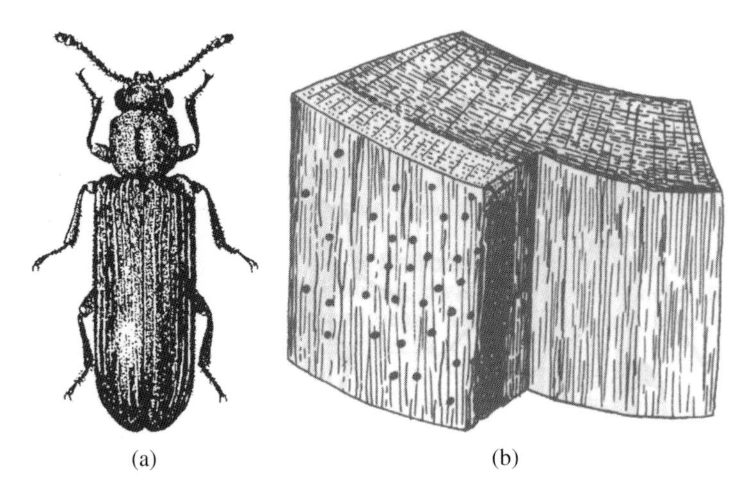

Figure 3.18 (a) European lyctus beetle, *Lyctus linearis* Goeze, and (b) its circular flight holes from oak wood

Source: Dominik, J. and Starzyk, R. (1983) – PWRL, Warszawa, Poland, 440 p

- *Adult:* This is 2.5–5 mm long, 1–2.1 mm wide, brown coloured, with fine hair on its body.

•• Lyctidae – powder-post beetles.

- Powder-post beetles are of size 2–7 mm. They are active in warmer areas, particular in Asia, Africa, America, Australia, but also in Europe, at a rather lower wood moisture of from 8 to 26%. They attack hardwoods with a minimum of 1% of starch substances; for example, sapwood of oaks and tropical species. In central Europe live the European lyctus beetle (*Lyctus linearis*) (Figure 3.18), the brown powder-post beetle (*Lyctus brunneus*), and the oak powder-post beetle (*Lyctus pubescens*). Circular galleries, under a thin intact surface skin of wood, have diameters from 1 to 1.5 mm, a random orientation, but mostly parallel to wood grains, and they are filled with cream-coloured fine bore flour.

3.3.2.2 Galleries of Hymenoptera insects

Approximately 10,000 insect species from the family Hymenoptera live in central Europe, mostly in a forest environment. They reproduce usually by fertilized eggs and they have complete metamorphosis (i.e. holometabolism). Two pairs of membrane wings are characteristic for adults.

- Siricidae – woodwasps.
  Larvae are cylindrical with minimally developed breast legs. They attack unhealthy trees and freshly felled logs of hardwoods and softwoods. Damage may be incorporated into buildings; however, new adult females are not

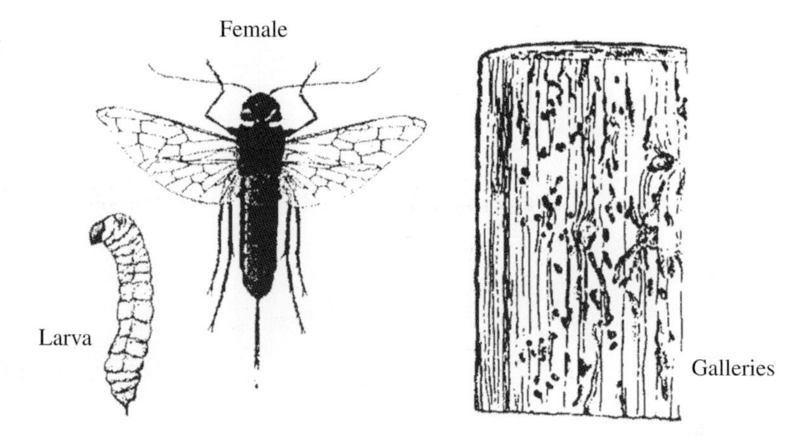

Figure 3.19 Giant wood wasp, *Urocerus gigas* L

Source: Novák, V., Hrozinka, F. and Starý, B. (1974) *Atlas of Insect's Pests of Forest Woods* (in Slovak). Príroda Bratislava, Czechoslovakia, 128 p

interested in dry unbarked timber. Galleries are of circular shape, rather long and tightly blocked with densely packed bore dust of wood colour. Flight holes are circular too, being 4–10 mm in diameter. Several wasps also spread from Europe in exported wood to other parts of world; for example, to Australia (Richardson, 1993). The most dangerous ones are the giant wood wasp (*Urocerus gigas* L.) (Figure 3.19) and the siricid wood wasp (*Sirex juvencus* L.).

- Formicidae – ants.

  Wood-damaging carpenter ants are harmful in Europe, America and other parts of the world. They attack wood (e.g. in living trees, logs, poles and buildings) for shelter rather than for food. In central Europe there are two well-known carpentry ants: *Camponotus herculeanus* L. and *Camponotus ligniperdus* Latr. In wood they create concentrically gnawed out tunnels having an interesting architecture: labyrinths with lengths of several metres (Figure 3.20).

  Carpenter ants attack mainly early wood of conifers used for utility poles, or in beams of ceilings, trusses and other parts of buildings with a poor structural protection when wood moisture is increased due to rain or condensation. Soft and wet wood suitable for ants is also present in decayed windows, balconies, and so on.

  The community of carpenter ants consists of a mother(s), female workers and males. Female workers are sexually undeveloped females and their main task is the care of offspring – bringing the food, larvae feeding and maintenance of cleanliness. Mothers are fertile females living for 10 years or more. Sexually developed males and mothers are hatched just in a certain period in summer; they are temporarily winged and they swarm outside the anthill. Males die shortly after insect swarming. The generation (from eggs

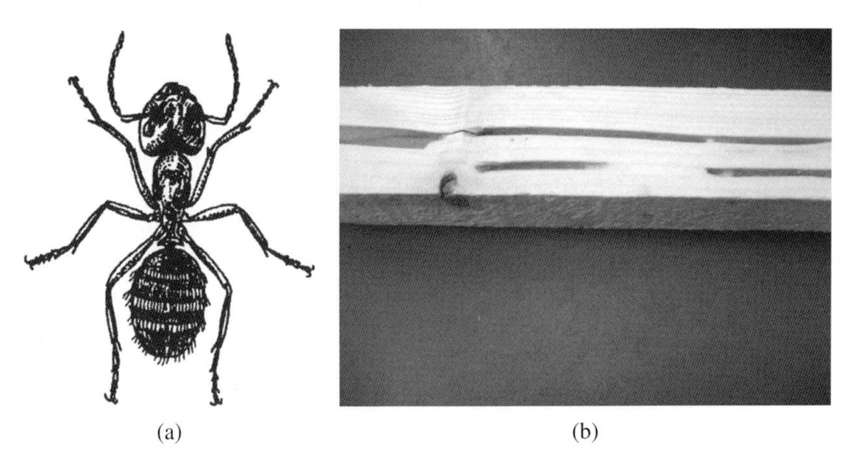

Figure 3.20 Typical labyrinth damage of early conifer-wood (b) by wood-damaging ants (a)

Source: (a) R., L. (2013) *Wood Protection*, Handbook, TU Zvolen, Slovakia, 134 p. Reproduced by permission of TU Zvolen

to adults) is affected by temperature; for example, it takes 1 month at 30 °C but 3 months at 15 °C.

3.3.2.3 Galleries of termites

Termites live in social communities termed colonies. They occur mainly in the tropics and subtropics (Richardson, 1993); however, several species also occur in Spain, France, Italy and even in Germany, Austria and Ukraine. Of the 2800 species of termites only 70–80 species are pests of wood; for example, in the USA there are about 56 species (Clausen, 2010). An environment with a temperature of 26–32 °C and a relative humidity of 70–90% is best for their life (Unger et al., 2001). In addition to wood, they also damage other materials that contain cellulose (i.e. various wooden composites, paper, cardboard and textiles), and also rubber, plastics or soft metals.

A colony of termites consists of a queen, king, larvae, nymphs (larvae with the wing bases), female workers, female pseudo-workers, soldiers and other forms with specific tasks. There are from 1000 to 2 million individuals living in a colony. Their metamorphosis is incomplete; that is, hemimetabolism without a pupa stage. The greatest one is a queen, approximately 120 mm long. The white body of larvae and female workers of termites is 3–10 mm long. Similar to ants, they prefer softer early wood, while creating interesting labyrinth formations. Their activity concentrates in particular in the internal zones of wooden beams – large tunnels are created in wood (Figure 3.21). In the case of internal damage, the wooden beams and other wood elements of the building seem to

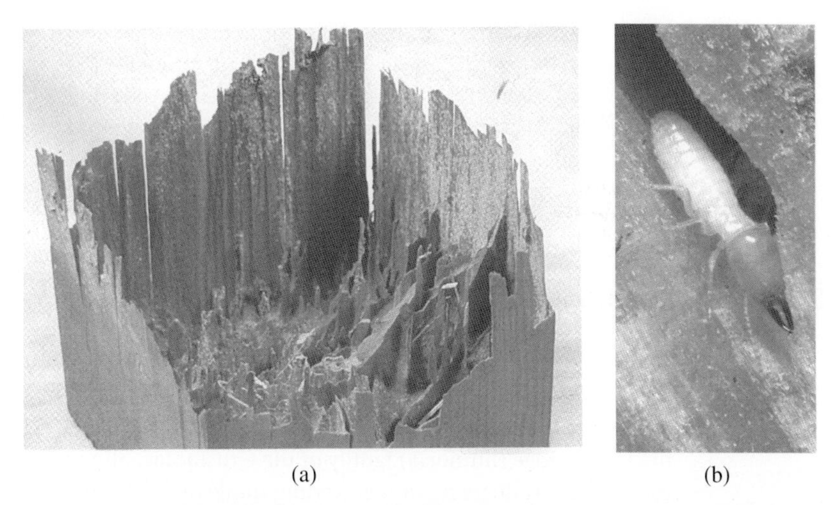

(a) (b)

Figure 3.21 Tunnel damage of wooden beam by termites (a) and termite soldier in a gallery (b)

be healthy from the outside, but the risk of collapse of elements and the whole building is increased.

Termites attacking wood are, from the standpoint of their activity, grouped into two main classes: ground inhabiting and wood inhabiting.

- Subterranean (underground or ground-inhabiting) termites; for example, Rhinotermitidae.
  - They build tunnels through soil or live in contact with the ground. Wooden structures can be attacked even from a distance of 20–100 m if there is no convenient insulation between the soil and wooden elements. The workers attack moist wood, and also dry wood when they have a constant source of moisture in the soil. They typically feed on the internal zones of beams and other wooden elements in buildings; that is, just an external 1–2 mm thin crust remains of the wooden element at the end (Figure 3.21).
  - The best-known termite in Europe is the Mediterranean termite (*Reticulitermes lucifugus*) acting in particular in Spain and Italy. *Reticulitermes santonensis* acts in the Paris territory in France, and *Reticulitermes flavipes* has been found in the Hamburg area in Germany for a long time. Subterranean termites seriously damage wooden houses in various regions of the USA (Peterson et al., 2006).
- Nonsubterranean (wood-inhabiting) termites; that is, Hodotermitidae and Kalotermitidae.
  - Damp-wood termites (Hodotermitidae) attack mainly damp and wet wood in contact with soil.

- Dry-wood termites (Kalotermitidae) attack air-dried wood, in particular in wooden houses. In beams or furniture they are able to move from one element to other. Well known is the yellownecked dry-wood termite (*Kalotermes flavicollis*). These termites damage wood just like the subterranean termites do; that is, just a thin crust remains. A colony is formed by a rather small number of individuals (1000–1500).

3.4 Wood damaged by marine organisms

Marine borers are organisms which in seas or salt water cause significant damage to wooden boats, ships, pilings or piers (Figure 3.22). The most dangerous species are found among the Teredinidae (shipworm) family of the Bivalvia, and the Limnoridae (limnoria) family of the Crustaceae. Shipworms cause damage to wooden products by tunnel feeding marks, while limnoria cause their erosion from surfaces (Richardson, 1993). Untreated wood – that is, without deep impregnation with creosote or other water-stable biocides – can be completely destroyed in a year or less by these organisms (Clausen, 2010).

3.4.1 Shipworms

Shipworms are molluscs of various species (*Teredo navalis* L., *Pholas dactylus* L., *Bankia* sp., etc.). They become imprisoned into wood and then in its structure create single blind-ending tunnels, precisely circular, length 100–1200 mm, and with a diameter of 5–15 mm (Figure 3.23). Shipworms usually secrete their own enzymes – cellulases; however, cellulolytic microbes in their body may act as well (e.g. in *Bankia setacea*). Shipworms feed mainly on cellulose and partially

Figure 3.22 Wooden pier damaged by marine organisms

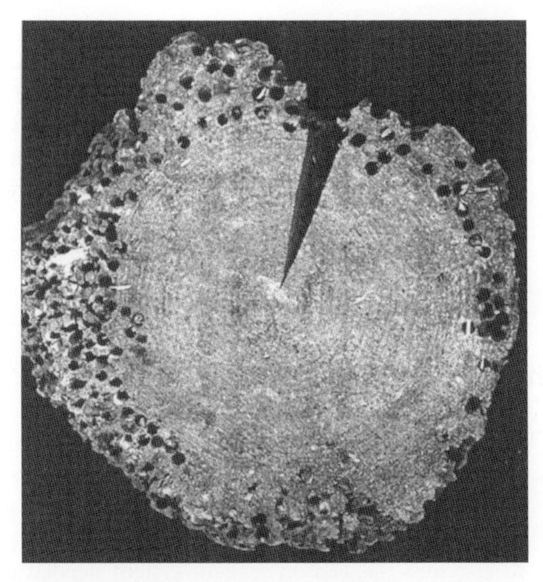

Figure 3.23 Wood damaged by *Teredo navalis*

Source: Unger, A., Schniewind, A. P. and Unger, W. (2001) *Conservation of Wood Artifacts*. Springer-Verlag Berlin Heidelberg, 578 p. Reproduced by permission of Springer

on hemicelluloses of wood (Kirk & Cowling, 1984). The entrance holes newer grow large, but the interior of wood may be completely honeycombed.

- *Teredo navalis* L.
  - Larvae are developed from eggs within 14 days, having a length of just 0.3 mm.
  - However, larvae can gradually grow to lengths of 0.1–1.2 m, and in wood bore long tunnels. They are able to damage sapwood and heartwood zones of broadleaves and conifers, making circular bore holes of diameter 6–8 mm. Wood in tunnels is all feed, and they leave a white layer of calcium in them. They are active in seawater containing ~0.9–3.5% salt (preferentially in warmer water).
  - A generation is 1–3 years.

3.4.2 Limnoria

Sea crustaceans (e.g. *Limnoria lignorum* L.) damage just the surface layers of wood by their mandibles, usually to a depth of not more than 50 mm; that is, they cause the erosion, shallow damage of wood, which gradually is markedly thinned.

- *Limnoria lignorum* L.
  - The individuals of *Limnoria lignorum*, having a length of 4–5 mm, live in pairs and in wood create holes having a diameter of 1.5–2 mm. They

damage wood of coniferous and broadleaves, mainly their early wood. Wood is damaged by erosion from surfaces gradually to a greater depth of 6–20 mm within 1 year. For their activity the seawater should have ~1.5% salt. *Limnoria lignorum* is a cold-water species, so the temperature of the seawater should be below 20 °C for life. Breeding begins when the temperature of the seawater reaches ~10°C (Eaton & Hale, 1993).

- The species has a 1-year generation.

3.5 Mechanisms of wood biodegradation

Bacteria, fungi, insects and marine borers decompose the structural components of wood cells by means of extracellular and intracellular enzymes (Table 3.8), together with help of low molecular weight agents functioning as precursors and co-agents. For example, brown-rot fungi produce in addition to enzymes the Fenton reagent system needed for cellulose decomposition, and low molecular weight oxidation substances are also helpful for white-rot fungi activity for deterioration of lignin and polysaccharide macromolecules (e.g. Koenigs, 1974; Goodell, 2003; Messner et al., 2003; Schmidt, 2006; Alfaro et al., 2014). By knowing the mechanism of rot and other biological damage of wood we can better understand the relations between its damaged molecular and anatomical structure and its properties, and also better propose the principles and technologies of its protection.

Enzymes are biocatalysts that accelerate and control biochemical reactions. In the literature, more than 3000 enzymes are described. The Nomenclature Committee of the International Union of Biochemistry and Molecular Biology group enzymes according to their catalytic function into six classes: (1) oxidoreductases catalyse oxidation and reduction reactions by transferring hydrogen and/or electrons (e.g. in lignin by their cooperation with hydrogen peroxide, H_2O_2); (2) transferases catalyse transmission of various functional groups (e.g. –OH, NH_2) from donors to acceptors; (3) hydrolases hydrolytically cleave bonds (e.g. in cellulose by incorporating molecules of water); (4) lyases catalyse non-hydrolytic cleavage or creation of bonds C–C, C–O, C–N, and so on; (5) isomerases catalyse reversible intramolecular transfers of atoms or groups of atoms (e.g. uridine diphosphate glucose to uridine diphosphate galactose); and (6) ligases catalyse synthesis of energetically rich covalent linkages in organic molecules with the help of compounds giving energy to the system (e.g. energy can be obtained at transformation of adenosine triphosphate to adenosine diphosphate).

Enzymes consist of protein (apo-enzyme) and a complex of organic molecules and/or metal-ions (cofactor). Enzyme dimensions range from 5 to 7 nm; therefore, their diffusion inside the cell walls of wood is often limited, mainly into micro-fibrils of cellulose. This means that for bio-damage of polysaccharides and lignin the enzymes need to cooperate with various

Table 3.8 Enzymes ensuring the biochemical decomposition of substrates in wood (modified from Reinprecht (1997))

Enzyme name	Substrate attacked	Catalysed reaction
endo-1,4-β-Glucanase	Cellulose	Statistically random hydrolysis of glucoside bonds in positions 1,4 in amorphous and crystalline cellulose
exo-1,4-β-Glucanase	Cellulose	Intended hydrolysis of glucoside bonds in positions 1,4 at the non-reducing ends of cellulose, leading to the splitting-off of cellobiose
1,4-β-Glucosidase	Cellobiose, cellodextrines	Hydrolysis of cellobiose and water-soluble cellodextrines to glucose
Cellobiose:quinone oxidoreductase (CBQ)	Cellobiose + quinones	Oxidation of cellobiose and other cellodextrines to lactones and acids + reduction of quinones and phenoxyl radicals (from lignin) to phenols
endo-1,4-β-Xylanase	Xylose	Statistically random hydrolysis of glycoside bonds in positions 1,4 in the frame of xylanes, while oligosaccharides, xylobiose and xylose are produced
1,4-β-Xylosidase	Xylobiose	Hydrolysis of xylobiose, and xylane oligosaccharides, to xylose
Acetyl(xylane)esterase	Xylanes	Hydrolysis splitting-off of acetic acid from xylanes (i.e. from C-2, C-3 acetylated carbohydrates)
endo-1,4-β-Mannanase	Mannanes	Statistically random hydrolysis of glycoside bonds in positions 1,4 in the frame of mannanes, while oligosaccharides, mannobiose and mannose are produced
1,4-β-Mannosidase	Mannobiose	Hydrolysis of mannobiose, and mannane oligosaccharides, to mannose
Lignin peroxidase	Lignin	Oxidation splitting of C–O–C and C–C bonds, splitting of aromatic ring while lactones are produced, oxidation of benzyl-alcohols, polymerization of phenols
Mn^{2+} peroxidase	Lignin	Oxidation splitting of C–O–C and C–C bonds, oxidation of phenols to phenoxyl radicals (and quinones)

(*continued*)

Table 3.8 (*Continued*)

Enzyme name	Substrate attacked	Catalysed reaction
Laccase	Lignin	Oxidation splitting of C–O–C and C–C bonds, oxidation of phenols to phenoxyl radicals (and quinones)
Aryl-alcohol oxidase	Aryl alcohols	Oxidation of aromatic C_α alcohols and production of H_2O_2, needed for peroxidases
Glucose-oxidase	Glucose	Oxidation of glucose to gluconolactone and production of H_2O_2
Methanol-oxidase	Glucose	Oxidation of methanol to formaldehyde and production of H_2O_2
Dioxygenase	Aryl alcohols	Splitting-off of aromatic ring in position 3,4 or 4,5, during which both atoms of the oxygen molecule are built into the substrate

Source: R., L. (1997) *Processes of Wood Degradation*, TU Zvolen, Slovakia, 137 p. Reproduced by permission of TU Zvolen.

smaller intermediaries; for example, metal ions, H_2O_2 and some organic low molecular weight substances.

The *catalytic efficiency of enzymes* is enormous; for example, one enzyme molecule can in ideal conditions catalyse 5×10^4 reactions per second. However, in the structure of the cell walls of wood, the catalytic potential of extracellularly acting enzymes decreases due to special steric hindrances: (1) in the cell walls of wood the various types of polysaccharides are covered or also cross-linked with phenyl-propane units of lignin; (2) the pores in the cell walls of wood are not always sufficiently filled with bond water, which is needed for diffusion of enzymes to polysaccharides and lignin. Many enzymes show a high specificity; that is, some types of reactions (hydrolysis, oxidation, isomerization, etc.) catalyse only in some types of macromolecules (pectin, mannanes, xylanes, guaiacyl lignins, etc.). This means that for the total biodegradation of wood a complex of more types of enzymes is needed to totally cleave the lignin–polysaccharide structures of cell walls, catalysing hydrolysis, oxidation, reduction and other reactions.

3.5.1 Biodegradation of cellulose

Despite more detailed studies related to the cellulose biodegradation by bacteria, fungi and other organisms, there is still some uncertainty as to how cellulose is damaged by pests (Frankova & Fry, 2013).

3.5.1.1 Degradation of cellulose by white-rot and soft-rot fungi

White-rot and soft-rot fungi use for decomposition of cellulose a multicomponent enzyme complex. It consists of cellulases, but also of polysaccharide monooxygenases and some other oxidative enzymes (Eriksson et al., 1990; Schmidt, 2006; Bey et al., 2013; Hori et al., 2013):

- *endo*-1,4-β-glucanase, or several types of *endo*-glucanases (C_X);
- *exo*-1,4-β-glucanase, or several types of *exo*-glucanases (C_1);
- 1,4-β-glucosidase;
- oxidation and oxidation–reduction enzymes for cellulose and cellobiose breakdown.

The *endo*-glucanases (C_X) hydrolyse 1-4-β-D-glucosidic linkages present inside of cellulose. The *exo*-glucanases (C_1) release cellobiose from non-reducing ends of cellulose. The final hydrolysis of cellobiose, and also of other oligosaccharides created before with the help of the C_X enzyme, is mediated by the 1,4-β-glucosidase (Figure 3.24). Cellobiose can also be destroyed by oxidation enzymes.

Hydrolase enzymes (celullases) synergetically transform cellulose to glucose (Figure 3.24). Then chitin and other substances needed for growth of white-rot

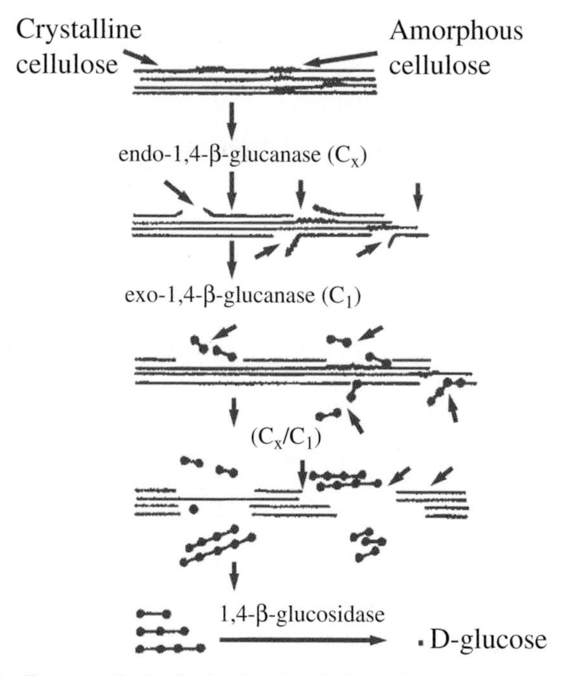

Figure 3.24 Enzymatic hydrolysis of cellulose by means of (C_X/C_1) *endo-/exo*-glucanases cooperation and 1,4-β-glucosidase

and soft-rot fungi are created from glucose. However, glucose can be decomposed to water and CO_2 as well. If the hydrolytic enzymes are not in equilibrium, or some of them are missing, the decomposition of amorphous cellulose will be slower and the decomposition of crystalline cellulose would not take place at all in the first stages of white rot (Hastrup et al., 2012).

3.5.1.2 Degradation of cellulose by brown-rot fungi

Brown-rot fungi use the following catalytic systems for the decomposition of cellulose (Eriksson et al., 1990; Goodell, 2003; Baldrian & Valášková, 2008; Arantes et al., 2012):

- *Hydrolase enzymes (usually incomplete complex without C_1 enzyme)*; that is, the *endo*-1,4-β-glucanase and 1,4-β-glucosidase, whilst the *exo*-1,4-β-glucanase is often missing. This system is able to impair cellulose just in its amorphous areas and just partially.
- *Non-enzyme oxidation system*; that is, the Fenton reagent system (Koenigs, 1974), $Fe^{2+}/H_2O_2/(COOH)_2$, from which is created the reactive hydroxyl radical ($OH^\cdot$) which is available to depolymerize the cellulose (Figure 3.25). However, in the non-enzyme system, other agents based on non-ferrous metals (e.g. Cu^{2+}, Ni^{2+}) can also be involved to create the $OH^\cdot$ radical on reaction with H_2O_2 (Halliwell, 2003). The aggressive $OH^\cdot$ radical can be created as well with the help and mediation of glycopeptide, which reduces Fe^{3+} to Fe^{2+} (Enoki et al., 2003), or with the help of quinone-type chelators (Shimokawa et al., 2004). All the low molecular weight oxidation systems mentioned are able to easily penetrate even between the micelles of crystalline cellulose and decompose it by oxidation reactions (Goodell, 2003).

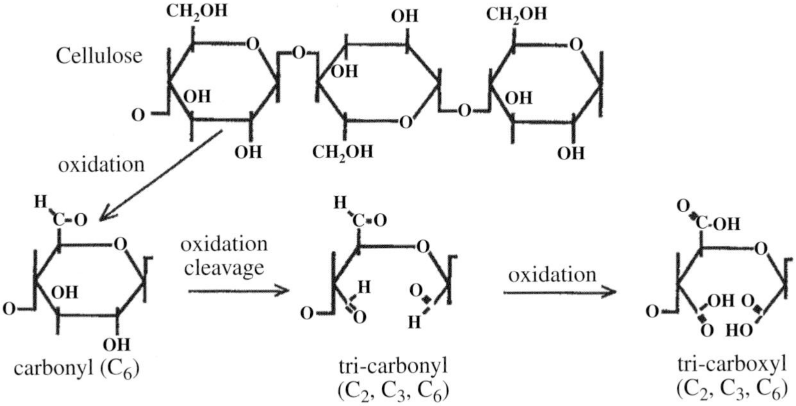

Figure 3.25 Non-enzymatic oxidation decomposition of cellulose by means of Fenton reagent, $Fe^{2+} + H_2O_2 \rightarrow Fe^{3+} + OH^- + OH^\cdot$

The Fenton oxidation system was proven in several brown-rot fungi acting inside of buildings; for example, in *S. lacrymans*, *Coniophora puteana* or *Oligoporus placenta* (Ritschkoff et al., 1992).

Brown-rot fungi, thanks to cooperation of the hydrolase enzymes and the low molecular weight aggressive oxidation systems, can quickly depolymerize cellulose. For example, Kirk and Cowling (1984) found that at 10% loss of wood mass caused by the brown-rot fungus *Poria monticola* the average degree of polymerization (ADP) of cellulose and hemicelluloses decreased by about five times (i.e. ~80%), while for the same 10% mass loss of wood due to action of the white-rot fungus *Trametes versicolor* the ADP of polysaccharides decreased only ~10%. This difference in the intensity of cellulose depolymerization by brown-rot and white-rot fungi is reflected in the mechanical properties of rotten woods (Table 3.9).

Generally, it can be said that brown-rot fungi successfully use the non-enzymatic oxidation systems due to the limitations of the hydrolytic enzymes to penetrate through lignified cell walls of wood. The brown-rot fungi can damage the three-dimensional spatial structure of lignin only to a certain degree. However, some brown-rot fungi are not able degrade pure cellulose substrates like standard filter paper or delignified wood; hence, this suggests an active role for lignin and its synergies in brown rot.

3.5.2 Biodegradation of hemicelluloses

The decomposition of hemicelluloses – xylanes in particular present in the wood of broadleaves in the range 15–35%, and mannanes in particular present in the wood of conifers – takes place by the action of a large spectrum of hydrolytic enzymes: the 'hydrolases'. These enzymes are produced by wood-decaying

Table 3.9 Effect of the degree of wood decay (determined by weight losses of rotten wood from 2 to 10%) on the percentage decrease of its mechanical properties

	Decrease of mechanical property (%)					
	Weight-loss with brown rot			Weight-loss with white rot		
Mechanical property	**2%**	**6%**	**10%**	**2%**	**6%**	**10%**
Impact bending strength	31–50	60–85	70–92	26	50	60
Compression strength	6–10	16–25	40	5	12–27	35
Bending strength	32	61	55–70	14	20–27	24
Hardness		20–28	35–45		18	25

Source: R., L. (2013) *Wood Protection*, Handbook, TU Zvolen, Slovakia, 134 p. Reproduced by permission of TU Zvolen.
Selected data from Wilcox (1978) and Reinprecht (1992, 1996).

fungi and some bacteria; they are also present in the digestive tracts of wood-damaging insets and marine borers (Eriksson et al., 1990; Saake et al., 2001; Schmidt, 2006; Schloss et al., 2006). Biochemical degradation of hemicelluloses in the cell walls of wood can occur either without simultaneous decomposition of their other macromolecular components (e.g. it takes place before decomposition of cellulose), or it takes place together with the decomposition of cellulose or also of lignin. Kim et al. (2015) by detailed microscopy and immunochemistry studies of early stages of rot, caused by brown-rot fungus *Poria placenta* in spruce and ash sapwood, found that hemicelluloses were first degraded in the S_3 layer and in the compound middle lamella of the cell walls.

The *xylanolytic complex of enzymes* is present in several bacteria and wood-decaying fungi and affects xylanes in the wood of broadleaves – hardwoods. It consists of the following enzymes: *endo*-1,4-β-xylanase, 1,4-β-D-xylosidase, α-D-glucuronidase, α-L-arabinofuranosidase and acetylxylanesterase.

The *mannanolytic complex of enzymes* provides for the decomposition of mannanes (galactoglucosomannanes and glucomannanes) in softwoods, but also in hardwoods. It is similar to the xylanotytic enzyme system with its effect; however, it consists of other types of enzymes: *endo*-1,4-β-mannanase attacks the frame of mannanes while producing oligomers, β-mannosidase, β-glucosidase and α-galactosidase, by which oligomer sugars hydrolyse to monosaccharides (i.e. to mannose, glucose and galactose).

Oxalic acid of brown-rot fungi can be involved first in the degradation of the side chains of the hemicelluloses, thus providing entrance to arabinose and galactose, and then depolymerize the main hemicellulose chain, or also amorphous cellulose (Green et al., 1991).

3.5.3 Biodegradation of lignin

Lignin is more resistant to biodegradation than polysaccharides are. This is implied by its three-dimensional and less polar macromolecular structure. Lignin is easily damaged just by white-rot fungi and some bacteria; however, for other organisms it is a barrier (Furukawa et al., 2014). For example, moulds and wood-staining fungi easily attack pure hemicelluloses, but they are difficult to reach in the cell walls of wood containing undamaged lignin.

The *ligninolytic complex of enzymes* is produced by white-rot fungi, soft-rot fungi and some bacteria. Brown-rot fungi attack only some parts of lignin using specific enzymes, or by Fenton-type reagents. White-rot fungi produce more types of oxidation and oxidation–reduction enzymes (Highley & Dashek, 1998). The primary phase of lignin decomposition thus need not be its depolymerization (as in the case of polysaccharide decomposition), but for example just splitting off of methoxyl groups. The decomposition of lignin in wood by the complex of ligninolytic enzymes is a complicated biochemical process, during which polysaccharides are also decomposed, as was described by many workers (e.g. Li, 2003; Messner et al., 2003; Fernandez-Fueyo et al., 2012; Floudas et al., 2012; Daniel, 2014). However, the classic view of lignin decomposition together with polysaccharides of wood was given by Eriksson et al. (1990) (Figure 3.26).

- Due to the impact of oligo-, di- and monosaccharides, created during the biodegradation of polysaccharides of wood by hydrolases, the creation of oxidoreductases (CBQ and XBQ) and oxidases (glucose-oxidase, xylose-oxidase) is supported.
- In the presence of synthesized oxidoreductases and oxidases, the preconditions for the reduction of quinones and phenoxyl radicals to phenols are created.
- The depolymerization of lignin itself is carried out in particular by phenoloxidases – lignin-peroxidases – by attacking C_α–C_β bonds or C_β–O–C_4 aryl–ether bonds while the methoxyphenyl derivatives are created, and by attacking the aromatic rings in relation to their splitting off and creation of lower molecular weight substances.
- Methoxyphenyl derivatives are further oxidized by laccase or Mn^{2+} peroxidase while phenoxyl radicals, quinones and methanol are created.
- Methanol is further oxidized by methanol-oxidase enzyme to formaldehyde while H_2O_2 is created, or further via formic acid to CO_2 and H_2O.
- The intermediate products with a split aromatic ring are incorporated in the cells of the pest organism in the Krebs cycle, while glucono- or cellobiono-δ-lactones are metabolized by the pentosane pathway.

Phenoloxidases (lignin-peroxidases, Mn^{2+} peroxidases, versatile peroxidases, laccases) are extracellular enzymes able to cooperate with H_2O_2 (peroxidases) or O_2 (laccases) to remove one electron from phenol-based substances, to which lignin belongs as well. Oxygen is in air, but even today it is not exactly known which enzyme plays the primary role in supplying the H_2O_2 (Li, 2003). Various species of white-rot fungi produce a plethora of ligninolytic phenoloxidases in different amounts (Fernandez-Fueyo et al., 2012; Floudas et al., 2012).

Phenoloxidases catalyse the following depolymerization and other reactions in lignin (Eriksson et al., 1990; Reinprecht, 1997):

- splitting of the C_β–O–C_4 aryl–ether bonds between two phenyl-propane units;
- splitting of the C_α–C_β bonds in propane sections of lignin, while derivatives of phenol nature are created;
- subsequent oxidation of the created derivatives of phenol nature to phenoxyl radicals that may be then transformed to quinones or recombined even to polymer structures;
- demethoxylation of lignin – that is, splitting off of methoxyl groups from its aromatic rings;
- splitting of the aromatic rings of the phenol section of lignin.

Fenton and similar reagents can also be involved in lignin degradation, because in their presence the hydroxyl radical $OH^{\cdot}$ (needed for oxidation

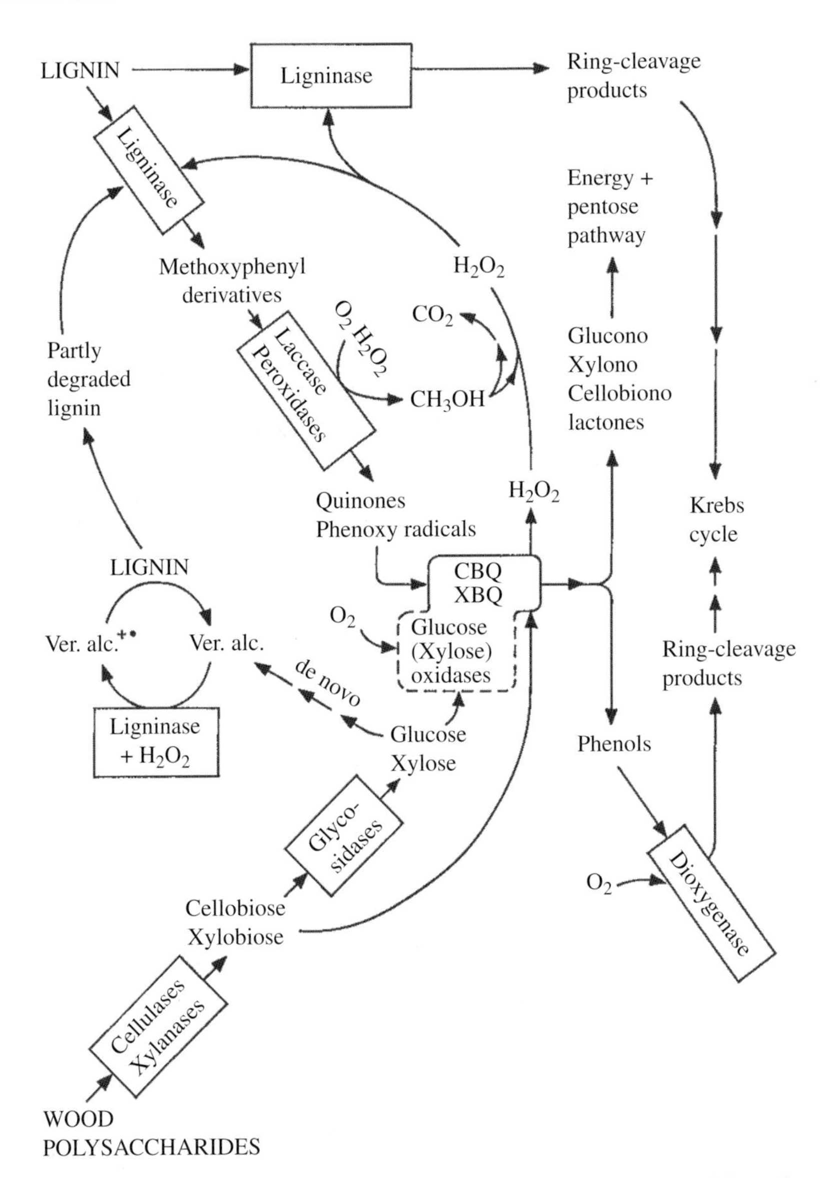

Figure 3.26 Fundamental principles at biochemical decomposition of lignin, cellulose and xylane in wood; CBQ: cellobiose:quinone oxidoreductase; XBQ: xylobiose:quinone oxidoreductase

Source: Eriksson, K-E., Blanchette, R. A. & Ander, P. (1990) *Microbial and Enzymatic Degradation of Wood and Wood Components*. Springer Verlag – Berlin Heidelberg, 407 p. Reproduced by permission of Springer

cleaving of covalent bonds in organic substances) is formed as well. For example, from the 'Mn^{2+} peroxidase/oxalate' system, in first step the relatively stable Mn^{3+} oxalate is produced, which in the second step oxidizes phenolic lignin compounds.

For a better understanding of the mechanisms of lignin biodegradation, reactions were also performed with the isotopes ^{14}C and ^{18}O incorporated into lignin or into its model molecular structures containing phenol-propane units (e.g. Umezawa & Higuchi, 1985). Progress in lignin degradation by pests was made (1) based on knowledge of the molecular genetics of lignin and (2) by the study of phenoloxidases produced, for example, by the white-rot fungus *Phanerochaete chrysosporium* that is often applied in laboratory experiments, using DNA and immune gold labelling techniques (Schmidt, 2006). From the investigations of Messner et al. (2003) and other researches, the important role of low molecular weight agents and extracellularly formed free radicals in lignin biodegradation, which are evidently smaller than enzymes and well diffusible in the cell walls of wood, has been ascertained.

3.6 Properties of biologically damaged wood

Wood-damaging pests cause various changes in the molecular, anatomical, morphological and geometry structural levels of attacked wood. Structural changes of bio-damaged woods were briefly mentioned in previous sections on the description of wood attacks by bacteria, fungi, insects or marine borers. With the structural changes are connected various physical and mechanical properties of damaged wood, and they are briefly mentioned now.

3.6.1 Properties of rotten wood

3.6.1.1 Physical properties of rotten wood

The physical properties of rotten woods change mainly in relation to the type of rot (brown, white, soft) and its degree and are commonly used in practice for inspection of damaged wooden products and constructions (see Section 7.3). The typical changes in rotten wood are as follows:

- *Density* is decreased, most significantly in the case of white rot.
- *Soaking (absorptivity)* is increased for water and other liquids; kinetics of soaking increase due to a better interconnection of wood cells impaired by rot, and the capacity of soaking increases due to the increased porosity of rotten wood.
- *Permeability and impregnability* are increased.
- *Hygroscopicity* is changed also in relation to the type of rot. In the case of brown rot, related to the degradation of polysaccharides, its reduction is observed in the majority of cases (Bech-Andersen, 1995). For example,

beech wood after decay with the brown-rot fungus *S. lacrymans*, characterized by a mass loss of 25%, had a moisture content of just 6.1% after conditioning (45 days, 20 °C, 45% relative humidity), whereas healthy beech wood had a higher moisture of 7.8% after the same mode of conditioning (Makovíny & Reinprecht, 1990). In contrast, in the case of white rot erosion type, related to the uniform degradation of polysaccharides and lignin, the hygroscopicity of wood does not usually change.

- *Electro-physical properties*, such as electric resistance R' (Ω), surface conductivity κ_p (S), bulk conductivity κ_o (S/m), relative permittivity ε', are changed specifically in relation to the type of rot. The conductivity and inductive capacity decrease with the decrease in density of conditioned (air-dryed) rotten wood. In contrast, an increase in permittivity is the consequence of an increased portion of free radicals in fresh rotten wood, as well as due to better the polarization ability of shorter depolymerized macromolecules of cellulose, hemicelluloses and lignin, or molecules created from them (Makovíny & Reinprecht, 1990).
- *Acoustic properties* are weakened. For example, the propagation of ultrasonic waves or pulse waves in rotten wood having cavities or a lower density is slowed down (Reinprecht & Hibký, 2011).

3.6.1.2 Mechanical properties of rotten wood

Mechanical properties of rotten wood decrease, evidently at higher degrees of rot and more in wet state of rotten wood:

- *Impact bending strength* decreases significantly in the first stages of rot. A more apparent decrease occurs in particular in wood damaged by brown rot, which relates to more significant depolymerization of cellulose (Table 3.9).
- *Static mechanical properties* of rotten wood (tension, bending, compression, hardness, etc.) are worsened, but in a more moderate way than the impact bending strength. The tensile and bending characteristics of rotten wood are reduced in a rather more significant way than the compression strength is, particularly due to brown-rot fungi (Table 3.9). Typical is their more apparent decrease in the wet state compared with in the air-dried or oven-dried state of rotten wood; that is, when the strengthening function of hydrogen and van der Waals bonds in the lignin–polysaccharide matrix of decayed wood is at a minimum (Reinprecht, 1992).

3.6.2 Properties of wood having galleries

The molecular structure of wood having galleries (feeding marks and bite-outs) created by insects and marine borers usually does not change (not taking into account the molecular structure of feeding marks in galleries). Therefore, some original physical properties of wood are not more apparently changed due to the presence of galleries (e.g. colour, hygroscopicity), particularly in the undamaged zones of wood. However, some physical properties of wood are significantly influenced (e.g. density, soaking kinetics and capacity, acoustic characteristics), and they can be used to identify galleries inside wooden materials.

An exemption to the lack of damage to the molecular structure of wood is the case when fungal mycelia develop near feeding marks and wood changes its colour and often also its strength and other properties. Insects transfer into wood spores of fungi on their wing-cases or other parts of body: (1) wood-decaying fungi (e.g. poplar borer, *Saperda carcharias*) and (2) wood-staining or ambrosia fungi (e.g. insects *Trypodendron lineatum* and *Scolytus intricatus*).

3.6.2.1 Physical properties of wood damaged by galleries

Physical properties of wood damaged by galleries change mainly in relation to the pest species and its feeding activity in wood, and in relation to the species of wood:

- *density* is decreased, particularly if the bore dust is removed from the galleries;
- *soaking (absorptivity)* is increased for water and other liquids, particularly its kinetics as a consequence of the increased portion of frontal areas in wood due to the presence of exit holes, or even its capacity may increase with the increase of empty pores or pores filled with less dense bore dust;
- *permeability and impregnability* are increased;
- *hygroscopicity* usually does not change, or it changes only in relation to changes in the chemical composition and in the water absorptivity of the bore dust;
- *acoustic and electro-physical properties* of wood impaired by galleries (by analogy with cracks or knots) change in a specific way; for example, the velocity of ultrasonic waves in damaged wood depends mainly upon the directional orientation of galleries with regard to the orientation of the ultrasonic waves.

3.6.2.2 Mechanical properties of wood damaged by galleries

The presence of insect and marine bore galleries has a negative influence on the strength and other mechanical properties of wood. The level of strength decrease depends not only upon the number and size of feeding marks, but significantly also upon their location with regard to the orientation of load forces.

The decrease in mechanical properties of wood damaged by a system of galleries can be hardly foreseen in practice. The reason is the non-homogeneous arrangement of galleries, while we even cannot see them from outside and their scope can be just assessed on the basis of the number of flight holes. The impact of the location of galleries and other macro-damage in wood on the decrease of its modulus of elasticity and strength in bending is greater when they are located from outside in the tension and compression zones of the wooden element.

References

Alfaro, M., Oguiza, J. A., Ramírez, L. & Pisabarro, A. G. (2014) Review – comparative analysis of secretomes in basidiomycete fungi. *Journal of Proteomics* 102, 28–43.

Ananthapadmanabha, H. S., Nagaveni, H. C. & Srinivasan, V. V. (1992) Control of wood biodegradation by fungal metabolites. In *IRG/WP 92*, Harrogate, England, UK. IRG/WP 92-1527.

Arantes, V., Jellison, J. & Goodell, B. (2012) Peculiarities of brown-rot fungi and biochemical Fenton reaction with regard to their potential as a model for bioprocessing biomass. *Applied Microbiology and Biotechnology* 94, 323–228.

Baldrian, P. & Valášková, V. (2008) Degradation of cellulose by basidiomycetous fungi. *FEMS Microbiology Reviews* 32(3), 501–521.

Bech-Andersen, J. (1995) *The Dry Rot Fungus and Other Fungi in Houses*. Hussvamp Laboratoriet ApS, Holte.

Becker, G. (1950) Zerstörung des Holzes durch Tiere. In *Handbuch der Holzkonservierung*, Malhke, F., Troschel, E. & Liese, J. (eds). Springer Verlag, Berlin.

Benko, R. (1992) Wood colonizing fungi as a human pathogen. In *IRG/WP 92*, Harrogate, England, UK. IRG/WP 92-1523.

Betina, V. (1990) *Mycotoxins: Chemistry, Biology, Ecology*. Alfa Bratislava, Slovakia (in Slovak).

Bey, M., Zhou, S., Poidevin, L., et al. (2013) Cello-oligosaccharide oxidation reveals differences between two lytic polysaccharide monooxygenases (family GH61) from *Podospora anserina*. *Applied and Environmental Microbiology* 79, 488–496.

Björdal, C., Nilsson, T. & Bardage, S. (2005) Three-dimensional visualisation of bacterial decay in individual tracheids of *Pinus sylvestris*. *Holzforschung* 59, 178–182.

Blanchette, R. A., Nilsson, T., Daniel, G. & Abad, A. R. (1990) Biological degradation of wood. In *Archaeological Wood – Properties, Chemistry, and Preservation*, Rowell, R. M. & Barbour, R. J. (eds). Advances in Chemistry Series, vol. 225. American Chemical Society, Washington, DC, chapter 6, pp. 141–174.

Cartwright, K. S. G. & Findlay, W. P. K. (1958) *Decay of Timber and its Prevention*. HMSO, London.

Clausen, C. A. (2010) Biodeterioration of wood. In *Wood Handbook – Wood as an Engineering Material*. USDA, Forest Product Laboratory, Madison, WI, pp. 14-1–14-16.

Creffield, J. W. (1996) *Wood Destroying Insects – Wood Borer and Termites*. CSIRO Publishing, Collingwood, Vic.

Daniel, G. (2014) Fungal and bacterial biodegradation: white rots, brown rots, soft rots, and bacteria. In *Deterioration and Protection of Sustainable Biomaterials*, Schultz, T., Goodell, B. & Nicholas, D. D. (eds). ACS Symposium Series, vol. 1158. American Chemical Society, Washington, DC, pp. 23–54.

Daniel, G. & Nilsson, T. (1986) Ultrastructural observations on wood degrading by erosion bacteria. In *IRG/WP 86*, Avignon, France. IRG/WP 1283.

Daniel, G. & Nilsson, T. (1998) Developments in the study of soft rot and bacterial decay. In *Forest Products Biotechnology*. Bruce, A. & Palfreyman, J. W. (eds). Taylor & Francis, London, pp. 37–62.

Despot, R. (1993) Improving the permeability of silver fir wood by bacterial action. *Drvna Industrija* 44(1), 5–14.

Dominik, J. & Starzyk, J. R. (1983) *Insects Damaging Wood*. Panstwowe Wydawnictwo Rolniczne i Lesne – PWRL, Warsaw (in Polish).

Duchesne, L. C., Hubbes, M. & Jeng, R. S. (1992) Biochemistry and molecular biology of defence reactions in xylem of angiosperm trees. In *Defence Mechanisms of Woody Plants against Fungi*, Blanchette, R. & Biggs, A. (eds). Springer, Berlin, pp. 133–146.

Eaton, R. A. & Hale, M. D. C. (1993) *Wood – Decay, Pests and Protection*. Chapman & Hall, London.

Efransjah, F., Kilbertus, G. & Bucur, V. (1989) Impact of water storage on mechanical properties of spruce as detected by ultrasonics. *Wood and Science Technology* 23(1), 35–42.

Enoki, A., Tanaka, H. & Itakura, S. (2003) Physical and chemical characteristics of glycopeptide. In *Wood Deterioration and Preservation: Advances in Our Changing World*, Goodell, B., Nicholas, D. D. & Schulz, T. P. (eds). ACS Symposium Series, vol. 845. American Chemical Society, Washington, DC, pp. 149–153.

Eriksson, K.-E. L., Blanchette, R. A. & Ander, P. (1990) *Microbial and Enzymatic Degradation of Wood and Wood Components*. Springer Series in Wood Science. Springer, Berlin.

Fassatiová, O. (1979) *Moulds and Microscopic Fungi in Technical Microbiology*. SNTL Praha, Czechoslovakia (in Czech).

Fernandez-Fueyo, E., Ruiz-Duenas, F. J., Ferreira, P., et al. (2012) Comparative genomics of *Ceriporiopsis subvermispora* and *Phanerochaete chrysosporium* provide insight into selective ligninolysis. *Proceedings of the National Academy of Sciences of the United States of America* 109, 5458–5463.

Floudas, D., Binder, M., Riley, R., et al. (2012) The Paleozoic origin of enzymatic lignin decomposition reconstructed from 31 fungal genomes. *Science* 336, 1715–1719.

Fojutowski, A. (2005) The influence of fungi causing blue-stain on absorptiveness of Scotch pine wood. In *IRG/WP 05*, Bangalore, India. IRG/WP 05-10565.

Frankova, L. & Fry, S. C. (2013) Biochemistry and physiological roles of enzymes that 'cut and paste' plant cell-wall polysaccharides. *Journal of Experimental Botany* 64 (12), 3519–3550.

Furukawa, T., Bello, F. O. & Horsfall, L. (2014) Microbial enzyme systems for lignin degradation and their transcriptional regulation. *Frontiers in Biology* 9(6), 448–471.

Gáper, J. & Pišút, I. (2003) *Mycology – System, Evolution and Ecology*. UMB Banská Bystrica, Slovakia (in Slovak).

Gogola, E. (1993) *Forest Entomology*. TU Zvolen, Slovakia (in Slovak).

Goodell, B. (2003) Brown-rot fungal degradation of wood – our evolving view. In *Wood Deterioration and Preservation: Advances in Our Changing World*, Goodell, B., Nicholas, D. D. & Schulz, T. P. (eds). ACS Symposium Series, vol. 845. American Chemical Society, Washington, DC, pp. 97–118.

Greaves, H. (1971) The bacterial factor in wood decay. *Wood Science and Technology* 5, 6–16.

Green, F., Larsen, M. J., Winandy, J. E. & Highley, T. L. (1991) Role of oxalic acid in incipient brown-rot decay. *Material und Organismen* 26, 191–213.

Gyarmati, B., Igmándy, Z. & Pagony, H. (1975) *Wood Protection*. MGK, Budapest (in Hungarian).

Halliwell, B. (2003) Free radical chemistry as related degradative mechanisms. In *Wood Deterioration and Preservation: Advances in Our Changing World*, Goodell, B., Nicholas, D. D. & Schulz, T. P. (eds). ACS Symposium Series, vol. 845. American Chemical Society, Washington, DC, pp. 10–15.

Harborne, J. B. (1988) *Introduction to Ecological Biochemistry*. Academic Press, London.

Hastrup, A.C.S., Howell, C., Larsen, F. H., et al. (2012) Differences in crystalline cellulose modification due to degradation by brown and white rot fungi. *Fungal Biology* 116, 1052–1063.

Hegarty, B., Buchwald, G., Cymorek, S. & Willeitner, H. (1986) Der Echte Hausschwamm – immer noch ein Problem? *Material und Organismen* 21, 87–99.

Henry, W. P. (2003) Non-enzymatic iron, manganese, and copper chemistry of potential importance in wood decay. In *Wood Deterioration and Preservation: Advances in Our Changing World*, Goodell, B., Nicholas, D. D. & Schulz, T. P. (eds). ACS Symposium Series, vol. 845, American Chemical Society, Washington, DC, pp. 175–195.

Highley, T. L. & Dashek, W. V. (1998) Biotechnology in the study of brown- and white-rot decay. In *Forest Products Biotechnology*, Bruce, A. & Palfreyman, J. W. (eds). Taylor & Francis, London, pp. 15–36.

Hori, C., Gaskell, J., Igarashi, K., et al. (2013) Genome wide analysis of polysaccharide degrading enzymes in eleven white and brown rot polyporales provides insight into mechanisms of wood decay. *Mycologia* 105, 1412–1427.

Huckfeldt, T. & Schmidt, O. (2005) *Hausfäule- und Bauholzpilze: Diagnose und Sanierung jetzt kaufen*. Rudolf Müller, Cologne.

Hůrka, K. & Čepická, A. (1978) *Reproduction and Development of Insects*. SPN, Prague, (in Czech).

Hutton, T. (1994) Non-destructive surveying the built environment, maintenance and monitoring. In *Conservation and Preservation of Timber in Buildings*, Drdácký, M., Palfreyman, J. W. & Singh, J. (eds). Aristocrat, Prague, pp. 16–20.

Jennings, D. H. & Lysek, G. (1999) *Fungal Biology*, 2nd edn. Bios Scientific, Oxford.

Kačík, F., Veľková, V., Šmíra, P., et al. (2012) Release of terpenes from fir wood during its long-term use and in thermal treatment. *Molecules* 17, 9990–9999.

Kern, V. D. & Hock, B. (1996) Gravitropismus bei pilzen. *Naturwissenschaftliche Rundschau* 49, 174–180.

Kim, J. S., Gao, J. & Daniel, G. (2015) Ultrastructure and immunocytochemistry of degradation in spruce and ash sapwood by the brown rot fungus *Postia placenta*: characterization of incipient stages of decay and variation in decay process. *International Biodeterioration & Biodegradation* 103, 161–178.

Kirk, T. K. & Cowling, E. B. (1984) Biological decomposition of wood. In *The Chemistry of Solid Wood*, Rowell, R. (ed.). Advances in Chemistry Series, vol. 207. American Chemical Society, Washington, DC, pp. 455–487.

Kirker, G. T. (2014) Genetic identification of fungi involved in wood decay. In *Deterioration and Protection of Sustainable Biomaterials*, Schultz, T., Goodell, B. & Nicholas, D. D. (eds). ACS Symposium Series, vol. 1158. American Chemical Society, Washington, DC, chapter 4, pp. 81–91.

Koenigs, J. W. (1974) Hydrogen peroxide and iron – a proposed system for decomposition of wood by brown-rot basidiomycetes. *Wood and Fiber* 6, 66–80.

Křístek, J. & Urban, J. (2004) *Forest Entomology*. Academia Praha, Czech Republic (in Czech).

Langendorf, G. (1988) *Holzschutz*. VEB Fachbuchverlag, Leipzig.

Li, K. (2003) The role of enzymes and mediators in white-rot fungal degradation of lignocellulose. In *Wood Deterioration and Preservation: Advances in Our Changing World*, Goodell, B., Nicholas, D. D. & Schulz, T. P. (eds). ACS Symposium Series, vol. 845. American Chemical Society, Washington, DC, pp. 196–209.

Liese, W. (1992) Holzbakterien und Holzschutz. *Material und Organismen* 27, 191–202.

Makovíny, I. & Reinprecht, L. (1990) Possibilities for determination of wood decay by its electro-physical properties. I: *Ochrana dreva '90 – Wood Protection '90*. DT Bratislava, Czechoslovakia, pp. 70–74 (in Slovak).

Messner, K., Facker, K., Lamaipis, P., et al. (2003) Overview of white-rot research: where we are today. In *Wood Deterioration and Preservation: Advances in Our Changing World*, Goodell, B., Nicholas, D. D. & Schultz, T. P. (eds). ACS Symposium Series, vol. 845. American Chemical Society, Washington, DC, pp. 73–96.

Mička, L. & Reinprecht, L. (1999) Rot of spruce wood due to the wood-destroying fungus *Serpula lacrymans* under stable and variable climatic conditions. *Drevársky Výskum (Wood Research)* 44(2), 28–40.

Milling, A., Kehr, R., Wulf, A. & Smalla, K. (2005) Survival of bacteria on wood and plastics: dependence on wood species and environmental conditions. *Holzforschung* 59(1), 72–81.

Mirič, M. & Willeitner, H. (1984) Lethal temperature for some wood-destroying fungi with respect to eradication by heat treatment. In *IRG/WP 84*, Ronneby Brunn, Sweden. IRG/WP 1229.

Novák, V., Hrozinka, F. & Starý, B. (1974) *Atlas of Insect's Pests of Forest Woods*. Príroda, Bratislava (in Slovak).

Palfreyman, J. W., Phillips, E. M. & Staines, H. J. (1996) The effect of calcium ion concentration on the growth and decay capacity of *Serpula lacrymans* (Schumacher ex Fr.) Gray and *Coniophora puteana* (Schumacher ex Fr.) Karst. *Holzforschung* 50, 3–8.

Pánek, M. & Reinprecht, L. (2008) Bio-treatment of spruce wood for improving of its permeability and soaking – part 1: direct treatment with the bacterium *Bacillus subtilis*. *Wood Research* 53(2), 1–12.

Pánek, M. & Reinprecht, L. (2011) *Bacillus subtilis* for improving spruce wood impregnability. *BioResources* 6(3), 2912–2931.

Peterson, Ch., Wagner, T., Mulrooney, J. E. & Shelton, T. G. (2006) *Subterranean Termites – Their Prevention and Control in Buildings*. Home and Garden Bulletin No. 64. USDA, Forest Service, Starkville, MS.

Phillips-Laing, E. M., Staines, H. J. & Palfreyman, J. W. (2003) The isolation of specific bio-control agents for the dry rot fungus *Serpula lacrymans*. *Holzforschung* 57, 574–578.

Rayner, A. D. M. & Boddy, L. (1988) *Fungal Decomposition of Wood – Its Biology and Ecology*. John Wiley & Sons, Ltd, Chichester.

Reinprecht, L. (1992) *Strength of Deteriorated Wood in Relation to its Structure*. Monograph VPA 2/1992. TU Zvolen, Slovakia.

Reinprecht, L. (1996) Structure and impact bending strength of beech and spruce degraded by brown-rot fungus *Serpula lacrymans* and white-rot fungus *Coriolus versicolor*. In *Biotechnology in the Pulp and Paper Industry*, Srebotnik, E. & Messner, K. (eds). Facultas-Universitätsverlag, Vienna, pp. 565–568.

Reinprecht, L. (1997) *Processes of Wood Degradation*. TU Zvolen, Slovakia (in Slovak).

Reinprecht, L. (2008) *Wood Protection*. Handbook. TU Zvolen, Slovakia (in Slovak).

Reinprecht, L. (2013) *Wood Protection*. Handbook. TU Zvolen, Slovakia.

Reinprecht, L. & Hibký, M. (2011) The type and degree of decay in spruce wood analyzed by the ultrasonic method in three anatomical directions. *BioResources* 6(4), 4953–4968.

Reinprecht, L. & Lehárová, J. (1997) Microscopic analyses of woods – beech (*Fagus sylvatica* L.), fir (*Abies alba* Mill.) and spruce (*Picea abies* L. Karst.) in various stages of rot caused by fungi *Serpula lacrymans*, *Coriolus versicolor* and *Schizophyllum commune*. In: *Drevoznehodnocujúce Huby '97* (*Wood-Damaging Fungi 97*), 1st Symposium. TU Zvolen, Slovakia, pp. 91–113 (in Slovak, English abstract).

Reinprecht, L. & Tiralová Z (2001) Susceptibility of the sound and the primary rotten wood to decay by selected brown-rot fungi. *Wood Research* 46(1), 11–20.

Richardson, B. A. (1993) *Wood Preservation*. E & FN Spon, London.

Ritschkoff, A-C., Pere, J., Buchert, J. & Viikari, L. (1992) The role of oxidation in wood degradation by brown-rot fungi. In *IRG/WP 92*, Harrogate, England, UK. IRG/WP 92-1592.

Robson, G. (1999) Hyphal cell biology. In *Molecular Fungal Biology*, Oliver, R. P. & Schweizer, M. (eds). Cambridge University Press, Cambridge, pp. 164–184.

Rypáček, V. (1957) *Biology of Wood Decaying Fungi*. N-ČSAV Praha, Czechoslovakia (in Czech).

Saake, B., Kruse, T. & Puls, J. (2001) Investigations on molar mass, solubility and enzymatic fragmentation of xylans by multi-detected SEC chromatography. *Bioresource Technology* 80(3), 195–204.

Schánĕl, L. (1975) Influence of environmental conditions on wood decay by fungi. *Drevársky Výskum (Wood Research)* 20(1), 59–79 (in Czech, English abstract).

Schánĕl, L. (2005) Influence of environmental conditions on wood decay by fungi – II. In *Sanace a Rekonstrukce Staveb 2005 (Restoration and Reconstruction of Constructions 2005), 27th Conference WTA*, Brno, pp. 31–38 (in Czech, English abstract).

Schloss, P. D., Delalibera, J, Handelsman, J. & Roffa, K. F. (2006) Bacteria associated with the guts of two wood-boring beetles: *Anoplophora glabripennis* and *Saperda vestita* (*Cerambycidae*). *Environmental Entomology* 35(3), 625–629.

Schmidt, O. (2006) *Wood and Tree Fungi – Biology, Damage, Protection, and Use*. Springer-Verlag, Berlin.

Schmidt, O. & Liese, W. (1994) Occurrence and significance of bacteria in wood. *Holzforschung* 48, 271–277.

Score, A. J., Bruce, A., King, B. & Palfreyman, J. W. (1998) The biological control of *Serpula lacrymans* by *Trichoderma* species. *Holzforschung* 52(2), 124–132.

Serdjukova, I. R. & Toskina, I. N. (1995) Some characters of biology and physiology of the common furniture beetle *Anobium punctatum* De Geer (Coleoptera, Anobiidae). *Russian Entomology Journal* 4(1–4), 35–43.

Shigo, A. L. (1979) *Tree Decay – An Expanded Concept*. Agricultural Information, Bulletin No. 419, USDA, Forest Service, Northeastern Forest Experiment Station, Durham, NH.

Shimokawa, T., Naklamura, M., Hayashi, N. & Ishihara, M. (2004) Production of 2,5-dimethoxyhydroquinone by the brown-rot fungus *Serpula lacrymans* to drive extracellular Fenton reaction. *Holzforschung* 58, 305–310.

Šimonovičová, A., Pangallo, D. & Chovanová, K. (2005) Microorganisms on wooden sculptures. In *Drevoznehodnocujúce Huby 2005 (Wood Damaging Fungi 2005), 4th Symposium*. TU Zvolen, Slovakia, pp. 95–97 (in Slovak, English abstract).

Singh, A. P., Nilsson, T. & Daniel, G. (1992) Resistance of *Alstonia scholaris* vestures to degradation by tunnelling bacteria. In *IRG/WP 92*, Harrogate, England, UK. IRG/WP 92-1547.

Terebesyová, M., Reinprecht, L. & Makovíny, I. (2010) Microwave sterilization of wood for destroying mycelia of the brown-rot fungi *Serpula lacrymans*, *Coniophora puteana* and *Gloeophyllum trabeum*. In *Wood Structure and Properties*, Kúdela, J. & Lagaňa, R. (eds). Arbora Publishers Zvolen, Slovakia, pp. 145–148.

Thörnqvist, T., Kärenlampi, P., Lundström, H., et al. (1987) *Vedegenskaper och mikrobiella angrepp i och på byggnadsvirke*. Swedish University of Agricultural Sciences, Uppsala (English abstract).

Tichý, V. (1975) Significance of synergetic and antagonistic effects at wood decay by fungi. *Drevársky Výskum* (*Wood Research*) 20(1), 37–57 (in Slovak, English abstract).

Umezawa, T. & Higuchi, T. (1985) Role of guaiacol in the degradation of arylglycerol-β-guaiacyl ether by *Phanerochaete chrysosporium*. *FEMS Microbiology Letters* 26, 123–126.

Unger, A., Schniewind, A. P. & Unger, W. (2001) *Conservation of Wood Artifacts*. Springer-Verlag, Berlin.

Urban, Z. & Kalina, T. (1980) *System and Evolution of Lower Development Plants*. SPN Praha, Czechoslovakia (in Czech).

Viitanen, H. & Ritschkoff, A.-C. (1991) *Mould Growth in Pine and Spruce Sapwood in Relation to Air Humidity and Temperature*. Swedish University of Agricultural Sciences, Uppsala.

Wakeling, R. (2006) New observation and interpretation for tunneling bacteria decay. In IRG/WP 06, Tromsø, Norway. IRG/WP 06-10579.

Ważny, H. & Czajnik, M. (1963) Zum Auftreten holzzerstörender Pilze in Gebäuden in Polen. *Folia Forestalia Polonica B-Series* 5, 5–17.

Wilcox, W. W. (1978) Review of literature on the effects of early stages of decay on wood strength. *Wood and Fiber* 9(4), 252–257.

4 Structural Protection of Wood

4.1 Methodology of structural protection of wood

The structural protection of wooden products and constructions is concerned with the aim to supress the danger of their damage by fungi, insects, marine organisms (Table 4.1), fire and weather factors in interiors and exteriors. This can be achieved by: (1) a selection of durable wood species and suitable types of wooden materials (see Sections 1.3 and 4.2), and (2) proper designs (Figure 4.1; Sections 4.3 and 4.4). Requirements of structural protection are applied to log cabins, half-timbered and assembled wooden houses, bridges, arbours, pergolas, carports, fences, garden furniture, balconies, trusses, ceilings, windows, doors, as well as to other wooden constructions and products, with the aim to prolong their service life (Table 4.2). The principles of structural protection of wood are addressed in several handbooks and publications (e.g. Feirer & Hutchings, 1986; Stalnarker & Harris, 1989; Lebeda et al., 1988; Žák & Reinprecht, 1998; Reinprecht & Štefko, 2000; Wilcox, 2000; Houdek & Koudelka, 2004; Štefko & Reinprecht, 2004; Underwood & Chiuini, 2007; Hugues et al., 2008).

The optimum transport, storage and assembly itself belong as well as to the significant tasks of the structural protection of wooden products. Their placing directly onto a ground without the protection against rainfall water is considered to be really inappropriate. The improper solution in exterior is also the placing of wooden elements under vapour-impermeable yet mechanically impaired foils, since there is an increased danger of their rot due to higher moisture.

4.2 Selection of suitable wood materials

Requirements for wood and other materials for wooden constructions relate to their input quality and resistance to pests, fire and weather.

Wood Deterioration, Protection and Maintenance, First Edition. Ladislav Reinprecht.

Table 4.1 Classes of wood use in conditions suitable for biological pests (modified from EN 335)

Use class	Wood-based product use conditions	Wood moisture w (%)	Occurrence of biological agents[a]					
			Wood-decaying fungi		Wood-staining fungi and moulds	Wood-damaging insects		Marine borers
			Bas.	Asc.		B	T	
1	Insider, without expectation of wetting	Permanently ≤20	—	—	—	E	L	—
2	Outside, undercover, wetting can occur	From time to time >20	E	—	E	E	L	—
3 3.1, 3.2	Outside, above ground, with weathering and wetting: 3.1: limited 3.2: prolonged	More frequently >20	E	—	E	E	L	—
4	Contact with ground and/or fresh water	Permanently >20	E	E	E	E	L	—
5	Contact with sea or brackish water	Permanently >20	E	E	E	E	L	E

[a] Bas.: Basidiomycota; Asc.: Ascomycota; B: beetles; T: termites; E: everywhere in Europe; L: local occurrence in Europe (e.g. termites just in the southern Europe).

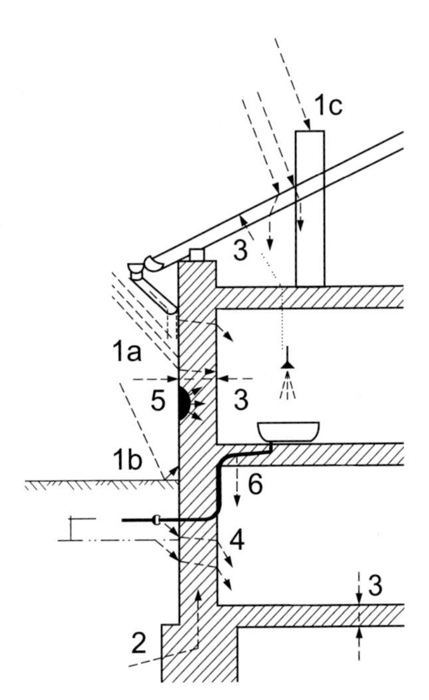

Figure 4.1 Zones with higher risks of moisture penetration to a building with the subsequent damage of wooden elements by wood-damaging fungi and insects: (1a) rainwater; (1b) splashing rainwater; (1c) rainwater entering through the chimney; (2) capillary water; (3) diffusion and condensation water; (4) water enteric by hydrostatic pressures; (5) hygroscopic water; (6) water from damaged pipes, radiators, and so on

Source: Lebeda, J., et al. (1988) *Reconstruction of Wet Masonry in Buildings* (in Czech). SNTL Praha, Czechoslovakia, 232 p

The *input quality of wood* is connected with use of wood products according to the relevant standards (in Europe EN); that is, products with unified dimensions, free from fungal rot and insect galleries, having required values of strength, hardness, and so on.

The *natural durability of wood* is its ability to resist the activities of wood-damaging fungi, insects and other pests, and in a broader view it also represents its resistance against weather factors and aggressive chemical agents. The data on the natural durability of wood in conditions convenient for its attack by decay-causing fungi and wood-boring beetles are stated in the EN 350-2 standard (Table 1.3).

Fire resistance of wood is low. The flammability of wood-based materials is dealt with by EN 13501-1 (Table 2.2). The flammability levels of wooden materials depend mainly upon their:

- density – that is, wood of hard broadleaves is not very flammable, whereas softer coniferous wood is medium flammable;
- constitution – that is, wood–cement particleboards are not very flammable, whereas the classic wood-particleboards are medium flammable.

Table 4.2 Basic methods of structural protection of wooden products and constructions

- The selection of suitable wood species, wooden materials and auxiliary materials
- The assurance of input quality of wood and all other materials
- The assurance of permanently low moisture of wood and the entire object in order to prevent the activity of biological pests:
    - shape optimization of wooden elements and the entire object
    - isolation of wooden elements and other materials from foundations, rainfall, splashing, condensed and service water
    - correct construction designs and suitable material composition of the object or its parts; e.g. the claddings and roof covers
    - regulation of climatic conditions within the object by correct ventilation, etc.
- The creation of mechanical barriers against ageing of wood caused by sun, water and other weather factors, as well as against the access of biological wood pests
- The creation of fire sections and observation of other fire safety principles

4.3 Design proposals for permanently low moisture of wood

Wood must have prescribed moisture contents for certain products. Wood must have less than 10% moisture where intended for interior use, less than 15% for bonded wooden elements, less than 25% for elements subjected to unprotected exposure if shrinkage is no defect, and no moisture is specified for elements exposed permanently to a humid environment. Input moisture content of wood and wooden materials is predetermined in particular by their exposure within an object; that is, the moisture of new wooden windows and claddings in central Europe should be less than 12%, of elements in trusses from 10 to 14% and floor friezes ~8% (Figure 4.2). Hence, one can avoid the shape deformation of wood elements and entire structural units (e.g. the shrinkage or swelling of parquet floors), the generation of cracks, and in particular the activity of fungi and insects (Chapter 3).

4.3.1 Estimated moisture of wood

Estimated moisture content of wood w_{est} is an important parameter for its structural protection. For individual wooden elements of the construction it can be calculated as follows:

$$w_{est}\ (\%) = w_{equilibrium} + (w_{input} - w_{equilibrium})a + w_{additional}a \qquad (4.1)$$

where $w_{equilibrium}$ is the equilibrium moisture content of wood for a particular exposure, w_{input} is the input moisture of wood at the time of building-in,

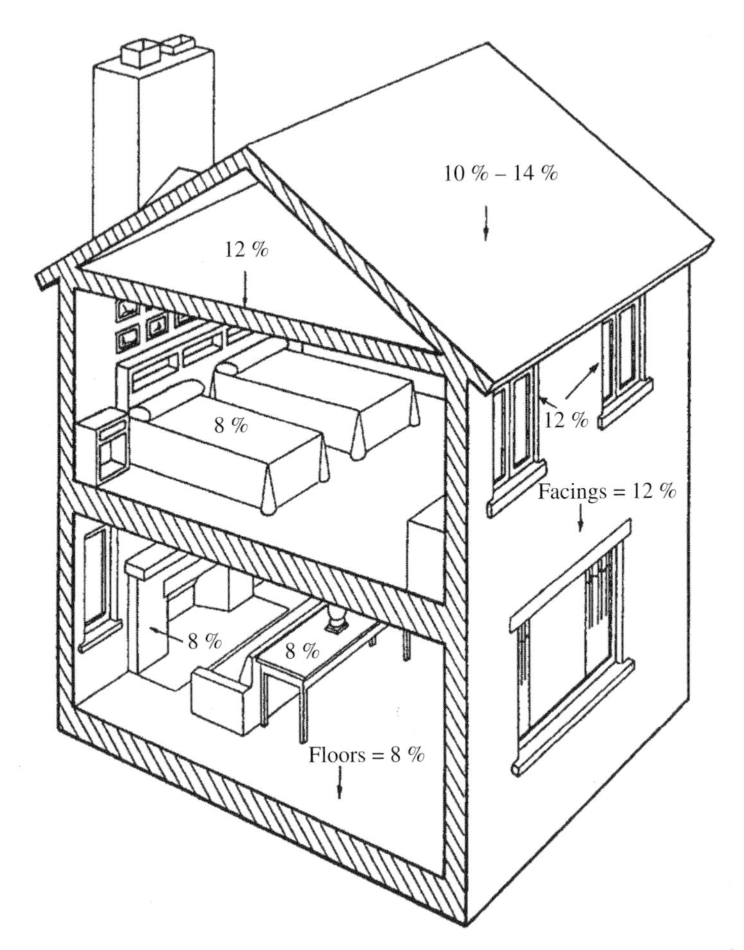

Figure 4.2 Typical moisture contents of wood at various locations of a residential house

Source: R., L. (2008) *Ochrana Dreva* (*Wood Protection*), Handbook, TU Zvolen, Slovakia, 453 p. Reproduced by permission of TU Zvolen

$w_{additional}$ is additionally increased moisture of wood (e.g. caused by rainfall or condensed water) with regard to its equilibrium moisture content (e.g. due to rainfall water it increases from 0% up to more than 60%), a is a dry-out factor of wood (from 0.1 to 0.9) that depends upon the dimensions, shape and roughness of the wood surface, upon its accessibility to ambient air and the overall climatic conditions within the object ($a = 0.1$ for less thick elements, $h < 50$ mm, accessible from all sides; $a = 0.9$ for thicker beams, treated by vapour-impermeable paints).

In central Europe, wooden elements of outdoor constructions in direct contact with ground, but sometimes also above ground, have moisture oftentimes above 30%. On the other hand, for correctly designed interior constructions (roof trusses, ceilings, stairways, etc.) it should not significantly exceed the limit of wood-boring insect activity ($w_{crit.min.insect}$) equal to 10%, and it should be permanently under the limit of fungal-decay activity ($w_{crit.min.fungi}$) equal to 20% (Reinprecht, 2008).

4.3.2 Shape optimizations for wood moisture reduction

4.3.2.1 Principles of shape optimizations of wooden elements and constructions for exteriors

The most important principles of structural protection of wood for exteriors are as follows.

- Restrict or absolutely eliminate the effect of water on the faces of solid wood and on all side areas of plywood and other wooden composites.
- Minimize the share of facial (frontal) areas with regard to other areas of wooden elements.
- Do not use elements with significant cracks or those prone to cracking:
  - do not use wood raw material with the pith that is prone to deformation and cracking (this is demanded mainly for elements with greater cross-section) (Figure 4.3);
  - elements with greater cross-section should preferably be made from layered wood in order to restrict cracking – an example is bonded beams (glulam and euro-beams – Figure 4.4);
  - facial areas of elements should be equipped with anti-splitting fixtures (e.g. gang-nail type plates, S-hooks, steel tapes).
- Wooden shingles, as well as other elements with permanent contact with rainfall water, should preferably be made by splitting alongside the fibre, not by sawing.
- The facial areas of utility poles, posts in fences, and so on should be protected against rainfall water from above by using a shelter (Figure 4.5) or they should be cut at an angle (Figure 4.6).
- Thin-walled profiles (e.g. claddings, floor boards) should be equipped with grooves from the back side in order to reduce stress and cracking (Figure 4.7).

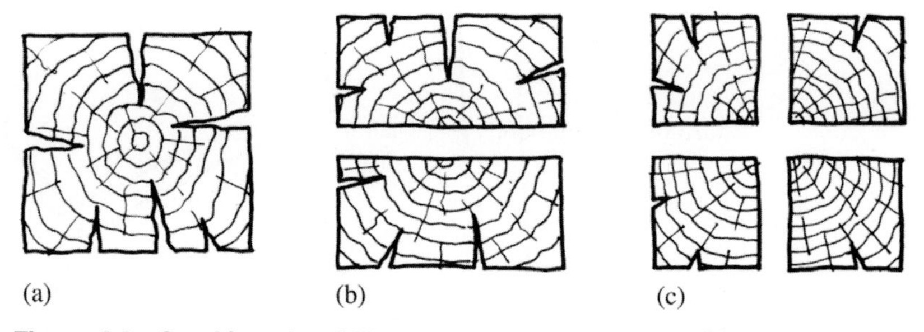

Figure 4.3 Cracking should be prevented by correct cutting of round wood; there should be no central pith in the elements with greater dimensions since it is usually the cause of the creation of large cracks (a), or the pith should be cut through (b, c) or cut out (c)

Figure 4.4 **Euro-beam – a window profile made of bonded wood lamellae has a great dimensional stability and resists cracking better than a massive beam does**

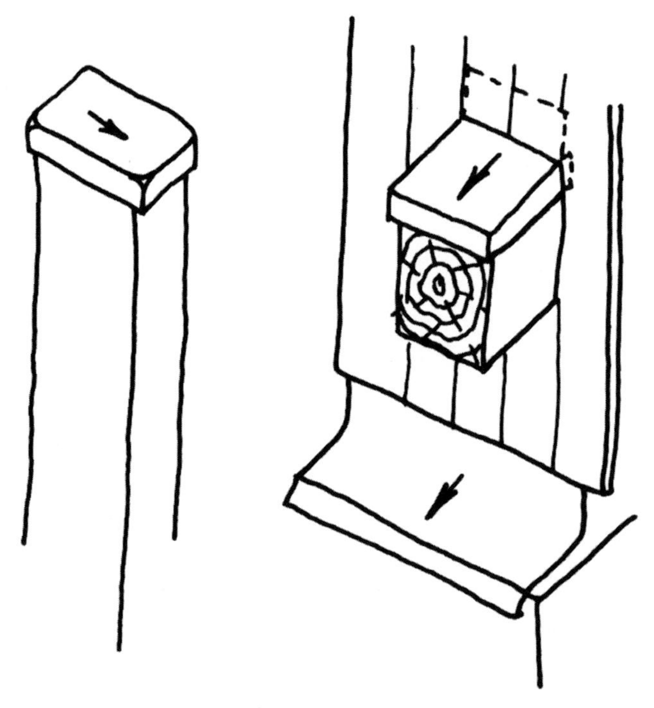

Figure 4.5 **Faces of wooden construction protected against rainwater using a shelter**

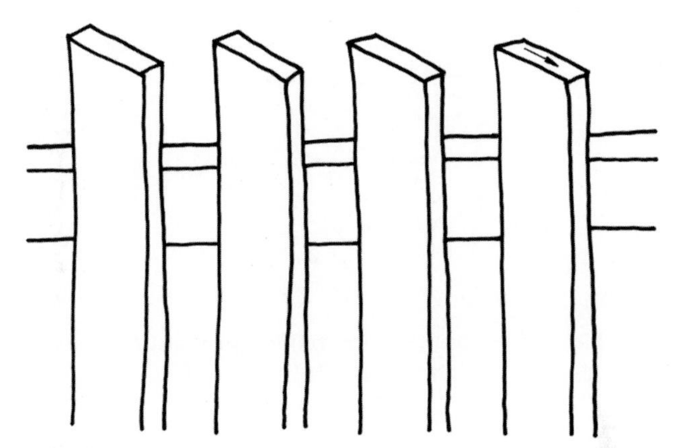

Figure 4.6 Faces of a wooden fence may be partially protected against rainwater by using an inclined cut

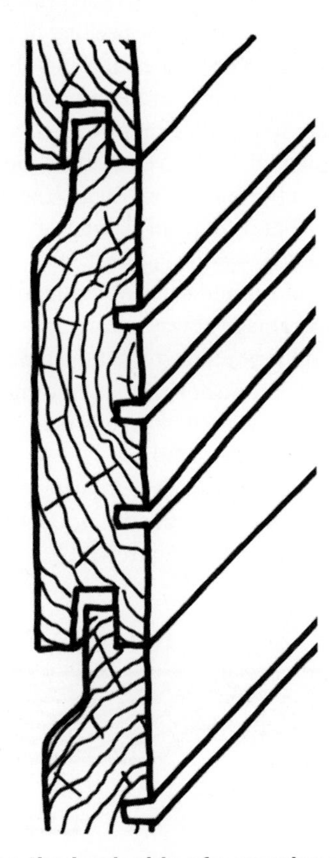

Figure 4.7 Grooves on the back side of a wooden panelling prevent its deformation and cracking as a result of climatic changes

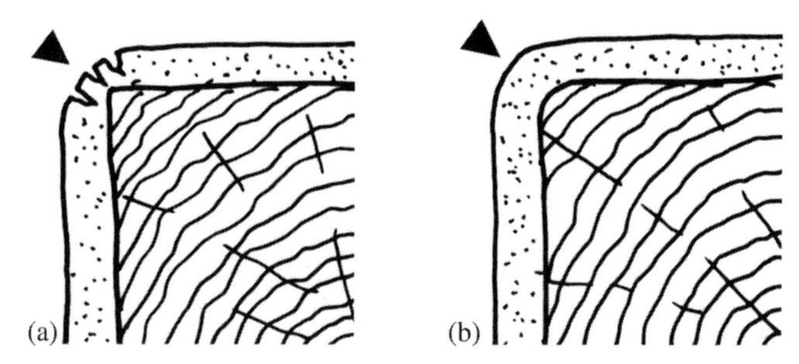

Figure 4.8 Paints adhere better to a rounded edge of a wooden element (b) than to a sharp edge (a)

- Horizontally placed elements (e.g. floor boards for terraces and balconies) should be slightly inclined so that water can run out down them, while it is suitable to leave gaps between the elements for water to run out as well as to account for possible swelling of wood.
- Do not use those types of joints that are easily soaked by water and from which water hardly evaporates. Moreover, it is suitable to smooth the joints in order to facilitate rapid water evaporation.
- The edges of wooden elements should be suitable rounded, which is particularly important before their treatment with paints against weather effects, with the aim to avoid flaking off of paints and then quick penetration of water to the wood (Figure 4.8).
- For windows and exterior doors it is desirable to apply details for provision of the drainage of water from gaps, against the condensation water generation and for outflow of water from structural details.
- For vertical and horizontal cladding panels with 'tongue and groove' joints it will be necessary to correctly design their orientation and structural details (Figures 4.9 and 4.10).
- Roofs should be dealt with using sufficiently long overlaps in order to protect the external walls of wooden houses, arbours and other structures against rainwater (Figure 4.11).

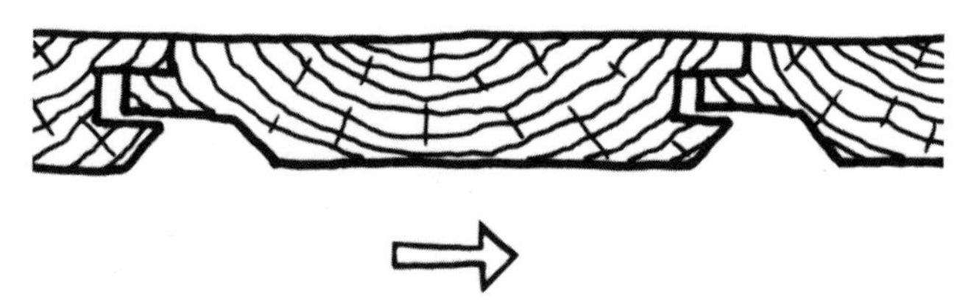

Figure 4.9 Correct orientation of 'tongue and groove' joint towards the direction of prevailing winds with rainwater in the case of vertically placed wooden panelling

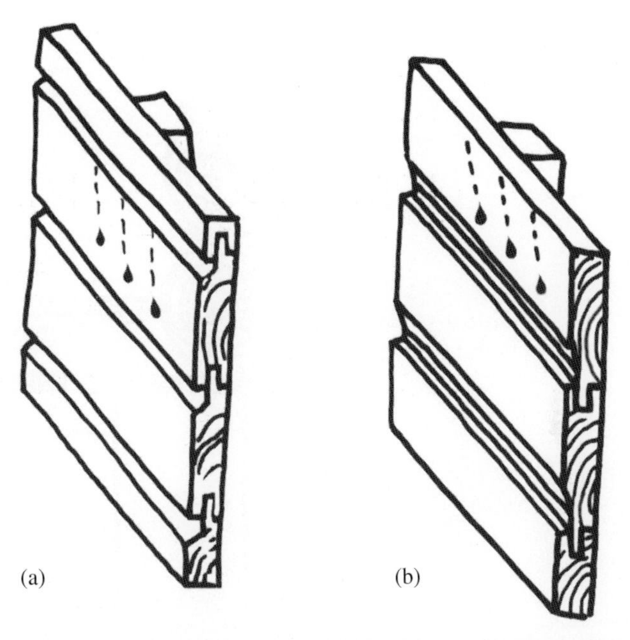

Figure 4.10 Correct (a) and incorrect (b) location of 'tongue and groove' joints towards the direction of rainwater in the case of horizontally placed wooden panelling

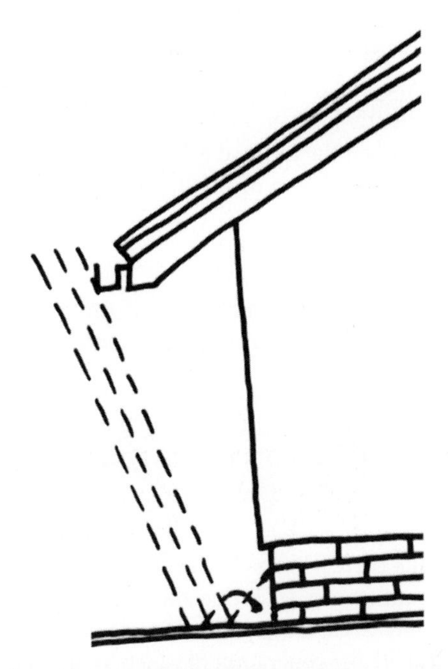

Figure 4.11 Roof overlap and height of concrete or rock bases affect the intensity of rainwater on wooden panels and facade elements (windows and doors) – see the sufficient roof overlap

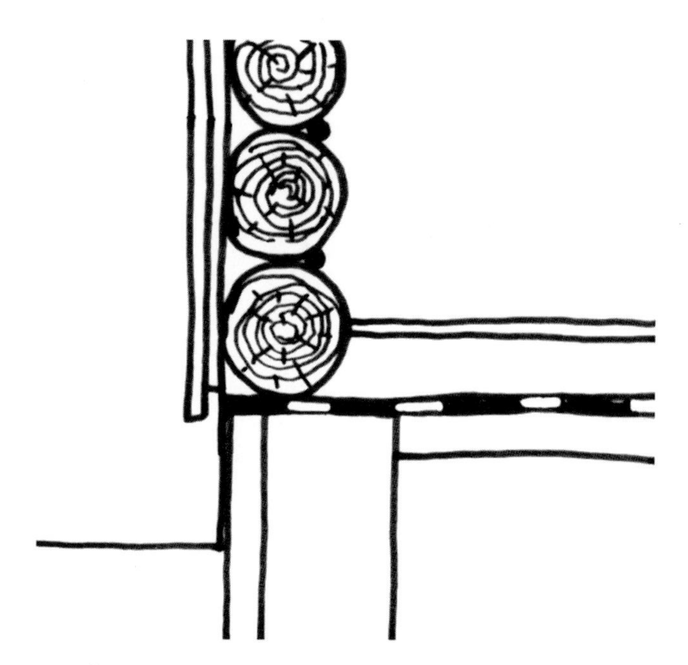

Figure 4.12 Old craftsmen have long known that the bottom side of a wooden panel should cover the hydro-insulation crevice between the base and wooden log-cabin construction

- Panelling of wooden walls should be designed so that the panels are sufficiently long in the bottom section and overlap the hydro-insulation gap between the mineral base and the wooden construction, and thus prevent the penetration of splashing water to the bottom part of the construction (Figure 4.12).

4.3.2.2 Principles of shape optimization of wooden elements against condensed water in interiors

The geometry of wooden and other elements of houses must be correctly designed with regard to providing a sufficient air flow in order to prevent the creation of zones with incorrect thermal fields. This is particularly important for zones around wooden windows, especially those in a roof:

- The correct shape of the lining and the location of the window are significant factors such that the heat transfer from the interior air to window will increase, then the surface temperature of the window will increase, and finally the generation of water condensate will be reduced (Figure 4.13). In the case where there is a suitably located heat source under the window the supply of dry air to the particular detail will be provided for, while the

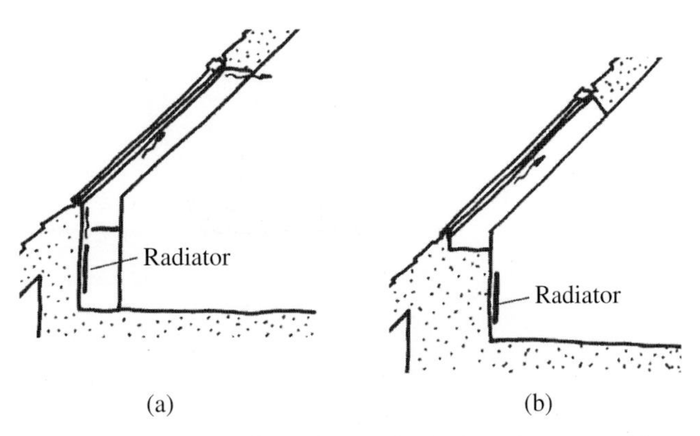

Figure 4.13 Geometry of lining of a roof window above a radiator will prevent (a) or will not prevent (b) condensation of water

precondition for this is also the correct shaping of window parapet and the area above the window.

- Venting gaps in the window design reduce water condensation.

4.3.3 Waterproofing and other isolations of wood and wooden composites from water sources

4.3.3.1 Isolation of wood in exteriors

Isolations from wet concrete or rock base, as well as from the ground and other water sources, are important mainly for wooden houses, bridges, children's playgrounds, pergolas, fences, pale fencing, noise-breaks and light-breaks next to roads. The shape of these objects must be properly constructed so that water cannot stay for a long time on them. Examples of this are shape optimizations of wooden elements and entire constructions (see Section 4.3.2). At the same time, it is necessary to create sufficient air or other isolations for them.

Air insulations are created between wood and wet mineral materials (concrete, masonry, etc.). They mainly prevent the penetration of water from porous mineral materials to wood by capillary forces (Figure 4.14). At the same time, they provide permanent aeration for wood and they stop rain water being reflected back from the terrain (Figure 4.15)

Hydro-insulations are inserted between the wooden element and other material. Hydro-insulations are made from suitable impermeable materials, using the barrier effect to prevent the penetration of water into the wood. Wooden constructions are isolated by hydro-insulations:

- from capillary water in the bases (e.g. by bitumen layer);
- from rainfall water on the roofs (e.g. by ceramic roofing and hydro-insulation foil, together with suitable ventilation with air gaps up to the frontispiece (Figure 4.16);

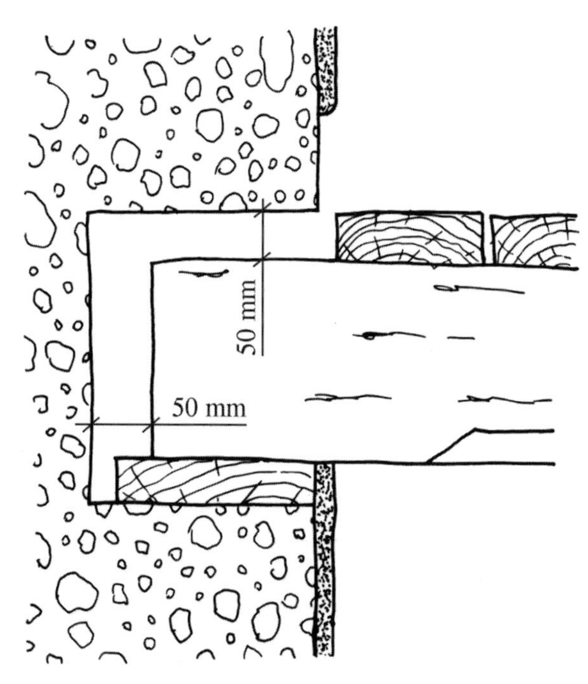

Figure 4.14 Heads of wooden elements must remain permanently vented; for example, it is necessary to leave at least 50 mm air gaps between the head of a beam and the masonry

- from rainfall water on the outside claddings (e.g. by waterproof vapour-permeable paints).

Waterproof paints prevent the access of both rainfall and splashing water to the building, and at the same time they slow down the transport of air humidity to the wood. However, paints should be sufficiently permeable to water vapour that no condensed and other water sources can accumulate under them. In practice, polyacrylate, acrylate-polyurethane, acrylate-alkyd, alkyd or other polymer film-forming coating systems are used. In contrast, the use of cover materials that are impermeable to vapour (e.g. oil paints, plastics, metals and other foils or panels) may be counterproductive, since wood in exteriors under vapour-impermeable barriers is often permanently moist and is attacked by fungi and other pests more frequently.

4.3.3.2 Isolations of wood in interiors

In interiors it is important to isolate wooden products (made from natural wood and wooden composites) from the sources of condensed water, and sometimes also from capillary water (Figure 4.14) or even rainfall water in the case of damaged roof covers. Also important is the correct installation, operation and

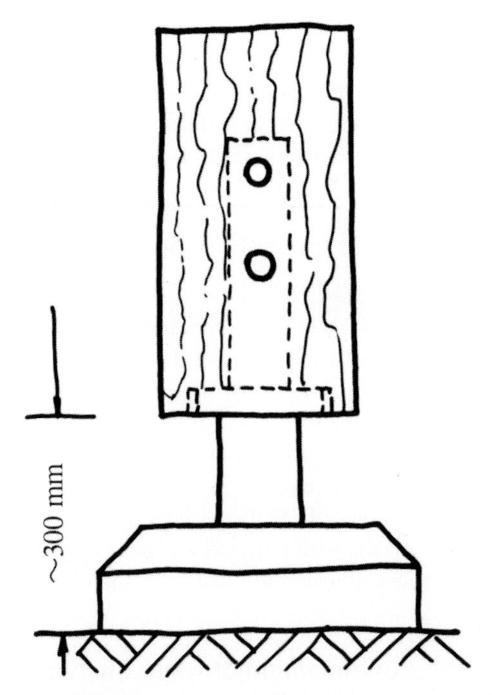

Figure 4.15 Air gap between the bottom part of a wooden product or construction (e.g. wooden pole) and the terrain should a minimum of 150–300 mm

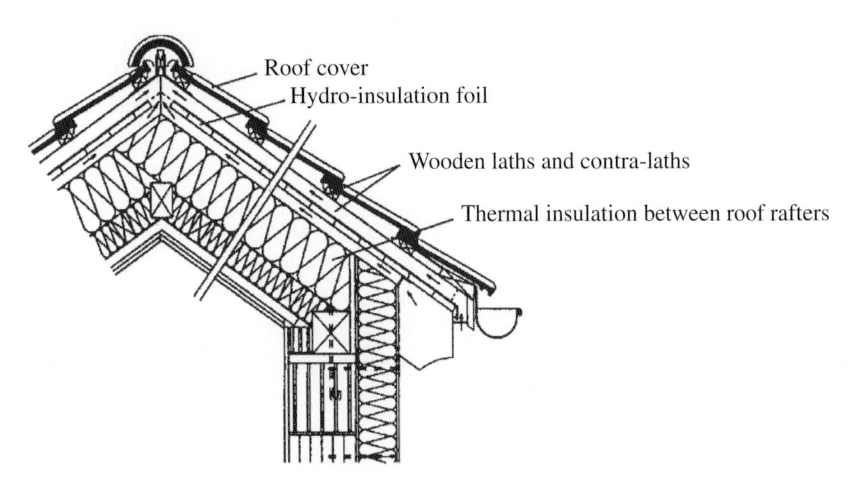

Figure 4.16 Correct ventilation of a three-layer roof will prevent an increase in moisture of wooden elements and insulation materials in the roof

maintenance of water, sewage and heating networks and various sanitary devices, such as basins, bath tubs, radiators, and so on. One should also not forget the problem of air insulations in the composition of roof covers and peripheral walls with the objective of their permanent ventilation and prevention of condensed water creation (Figure 4.16; see Section 4.3.4).

4.3.4 Structural design to prevent condensed water generation

Condensed water is generated in badly designed structural and material objects; for example, with a bad composition of peripheral walls and roof covers (Australian Building Codes Board, 2011). It occurs in particular in winter upon the diffusion of water vapour from the interior to the exterior in places of thermal bridging. In order to prevent the generation of condensed water, it is necessary to observe the principles of construction physics and aerodynamics. From the point of view of construction physics, it is important to realize such design and material solutions that significantly eliminate the risk of water vapour condensation on the inside surfaces of peripheral walls and roof covers – the best perform thermal insulation from the outside of the building.

4.3.4.1 Condensation of water on interior walls in wooden constructions

Condensation of water on surfaces of interior's walls in wooden and other buildings occurs when the surface temperature T_S drops below the dew point T_{DP} ($T_{DP} > T_S$). The interior surface temperature T_S depends mainly upon the thermal and insulation properties of walls. The dew point T_{DP} on the wall surface depends on the air temperature T_A and interior relative humidity (RH); for example, for T_A of 20 °C and RH of 55% the dew point T_{DP} is ~10 °C.

Condensation of water on the surfaces of interior cold walls can be eliminated by a decrease in the air's RH (by air conditioning or a correct ventilation) or by increasing the thermal resistance of the peripheral walls in the construction; for example, by increasing the thickness and thermal resistance of individual materials in the peripheral wall. A high risk of condensation occurs in particular at the location of thermal bridges, corners and weakened spots. Thermal bridges in the peripheral log-cabin wall or sandwich wooden wall are auxiliary materials with higher thermal conductivity crossing the wall (e.g. steel connecting bolts or steel beams). They can be determined by thermography cameras.

4.3.4.2 Condensation of water inside peripheral walls or roof covers

Condensation of water inside peripheral walls depends upon the temperature of the particular place of a given construction and the distribution of diffusion resistances of the individual layers of that construction, alongside its thickness. When diffusing via pores of the individual materials and construction crevices,

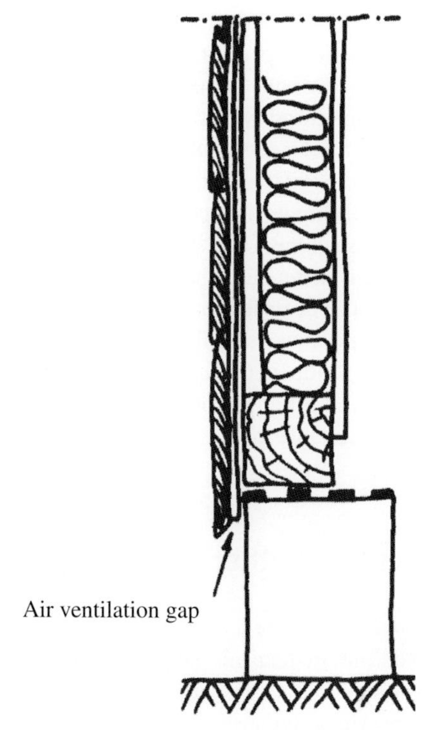

Figure 4.17 Air ventilation gap between the facade panel and bearing wall of a wooden construction

water vapour condenses when it hits a layer with a large diffusion resistance and which at the same time has a low temperature.

In order to prevent such condensation, it is necessary to locate materials with higher diffusion resistances from the inside of the construction (e.g. vapour-barrier foils and similar layers with high diffusion resistance preventing the penetration of water vapour from the interior to the peripheral walls and roof covers), or not to use such materials in construction composition at all. The suppression of generation of condensed water and the speeding up of its release from wooden houses and other constructions may also be achieved by the suitable location and sufficient thickness of thermal insulation, and also by the suitable location of ventilation gaps (Figure 4.17). Thermal insulation based on polystyrene or mineral fibres must be placed from the external side of buildings. Their thickness in the roof should be at least 200–240 mm in central Europe.

4.3.5 Regulation of climatic conditions in interiors

The regulation of climatic conditions in interiors is an important task in the structural protection of objects based on wood. This is implied by the fact that the physical processes in the construction materials are affected by the internal climate (i.e. RH, temperature, pressure and circulation of air). Correct amendments to the interior climatic conditions can in particular prevent the generation of condensed water. If condensed water is generated, the quickest way to remove it is by a suitable air conditioning. Air conditioning also regulates other phenomena that could lead to the damage of wood and other materials. The internal climate in depositories, hospitals, industrial halls, houses and some other places is often regulated using air conditioners.

A general principle regarding the need for correct and regular ventilation of interiors is in order to provide for a suitable RH of air (usually 45–55%) and its temperature (usually 18–22 °C). This will ensure the shape permanence of wooden products (parquets, furniture, etc.), without the creation of cracks, shrinkage and other physical defects, and at the same time provide conditions that are unsuitable for the activity of biological pests.

4.4 Fire sections and other fire-safety measures

The *Code of Practice for Fire Safety in Buildings 2011* (Buildings Department, 2012) defines in detail all the rules for safety measures in constructions, including wooden ones, as well as in constructions having wooden walls, ceilings or floor coverings, and also in constructions with wooden openings in walls, such as doors or windows. The creation of fire safety sections in buildings is very important.

A *fire section* is an individual room, living or working space in the construction. It should be created from fire-resistant walls and ceilings; it should also contain fire safety openings and suitable fire equipment. If a fire occurs in one fire section, it should be isolated so as not to spread to other parts of the construction (Hietaniemi & Mikkola, 2010).

The term *fire risk of a construction* defines the probable intensity of a possible fire in the construction or in its assessed fire sections. Studies, such as those of Frangi and Fontana (2010), Fragiacomo et al. (2012) or Barber and Gerard (2015), have thoroughly analysed the proposed new fire design concepts for safety wooden elements and whole multistorey tall-wood buildings.

Fire risks of wooden constructions may be calculated according to several models (Hurley & Bukowski, 2003; Jeon et al., 2010; Xin & Huang, 2013) and standards. These risks derive from the material composition, the permanent and accidental fire loads, and the surrounding environment. Fire safety is determined from the fire risk of a given construction. Subsequently, the requirements for wood species and type of wooden and other materials in the construction have to be confirmed or newly assessed while considering mainly their fire resistance and dimensions (Buchanan et al., 2014). The lack of sufficient fire resistance of wooden and other materials used in buildings can be improved by their treatment with fire retardants (Section 5.2.4).

Fire protection of a wooden construction should be dealt with in a complex way at the following levels:

- selection of convenient, more fire-resistant woods and wooden and other materials – that is, with the aim of using materials having a lower flammability level or treatment with a fire retardant;
- proper design draft – that is, so that the individual materials and components of the construction are of a suitable shape and of sufficient fire resistance; so that (1) the construction as a single unit is sufficiently fire safe; (2) the opportunities for further spread of a new fire in an already fire-attacked construction is limited to the maximum level; (3) the risk of loss of human life in the case of a fire developing is eliminated or limited.

References

Australian Building Codes Board (2011) *Information Handbook: Condensation in Buildings*. ABCB, Australian Institute of Architects, Canberra.

Barber, D. & Gerard, R. (2015) Summary of the fire protection foundation report – fire safety challenges of tall wood buildings. *Fire Science Reviews* 4, 5. doi 10.1186/s40038-015-0009-3.

Buchanan, A., Östman, B. & Frangi, A. (2014) Fire resistance of timber structures: a report for the National Institute of Standards and Technology. Draft report, 31 March. http://www.nist.gov/el/fire_research/upload/NIST-Timber-Report-v4-Copy.pdf (accessed 4 May 2016).

Buildings Department (2012) *Code of Practice for Fire Safety in Buildings 2011*. Based on Ove Arup & Partners Hong Kong Ltd consultancy study. Buildings Department, Hong Kong. http://www.bd.gov.hk///code/_code2011.pdf (accessed 4 May 2016).

Feirer, J. L. & Hutchings, G. R. (1986) *Carpentry and Building Construction*. Glencoe Publishing Company, Peoria, IL.

Fragiacomo, M., Menis, A., Clemente, I., et al. (2012) Fire resistance of cross-laminated timber panels loaded out of plane. *Journal of Structural Engineering* 139(12), 145–159.

Frangi, A. & Fontana, M. (2010) Fire safety of multistorey timber buildings. *Structures and Buildings* 163(4), 213–226.

Hietaniemi, J. & Mikkola, E. (2010) *Design Fires for Fire Safety Engineering*. VTT, Working Papers 139.

Hugues, T., Steiger, L. & Weber, J. (2008) *Timber Construction: Details, Products, Case Studies*. Architektur-Documentation GmbH & Co. KG, Munich.

Hurley, M. J. & Bukowski, R. W. (2003) Fire hazard analysis techniques. In *Fire Protection Handbook*, 19th edn. NFPA, Quincy, MA, section 3, chapter 7, pp. 121–134.

Houdek, D. & Koudelka, O. (2004) *Log Cabins from Round Wood*. ERA, Brno (in Czech)

Jeon, H.-K. Choi, Y.-S. & Choo, H.-L. (2010) A study for the fire hazard evaluation through the fire simulation of an apartment fire accident. *Journal of Korean Institute of Fire Science and Engineering* 24(4), 69–78.

Lebeda, J., et al. (1988) *Reconstruction of Wet Masonry in Buildings*. SNTL, Prague (in Czech).

Reinprecht, L. (2008) *Wood Protection*. Handbook. TU Zvolen, Slovakia (in Slovak).
Reinprecht, L. & Štefko, J. (2000) *Wooden Ceiling and Trusses – Types, Failures, Inspections and Reconstructions*. ABF–ARCH, Prague, Czech Republic (in Czech).
Stalnarker, J. J. & Harris, E. C. (1989) *Structural Design of Wood*. Structural Engineering Series, Springer Science + Business Media, New York.
Štefko, J. & Reinprecht, L. (2004) *Wooden Buildings – Constructions, Protection and Maintenance*. Jaga Group, Bratislava (in Czech).
Underwood, R. & Chiuini, M. (2007) *Structural Design: A Practical Guide for Architects*. John Wiley & Sons, Inc., Hoboken, NJ.
Wilcox, W. W. (2000) The influence of building design on wood decay. In *IRG/WP 00*, Kona, HI, USA. IRG/WP 00-10339.
Xin, J. & Huang, Ch. (2013) Fire risk analysis of residential buildings based on scenario clusters and its application in fire risk management. *Fire Safety Journal* 62-A, 72–78.
Žák, J. & Reinprecht, L. (1998) *Protection of Wood in Buildings*. ABF–ARCH, Prague (in Czech).

Standards

EN 335 (2013) Durability of wood and wood-based products – Use classes: definitions, application to solid wood and wood-based products.
EN 350-2 (1994) Durability of wood and wood based products – Natural durability of solid wood – Guide to natural durability and treatability of selected wood species of importance in Europe.
EN 13501-1+A1 (2009) Fire classification of construction products and building elements – Part 1: Classification using data from reaction to fire tests.

5 Chemical Protection of Wood

5.1 Methodology, ecology and regulation of chemical protection of wood

The chemical protection of wood is performed with chemical protective agents called 'wood preservatives'. The best known are bactericides (against bacteria), fungicides (against fungi), insecticides (against insects), fire retardants (against fire), anti-weathering and anti-corrosive agents (against atmospheric and aggressive-chemical effects).

Wood should be adequately protected with preservatives in consideration of its intended use, defined, for example, by EN 335 (see Table 4.1) or American Wood Protection Association (AWPA) Commodity Standards. Some of the active substances of preservatives, including commercial preservatives, can be withdrawn in the course of time and replaced with new ones that are more effective and that have less negative impacts on the environment.

Chemical protection is intended in particular for long-term preventive protection of wooden products in the exterior environment; for example, railway sleepers, utility poles, bridges, terraces, pergolas, fences or windows. However, it is occasionally also used in interiors; for example, for log cabins and other wooden houses, ceilings or trusses. Moreover, it is applied in short-term protection of round wood and sawn timbers during their storage and transport. It is important also for sterilization disposal of biological pests in transported wooden materials and infected wooden products.

The quality of chemical protection of wood depends upon (1) the sufficient efficiency, distribution and stability of the preservative in the wood and (2) the minimal risk of the preservative to people and the environment. More of these factors are directly influenced as well as by the wood structure (dimension, porosity, permeability, etc.), the application properties of the preservative, and the technology using of the preservative.

Wood Deterioration, Protection and Maintenance, First Edition. Ladislav Reinprecht.

5.1.1 Methodology and legislation of chemical protection of wood

The methodology and legislation of chemical protection of wood are in principle similar in Europe, the USA, Japan, Australia and other world regions, though with some specific differences for each one.

In Europe, the chemical protection of wood is based on EN standards, resulting in the following procedures and principles:

- *A suitable wood species and/or the type of wooden material is first selected for the wooden product* based also on its natural durability:
  - Classes of wood durability in the presence of biological pests according to EN 350-2 (Table 1.3).
  - Classes of wood reaction to fire according to EN 13501-1 (Table 2.2).
- *The expected damage to the wooden product is identified in the chosen environment:*
  - Action of biological pests – that is, according to EN 335 (Table 4.1) in relation to EN 351-1 (Figure 5.1).
  - Fire hazard – that is, from a heating load of the individual sections of the wooden structure.
  - Weather impacts – that is, according to exposure conditions and structural protection of wooden elements.
- *The natural resistance of the wooden material is assessed in the particular environment:*
  - If its resistance is not sufficient, it shall be improved by use of preservatives (Section 5.2), or alternatively by thermal, chemical or other modification processes (Chapter 6).
- *The particular chemical protection required for the wooden product is elaborated by* focusing on the selection of the preservative and the technology of its use:
  - For example, chemical protection of wood with a fungicide against fungal decay should be performed (1) within EN 460 to decide whether such protection is needed at all (Table 5.1), (2) then within EN 599-1 and EN 599-2 for selection of a specific effective type of preservative and its concentration and (3) finally determining the class of wood impregnability by EN 350-2 together with EN 351-1 and proposing a suitable treatment technology to achieve the required penetration class and retention class (Figure 5.1).
- *The wood preservative is selected* – within a required protection effect(s) in a particular class of use (e.g. determined by EN 335). A suitable concentration of the preservative is determined from its efficacy, the technology of its application and the exposure conditions of the wooden product. The load-bearing efficacy of all wooden preservatives have to be marked after an official approval by state institutions (e.g. for industry, buildings, risk assessment and environment). For example, in Germany and Slovakia the following marks are used for wood biocides: I_V, preventively effective against wood-damaging insects; P, effective against brown- and white-rot fungi; B,

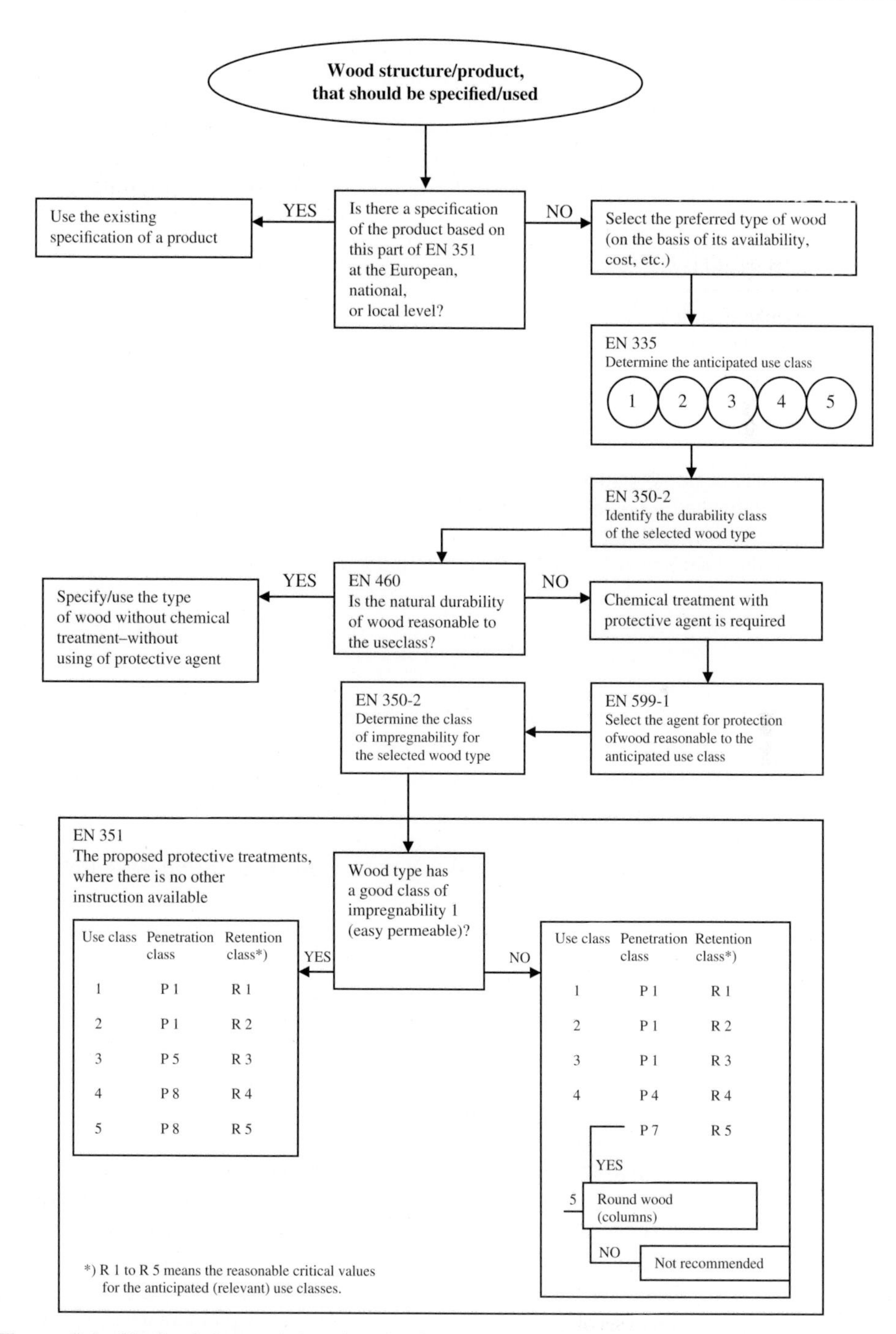

Figure 5.1 Methodology of the chemical protection of wooden structures (in accordance with EN 351-1)

Table 5.1 Requirements for the chemical protection of wood-based products with fungicides efficient against wood-decaying fungi in relation to its moisture content *w* (in accordance with EN 460)

Use class of wood (EN 335)	Wood natural durability against rot according to class (EN 350-2)				
	1	2	3	4	5
1 (permanently: $w < 20\%$)	I	I	I	I	I
2 (from time to time: $w > 20\%$)	I	I	I	(I)	(I)
3 (more frequently: $w > 20\%$)	I	I	(I)	(I)–(X)	(I)–(X)
4 (permanently: $w > 20\%$)	I	(I)	(X)	X	X
5 (permanently: $w > 20\%$ in salty environment)	I	(I)	(X)	X	X

I: natural durability of wood is sufficient and chemical protection of wood with a fungicide is not needed; X: wood has to be treated with a fungicide as its natural durability is not sufficient; (I) or (X): choice of fungicidal treatment of the wooden element (structure) also depends on other factors (e.g. on its static function or on its lifetime requests).

effective against wood-staining fungi and moulds; E, effective against soft-rot fungi in extreme wet conditions; M, effective against marine organisms; W, the protective agent is also suitable for exterior use without ground contact where the active component shall not be leached away from the wood by rainwater.

- *The technology for chemical protection of the wood is proposed,* preferably observing the following principles:
  - Only mechanically processed wood should be subject to chemical protection. If a layer of chemically treated wood is removed by subsequent sawing, planning, boring, and so on, then this section of wood has to be protected again.
  - The impregnation of poorly permeable 'refractory' wood species (e.g. spruce and fir) can be improved by mechanical and laser perforations, or by biological and chemical pretreatments.
  - The optimum moisture of wood prior to protection depends upon the technology used (pressure or diffusion driving forces).
  - The particular technology of protection is given by the requirement of the depth of penetration and retention of the preservative in the wood.
  - The technology of chemical protection of wood should be selected on the basis of its class of use – depending on danger of attack by biological pests (EN 335 – see Table 4.1), based on the principles of EN 351-1 (Figure 5.1):

Class of wood use (EN 335)	Technology of the preservative application
1	Not predetermined in detail (painting, spraying, etc.)
2	Not predetermined in detail (painting, spraying, etc.)
3	Minimally dipping (extraordinary painting, spraying), but optimally vacuum-pressure technologies
4	Vacuum-pressure technologies
5	Vacuum pressure technologies

- The depth of penetration of a preservative into the wood (for biocides prescribed by EN 351-1: from P1 to P9) depends upon the permeability of the wood as well. For example, in the case of use class 3, the biocide should penetrate from lateral areas of the well-permeable wood species (beech, hornbeam, lime, etc.) at least to 6 mm → class P5; on the other hand, for the poorly impregnable species (spruce, fir, heart of oak, etc.) there are no demands to depth of penetration → class P1.
- The technology of chemical protection of wood by fire retardants depends upon their type (i.e. intumescent paints just on the wood surface; other types also into a depth of wood by soaking or vacuum-pressure technologies).
- The technology of chemical protection of wood surfaces against weather conditions is performed usually by one to three coating layers, using immersion, spraying or painting.
- Wood composites (lamella wood, plywood, laminated veneer lumber (LVL), particleboard (PB), oriented strand board (OSB), etc.) can be preserved by spraying, painting or other post-manufacture treatment technologies, possibly as well as by special in-process treatment technologies during their manufacture.
- The fixation of a preservative in wood is important for those products intended for exterior use.

5.1.2 Toxicological and ecotoxicological standpoints of chemical protection of wood

For chemical protection of wood, the types of preservatives that can be used are those which are nontoxic or only minimally toxic to people, mammals, fish and plants and which are environmentally friendly (Stook et al., 2005; Kües et al., 2007). The environmental standpoint is important not only in the technology of wood treatment, but also during its exposure and disposal of the treated wood after its useful life by incineration, composting, and so on (Solár & Reinprecht, 2004; Lahary et al., 2006; Townsend & Solo-Gabriele, 2006; Zhang et al., 2015).

The *toxicity* of preservatives for warm-blooded organisms is assessed first of all by the acute toxicity. This toxicological question is characterized by the

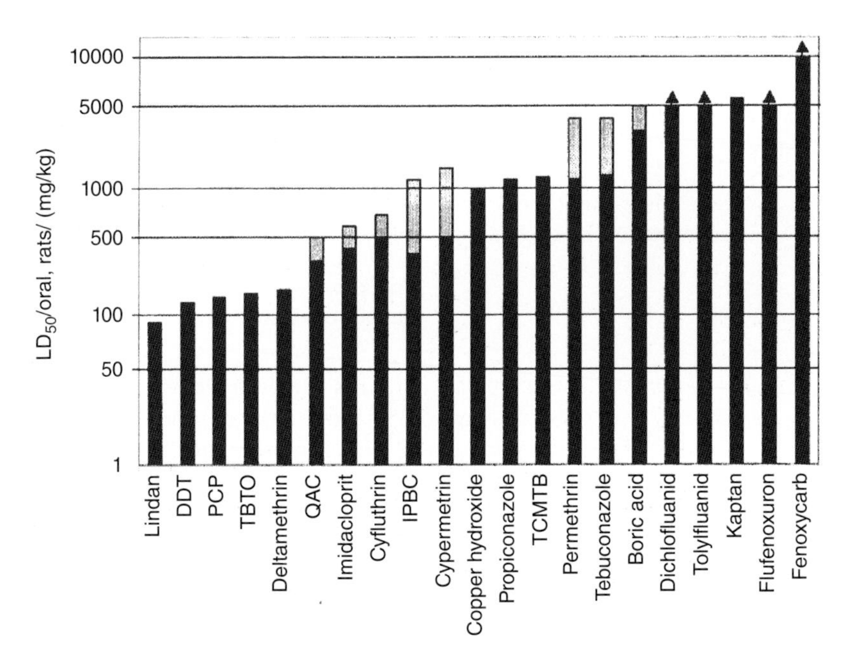

Figure 5.2 Acute toxicities $LD_{50\ (orally\ rats)}$ of some biocides for wood protection

Source: R., L. (2008) *Ochrana Dreva* (*Wood protection*), Handbook, TU Zvolen, Slovakia, 453 p. Reproduced by permission of TU Zvolen

lethal dose (LD), and usually by the LD_{50} (mg/kg, ppm). The LD_{50} is the dose of chemical substance in milligrams per kilogram of live weight of laboratory animals (e.g. rats) that results in 50% mortality occurring in the test population. Preservatives with higher values of the LD_{50} are less toxic, both for animals and human (Figure 5.2).

In Europe, biocide use, including wood preservatives, is regulated by the Biocidal Product Directive 98/8/EC. Wood preservatives used now should be relatively harmless (LD_{50} >15,000 mg/kg) to slightly toxic (LD_{50} >500 mg/kg). In addition to acute toxicity, the protective agent may not show a chronic toxicity, mutagenic, carcinogenic or teratogenic effects.

The *ecotoxicity* of preservatives defines their impact on the environment; that is, on the air, rivers, oceans, soils and plants (Aschacher & Gründlinger, 2000). From the point of view of ecology of preservatives, but also their efficacy, it is very important also that they are stable in the preserved wood. The main problem associated with the development of environmentally friendly organic preservatives, for high hazard end uses, is the observed biotransformation of these chemicals. Certain strains of proteobacteria (e.g. *Alcaligenes*, *Enterobacter*, *Pseudomonans*, *Ralstonia*) are able to degrade some organic biocides; for example, 3-iodo-2-propynyl-butyl-carbamate (IPBC), propiconazole and chlorothalonil (Wallace & Dickinson, 2006). Propiconazole and tebuconazole are also partly degradable by the black-stain fungus *Epicoccum purpurascens* (Stirling & Morris, 2010). According to these authors, if it proves to be possible to disrupt the mechanism of detoxification processes of organic fungicides, this could herald a new generation of environmentally friendly wood preservatives. On the other hand, intentional detoxification of organic fungicides can be important at reconstruction of old heritage buildings (e.g. preserved in past with pentachlorophenol or dichlorodiphenyltrichloroethane (DDT)), and at remediation of carbon-based preservative-treated wood at the end of its service life.

Volatile organic compounds (VOCs) are organic chemicals (solvents, active substances of preservatives, etc.) which evaporate from the preserved wood during its treatment and long-term exposure. The amount of VOCs (i.e. air pollutants) is limited by standards; for example, in Europe for coatings by the Paints Directive 2004/42/EC and EN 16402, or in the USA by the Environmental Protection Agency (EPA).

5.1.3 Regulation of chemical protection of wood

The regulation of preservatives for the chemical protection of wood is a very important for their proper and safe implementation in practice.

In Europe, as a result of the health and environmental demands on biocidal products determined in Directive 98/8/EC of the European Parliament and of the Council of 16 February 1998, and also in other Commission Regulation reports of this directive (e.g. No. 2007/1451/ES), only a limited number of active substances in wood preservatives should be permitted in the European market in years 2014–2016 (Krajewski & Strzelczyk-Urbańska, 2009). In Annex I to Directive 98/8/EC, there are a limited number of active substances from PT 8 (wood preservatives); for example, IPBC, potassium salt of cyclohexylhydroxydiazene 1-oxide, dialkyldimetylammonium compounds (DDACs), propiconazole, tebuconazole, thiabendazole and dichlofluanid from the fungicides.

In the USA, the EPA is responsible for regulation of biocides. Before registering a new biocide or new use for a registered preservative, the EPA must first ensure that the preservative can be used with a reasonable certainty of no harm to human health and without posing unreasonable risks to the environment (Mai & Militz, 2007; Leithoff et al., 2008; Lebow, 2010).

Today, the International Council-Evaluation Service (ICC-ES) issues Evaluation Reports that provide evidence that a building product complies with building codes – Acceptance Criteria (AC). Performance criteria of products containing wood preservatives are set out in AC326, arising from AWPA, ASTM and EN standard test methods (Lebow, 2010).

As a result of regulations for wood preservatives, the registration of new biocides and commercial products is under constant review by responsible organizations. In future, there probably will be a reduced interest for investigation of new biocidal (e.g. fungicide and insecticide) substances, mainly those on the basis of heavy metals (not only chromium, tin or arsenic, but also copper, etc.), the usage of which is already limited today.

According to Leithoff and Blancquaert (2006), the price for notification of existing active substances of preservatives varies from €3.3 million to €6.0 million, of which a substantial 85–90% of this price is connected with toxicology and ecotoxicology studies. Lower prices are needed for registration of existing preservatives (€0.2 million to €0.5 million) or new preservatives based on known active substances (€0.3 million to €1.4 million). However, the highest prices (from ~€20 million to €30 million) are needed for evaluation of new active substances – synthesis and analysis, their biological screening tests, and delimitation of optimal technologies for their application in practice and their verification (Reinprecht, 2008).

5.2 Preservatives for wood protection

Preservatives for wood protection comprise:

- one or several active substances;
- one or several accompanying components (i.e. solvent, stabilizer, emulsifier, fixative, pigment, colouring, etc.).

The active substances in commercial wood preservatives have various protecting effects:

- bactericides, effective against bacteria;
- fungicides, effective against wood-decaying fungi, staining fungi and moulds;
- insecticides, effective against wood-damaging insects;
- fire retardants, reducing the flammability of wood;
- inhibitors of weather corrosion, improving the resistance of wood against UV radiation, water, oxygen, and so on;
- inhibitors of chemical corrosion, improving the resistance of wood against aggressive chemicals.

5.2.1 Bactericides

Bactericides eradicate bacteria that attack of wood, concrete, stone, plastics or other materials. New types of bactericides contain silver and its compounds; for example, silver chloride, silver chloride mixed with titanium dioxide (TiO_2) and sodium sulfosuccinate, or silver mixed with copper oxide and zinc silicate (Wasserbauer, 2000). Silver exhibits a broad range of antimicrobial activity against Gram-positive and -negative bacteria, but also against yeast and moulds. However, more fungicides for wood protection (e.g. quaternary ammonium compounds (QACs); see Section 5.2.2) are applied as bactericides as well (Ash & Ash, 2004).

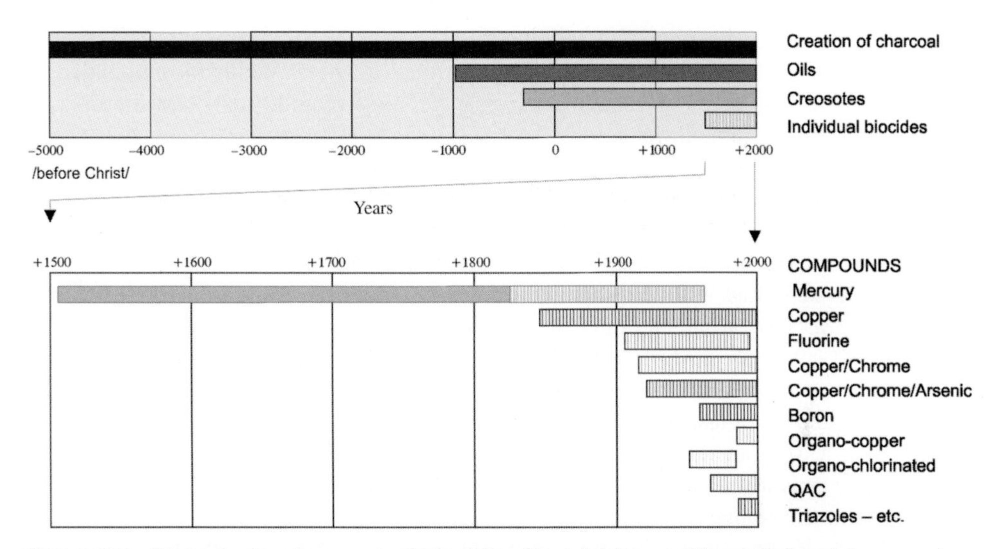

Figure 5.3 Historic development of biocides (fungicides and insecticides) for wood protection

Source: Pallaske, M. (2004) Chemical wood protection: improvement of biocides in use. In COST Action E22 – Environmental Optimization of Wood Protection,11 p. Lisbon, Portugal. Proceedings of the Final Conference; Estoril, Portugal

5.2.2 Fungicides: for decay, sap-stain and mould control

Fungicides for wood protection (Figure 5.3) are classified according to their chemical structure and origin as:

- inorganic
  - water soluble, able to be leached out from wood;
  - water soluble, fixable in wood;
- organic
  - creosote oils (prepared by distillation of black-coal tar), containing aromatic hydrocarbons;
  - synthesized substances (e.g. organometals, QACs, carbamates, heterocycles), but also other types, usually applicable in organic solvents, water emulsions, or some of them also in aqueous solutions;
  - natural substances (e.g. chitosan).

5.2.2.1 Inorganic fungicides

Inorganic fungicides comprise active substances based on the following chemical elements: copper (Cu^{2+}), zinc (Zn^{2+}), antimony (Sb^{3+}), silver (Ag^{+}), cadmium (Cd^{2+}), cobalt (Co^{2+}), nickel (Ni^{2+}) and boron (B^{3+}). Those used in the past also contained arsenic (As^{3+} and As^{5+}), chromium (Cr^{6+}), and also mercury (Hg^{2+}) long ago.

Inorganic fungicides without fixable substances (e.g. without Cr^{6+}) are often easily leached out from wood; that is, they are preferentially recommended for the protection of wooden products in buildings or under a shelter without exposure to rainfall water.

Copper (Cu^{2+}) is commonly applied as copper oxide (CuO), copper hydroxide ($Cu(OH)_2$), copper carbonate ($CuCO_3$), or copper hydroxide-carbonate ($Cu(OH)_2 \cdot CuCO_3$). In the past it was used as pentahydrate of copper sulphate ($CuSO_4 \cdot 5H_2O$). In the recent past the copper was very effectively combined with chromium and arsenic – both these chemical elements were used as a biocide and also as a fixative agent. The copper/chromate/borate salts were used mainly in Europe. The copper/chromate/arsenic (CCA), acid/copper/chromate, ammoniacal/copper/arsenate and ammoniacal/copper/zinc/arsenate salts were used, or some of them are still used, mainly in the USA, Canada and Australia. However, preservatives containing chromium and arsenic are now restricted-use biocides, including those used in combinations with copper.

Copper denatures the enzymes of fungi. It is active in particular against soft-rot fungi (Ray et al., 2010). On the other hand, more brown-rot fungi (e.g. *Serpula lacrymans, Serpula himantioides, Antrodia radiculosa, Oligoporus placentus, Fomitopsis palustris*), which use the Fenton reagent to depolymerize cellulose, are tolerant towards copper-based wood preservatives as a consequence of the creation of non-active copper oxalate crystals (Hastrup et al., 2005; Schilling & Inda, 2010). Copper oxalate is insoluble in water, and copper in this form has a greatly reduced inhibitory effect on fungal growth (Humar et al., 2001). Therefore, broad-spectral preservatives require the addition of a suitable co-fungicide (e.g. boron compound) to protect wood against copper-tolerant fungi.

Copper compounds are mainly used in water, amine or organic solutions, but in recent years also in the form of nano-copper – micronized copper quat (MCQ) particles, whose dimension is usually 10–700 nm (Matsunaga et al., 2007; McIntyre, 2010). $CuCO_3$ used in the form of MCQ is fixed to wood, creating octahedral complexes with six oxygen atoms surrounding the central copper (Xue et al., 2010). Today it is known that in the presence of Cu^{2+} there is an ongoing oxidation process in polysaccharides, and also in guaiacyl-lignin to quinone-methides, which leads to complex formation of copper with all wood components. Wood treated with copper fungicides has a green colour, and this can be a disadvantage for some products.

Zinc (Zn^{2+}) is today applied as zinc chloride in water solutions, or in the form of nano-systems as zinc borate ($Zn(BO_2)_2$) or zinc oxide (ZnO) (Reinprecht et al., 2015).

Silver (Ag, Ag^+) has become very interesting fungicide for wood protection recently. Its compositions in the form of Ag^+ (e.g. silver chloride and silver iodide) or in the form of Ag metal as nano-dispersions or micro-colloid systems have proved to be highly efficient against wood-decaying fungi – for example, *Gloeophyllum trabeum* (Ellis et al., 2007) and termites (Green & Arango, 2007) respectively.

Boron (B^{3+}) in the form of boric acid (H_3BO_3), sodium tetraborate (borax, $Na_2B_4O_7 \cdot 10H_2O$), and sodium octaborate (timbor – polyboron) is normally used in aqueous solutions (Lin & Furuno, 2001). All these inorganic boron compounds are characterized by combined fungicide, insecticide and partly fire-retardant effects (Lloyd, 1998). Their biocide effect lies in the creation of stable complexes with the structural components of cells of fungi and insects, affecting the transport processes between living cells, and reduction of enzyme production. However, the antimould effect of boron compounds is a rather weak one (Reinprecht, 2007). Boron under standard conditions is not fixed in the molecular structure of wood; that is, without special fixation it is suitable only for protection of wood products in interiors. Inorganic compounds of boron have a satisfactory biocide effect in wood at retentions from 3 to 20 kg/m^3 (Pallaske, 2004; Lyon et al., 2009).

Inorganic compounds of boron may also be used in exteriors where this no contact of treated wood with soil, but only in such situations, when boron is fixed in wood either by suitable additives or by waterproof paints (Lloyd et al., 2001; Mazela et al., 2007). Fixation of boron in wood can be performed with the following additives:

- colloidal silicone acids (Yamaguchi, 2001);
- calcium chloride, zinc sulphate or lead acetate, in whose presence are produced less-water-soluble metal borates $M(BO_2)_2$ (M = Ca, Zn, Pb); zinc borates are also applied in the protection of wood composites (Kirkpatrick & Barnes, 2006);
- animal proteins producing gel structures with boron upon increased temperature (Thevenon & Pizzi, 2003),
- tannins (Pizzi & Baecker, 1996; Tondi et al., 2012);
- monoglycerides, creating stable complex with borate ions (Mohareb et al., 2010);
- glycerol/glyoxal system (Toussaint-Dauvergne et al., 2000).

A better transport of boron into wood and wooden composites is obtained by using the gaseous azeotrope mixture of tri-methyl borate (TMB), a colourless liquid having a boiling point of 68.7 °C, with methanol (Murphy, 1994). TMB is then hydrolysed in wood with molecules of bonded water to H_3BO_3:

$$B(OCH_3)_3 + 3H_2O \rightarrow H_3BO_3 + 3CH_3OH \qquad (5.1)$$

Organic esters of boron – for example, trihexylene-glycol-biborate soluble in organic solvents, tetramethyl-ammonium bis(salicyl)borate and tetramethyl-ammonium bis(2-hydroxymetylphenyl)borate (Humphrey et al., 2002) or ammonium borate oleate (Lyon et al., 2009) – are usually more stable against leaching with water than its inorganic compounds. The newly developed didecyl-dimethyl-ammonium tetrafluoroborate (DBF) is acceptably stable and also effective against white-rot and brown-rot basidiomycota (Kartal et al.,

2005) and against soft-rot fungi (Kartal et al., 2006). DBF, with or without incorporation of acryl-silicon-resin emulsion, has a good decay resistance against basidiomycota even after severe weathering (Kartal et al., 2004, 2006). The fixation reaction of boronic vinyl ester (synthetized from vinyl acetate and 4-carboxyphenylboronic acid) shows bonding to wood –OH groups and secured sufficient efficacy against brown-rot fungi at a weight-percent-gain (WPG) as low as 3% (Jebrane, 2015):

Boronic vinyl ester — Boronated wood

Wood—OH + [boronic vinyl ester] ⇌ Wood–[boronated ester] + vinyl alcohol (OH) ⇌ acetaldehyde (O)

(5.2)

5.2.2.2 Creosotes

Creosotes are oils obtained by fractional distillation of black-coal tar; for example, the older classic creosote 'tar oils' with a boiling temperature range of 200–400 °C (Hunt & Garratt, 1967; Richardson, 1993). Creosotes, as effective preservatives against biological pests, were patented in 1836 by German chemist Franz Moll and first used for wood impregnation by John Bethell in 1838. In the 19th and 20th centuries, creosotes were the most commonly used wood preservatives throughout the world, with a worldwide production of ~16 × 10^6 t per year (Eaton & Hale, 1993). Creosotes contain ~96% of various nonpolar polycyclic aromatic carbohydrates (fluorene, phenanthrene, anthracene, etc.) and ~4% tar phenols (cresol, naphthol, etc.) and tar bases (quinolone, acridine, etc.). In the past, the creosotes were applied mainly for industrial protection of wooden sleepers, utility poles, bridges and other structures for exterior use (Richardson, 1993). Now, they are used to a lesser extent, substantially only for industrial treatment of special sleepers and poles, and only by vacuum-pressure technologies at temperatures above 100 °C.

In Europe, the composition and usage of creosotes is controlled by the West European Institute for Wood Impregnation (WEI). The WEI type C creosote, which contains mainly acenaphthene, fluorene, phenanthrene and anthracene, is from the environmental point of view a more friendly than older types (WEI type A and WEI type B), because this creosote has only compounds with tapered boiling points, from ~280 to 360 °C; that is, without easily volatile naphthalene derivatives and also without higher condensed aromatics. For example, the maximum content of benzo(*a*)pyrene, whose boiling point is above 450 °C, and which has a strong carcinogenic effect, had to be less than 50 ppm (Crawford et al., 2000); for example, in the WEI type C maximally from 5 to 50 ppm.

An advantage of creosotes is their stability in wood (low water leachability and low volatility), no corrosive effects on wood and metals, and affordable price. However, chemical analyses of old impregnated sleepers prove that after 25–30 years of exposure in railway routes just one-third of the original amount of oils remains in them. The disadvantages of creosotes are the dark colour, the smell, the variable chemical composition, the high viscosity (and thus also the need to heat oils for wood impregnation to temperatures of 100–140 °C), the moderate increase of wood flammability and particularly the health-damaging effects on humans.

The durability of wooden sleepers can be increased after pressure impregnation by creosotes three times (oak) to ten times (beech); that is, a non-impregnated beech sleeper has a service life of ~3 years and an impregnated one ~30 years. Retention of 30 kg/m^3 (30 kg of oil per 1 m^3 of wood) is required against wood-decaying Basidiomycota fungi causing brown and white rot, or against wood-damaging insects. However a substantially higher retention of 120 kg/m^3 is required against wood-decaying Ascomycota fungi, which cause soft rot.

Creosotes dissolved in mineral oils (carbolineum) or emulsified in water (pigmented emulsified creosote) can be used at ambient temperature (Mai & Militz, 2007) by painting or other non-autoclave technologies. But it should be emphasized that not even these forms of creosotes may be used for the protection of wood in interiors or children's playgrounds. Generally, now and mainly in the future, application of creosotes will be significantly restricted because of their negative influences on the environment.

5.2.2.3 Synthetized organic fungicides

In the second-half of the 20th century the successful chemical protection of utility poles, log cabins and other wooden products against the activity of wood-damaging fungi was performed with pentachlorophenol and the sodium salt of this fungicide. Also well known were the organic fungicides such as tri-butyl-tin-naphthenate, bis-(tri-butyl-tin-oxide), 2-thiocyanomethylthio-benzothiazole and phenylsulphamides (dichlofluanid, tolylfluanid) (Reinprecht, 1996).

At present, the preservation of wooden products should be performed with more environmentally friendly organic fungicides, from which the following are the most important.

N-Organodiazeniumdioxy-metals are organometal substances in which an atom of metal is bound to an oxygen atom present in one or more organodiazenium molecules. The organic part of these substances is, for example, cycloalkyl, alkyl or aryl. The metal is usually potassium (K), copper (Cu), zinc (Zn), nickel (Ni), cobalt (Co) or aluminium (Al). These fungicides are applied in water solutions with the presence of suitable additives for their dissolution, or in organic solvents.

Bis-[*N*-cyclohexyldiazeniumdioxy]-copper (Cu-HDO) is efficient against more wood pests, in particular against brown-rot fungi (specifically against species from the *Antrodia* genus) and white-rot fungi, but also against soft-rot fungi and wood-damaging insects (Göttsche & Marx, 1989). Its biocidal effect lies in the denaturation of proteins and enzymes, and also in the damage of mitochondria in the cells of fungi. Cu-HDO is stable in wood; that is, it is not soluble in water and is not volatile. Cu-HDO is applied in aqueous solutions in the presence of additives, usually organic acids (e.g. lactic acid), alkaline amines (e.g. diethylenetriamine), or complexing polymer amines (e.g. polyethyleneamines). These additives allow temporary transformation of Cu-HDO to a water-soluble complex (at pH lower than 4 or higher than 9.5), after which it can be fixed in the wood using neutralization processes. Several commercial products based on Cu-HDO, also containing other biocides (e.g. copper hydroxide and H_3BO_3), are used for the industrial impregnation of shingles and various elements of wooden structures, mainly in above-ground applications.

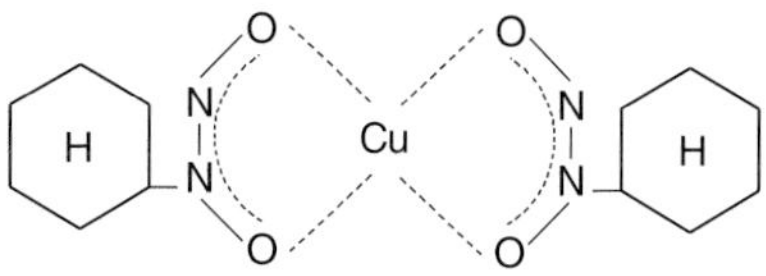

Cu-HDO

Other organometal fungicides are in several cases based on copper. The best known are the copper naphthenates (CuNs), copper-8-quinolinolate, copper chelates, ammonium copper carboxylates with the structure R–COO–Cu–OOC–R (e.g. ammonium copper citrates), alkaline copper quat (ACQ) formulations (e.g. containing CuO and the QAC DDAC), copper azoles consisting of ~96% of amine copper and ~4% of tebuconazole or propiconazole, or copper bis-dimethyldithiocarbamate (CDDC) formed directly in wood by reaction of two separate compounds (Eaton & Hale, 1993; Pankras et al., 2009; Lebow, 2010). CDDC is created in the wood by its first treatment with a solution containing a maximum of 5% of bivalent copper ethanolamine and then with a solution containing a minimum of 2.5% of sodium dimethyldithiocarbamate.

It should be highlighted, that CuN (Cuprinol) was used from the beginning of 20th century in oils, but now it is used as a water-borne solution in a mixture of water and ethanolamine as well (Lebow, 2010).

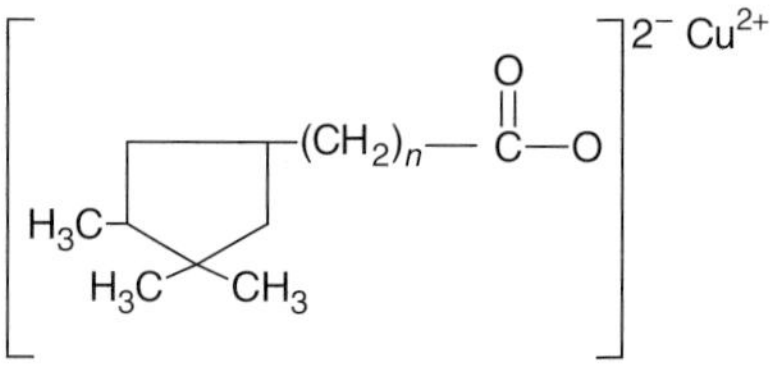

CuN

QACs are efficient in particular against bacteria, moulds and staining fungi (Micales-Glaeser et al., 2004). They are less active against wood-decaying fungi, at which they have a certain efficacy even against wood-boring beetles and termites. Biocide activity of QAC can by increased by inorganic compounds of copper. Their fungicidal efficiency also depends on the type of alkyl or aryl groups (R_1, R_2, R_3, R_4) bounded to the nitrogen atom, and on the type of anion (X^-; e.g. chloride, tetrafluoroborate, nitrate or acetate) (Zabielska-Matejuk & Skrzypczak, 2006).

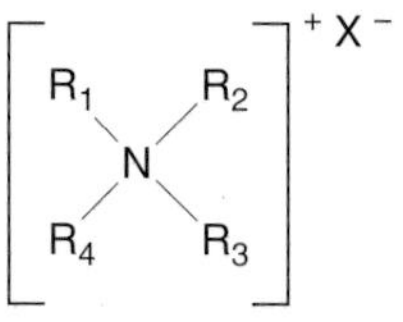

$$\left[N R_1 R_2 R_3 R_4 \right]^{+} X^{-}$$

QAC

R_1, R_2: alkyl groups with 1 to 6 atoms of carbon; R_3: alkyl groups with 1 to 20 atoms of carbon, or benzyl group; R_4: alkyl groups with 8 to 22 atoms of carbon; X^-: anion of acid (e.g. Cl^-, Br^-), hydroxyl (OH^-), etc.

QACs are well soluble in water and miscible in alcohols. They are often added to paints (e.g. into vinyl acetate or acrylate latexes). Several types of QACs (e.g. the benzylalkyldimethylammonium compounds, DDACs and alkyltrimethylammonium compounds) are used for wood protection. In wood they are fixed on the basis of ion-exchange reactions with carbonyl groups of lignin and hemicelluloses, and also by interactions with –OH groups of all wood components (Nicholas et al., 1991). Their fungicide activity lies in the dehydration of cellular walls of fungus, in which are created holes, and subsequently the fungus dies. Nowadays, QACs are used for treatment of structural timbers in interiors, but sometimes also in exterior above-ground exposures. Owing to their lower stability in the environment, their rapid fixation close to the wood surface and their influence on higher absorption of water by wood from the environment, QACs are not convenient for treatment of wood in contact with the ground. Commercially, they are usually used in combination with copper compounds (e.g. $CuCO_3 \cdot Cu(OH)_2$), boron compounds (e.g. H_3BO_3) or triazoles (e.g. propiconazole) (Härtner & Barth 1996; Reinprecht, 2010).

Carbamates are well known biocides. IPBC is efficient against various wood-damaging fungi; however, in practice, it is mainly used against moulds and staining fungi. IPBC can be applied in organic solvents (e.g. in acetone or xylene) and also in water emulsions. Biocidal activity of IPBC against fungi can be enhanced with borates (Cassens & Eslyn, 1981), or against the true dry-rot fungus *Serpula lacrymans* about ~50% in the presence of 2 g/l of α-aminoisobutyric acid (Bota et al., 2010). The methylbenzimidazole-yl-carbamate (Carbendazim) is efficient against staining fungi.

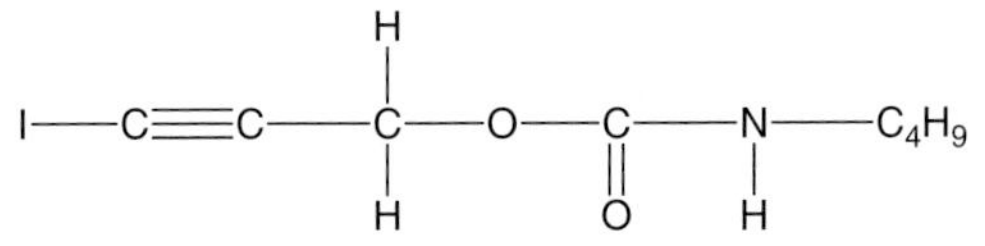

IPBC

Derivatives of 1,2,4-triazole are active against moulds, staining fungi and wood-decaying fungi. The best known are azaconazole, cyproconazole, propiconazole and tebuconazole. They suppress biosynthesis of sterole, which is important for the function of membranes of cells in fungal organisms. Their limit of efficiency against decay-causing fungi is 0.15–1.2 kg/m^3 for propiconazole (1-[[2-(2,4-dichlorophenyl)-4-propyl-1,3-dioxolan-2-yl]-methyl]-1*H*-1,2,4-triazole) or 0.05–0.5 kg/m^3 for tebuconazole (α-[2-(4-chlorophenyl)-ethyl]-α-(1,1-dimethylethyl)-1*H*-1,2,4-triazole-1-ethanol).

The 1,2,4-triazoles are applicable in organic solvents (e.g. acetone, toluene, ethyleneglycol and other glycols) or water emulsions. They are stable in exteriors and have only a low toxicity to animals ($LD_{50\ (\text{orally rats})}$ of 1500–4000 ppm). Today, they belong to the most frequently used fungicides for the protection of construction and joinery products, such as windows, exterior doors, and so on. The 1,2,4-triazoles are efficient in particular against the true dry-rot fungus *Serpula lacrymans* and other species of Basidiomycota fungi. Their fungicide efficiency can be increased synergetically by the addition of antioxidants and chelates of metals (Bakhsous et al., 2006). A typical natural antioxidant is caffeine (1,3,7-trimethylxanthine), which induces a strong alteration of cell wall architecture of fungi, inhibiting their growth (Lekounougou et al., 2007).

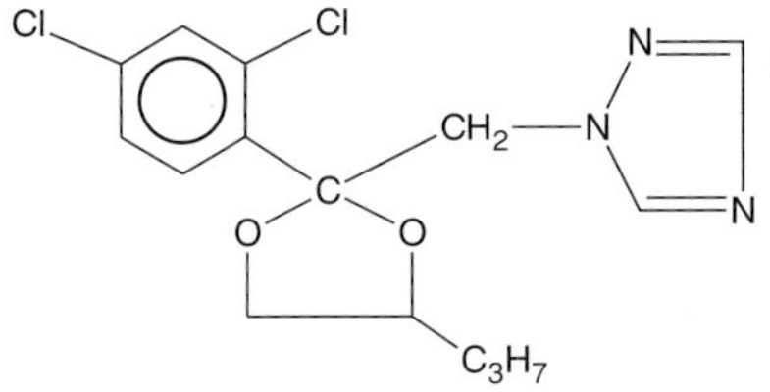

Propiconazole

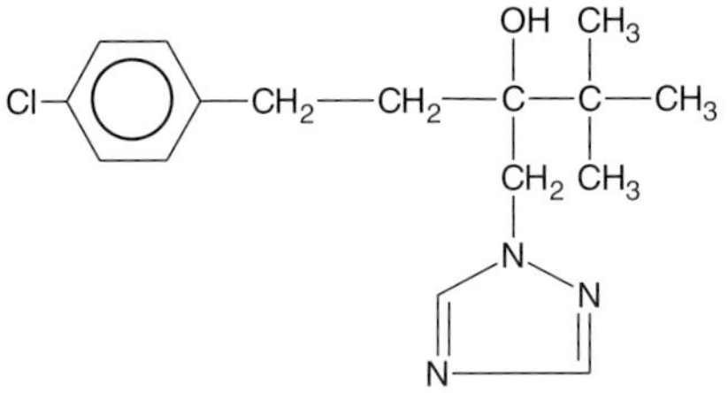

Tebuconazole

Metal azole complexes have been known from the 1970s. Now, after restriction of CCA salts, the use of copper in combination with organic co-biocide has become of increasing interest (Evans et al., 2008). Metal-centred azole complexes formed from propiconazole or tebuconazole with copper acetate (e.g. $Cu(tebuconazole)_2(OAc)_2$) or other metal substances create crystals which have been found to have a higher efficiency against fungi in comparison with the original azole and metal compounds.

Derivatives of isothiazolone (ITA) – for example, 4,5-dichloro-2-*n*-octyl-4-isothiazol-3-one (DCOIT), 2-*n*-octyl-4-isothiazol-3-one (Kathone) or 5-chloro-2-methyl-4-isothiazol-3-one (Kathone CG) – are applied for protection of wood against bacteria and wood-decaying fungi. Some of them are also efficient against termites. The action mechanism of ITAs is disruption of the metabolic pathways in organisms involving dehydrogenase enzymes (Williams, 2007). They can be applied in organic solvents (e.g. in toluene, xylene and ethanol) or after the addition of emulsifiers also as a water dispersion. DCOIT is efficient against wood-decaying fungi in doses of 0.15–1.28 kg/m^3. It has a similar stability in wood as other ITA derivatives (Hegarty et al., 1997) and has acceptable toxicological and ecotoxicological properties.

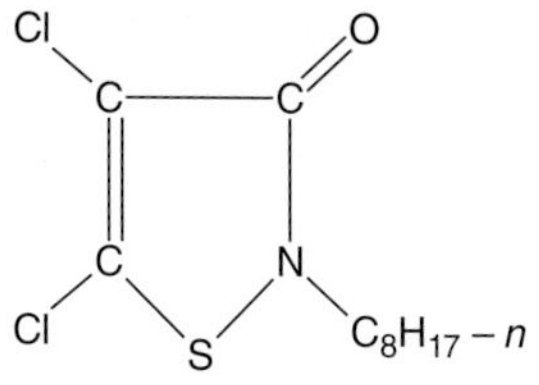

DCOIT

5.2.2.4 Natural organic fungicides

Also suitable for wood protection are some natural substances having bactericidal, fungicidal and/or insecticidal effects. Among the best known are: (1) essential vegetable oils (e.g. cinnamon oil, clove oil, geranium oil, neem oil, oregano oil, tea tree oil, thyme oil) and their effective compounds (e.g. 1,8-cineole, α-cadinol, α-terpinene, α-thujone, elemol, eugenol, limonene, *p*-cymol, thymol) obtained from convenient plant species (e.g. Dhyani & Tripathi, 2006; Yang & Clausen, 2006; Li et al., 2007; Chittenden & Singh, 2011; Mohareb et al., 2013; Pánek et al., 2014); (2) wood extracts (e.g. tannins, flavonoids, terpenoids) obtained from the most durable wood species (e.g. Amusant et al., 2005; Asamoah & Antwi-Boasiako, 2007); (3) extracts from chilli with the effective compound capsaicin (Singh et al., 2006); (4) propolis from beeswax with an antibiotic effect (Budija et al., 2008); and (5) chitosan (Eikenes et al., 2005).

Essential vegetable oils are obtained from suitable species of plants by distillation with water vapour or by extraction with water or organic solvents. Well known are cedar oil, Egypt geranium oil (from *Pelargonium graveolens*), thuja oil (from *Thuja occidentalis*), thyme oil (from *Thymus zygis*), hiba oil (from *Thujopsis dolabrata*), tung oil (from *Aleuritis fortii* and *Aleuritis cordata*), lavender oil (from *Lavandula angustifolia*), neem oil (from *Azadirachta indica*) and clove oil (from *Syzygium aromaticum*), but also linseed, citrus, olive and several other types of oils. These oils are mixtures of various alcohols, aldehydes, ketones, esters, lactones, carbohydrates, terpenes and heterocycles (Wang et al., 2005, Yang & Clausen, 2006; Pánek et al., 2014). They are less soluble in water, but well-dissoluble in alcohols and aromatic carbohydrates. Several of them are partially volatile; that is, they act also as repellents to wood-damaging insects.

Essential oils were applied even 2000 years ago, either as concentrates or in organic solvents for the protection of construction timber. Essential oils with phenolic or aldehyde and ketone components, like thymol, eugenol, carvone and carvacrol, show better efficiency against wood-decaying basidiomycetes than those with various terpenes (Amusant et al., 2009). However, Yang and Clausen (2006) note that oils containing thujaplicins and some other terpenes (e.g. thyme oil and geranium oil) have good fungicide properties. Carvone, citronellol, geraniol, thymol and borneol are effective inhibitors of mould spore germination. Essential oils are effective also against bacteria (Papadopoulos et al., 2006; Wong et al., 2008), and they can be used for protection of wooden products against the accumulation and growth of disease-causing and harmful bacteria in interior areas such as kitchens, dining rooms and hospitals.

Specific disadvantages of several essential oils are high volatility, unstable concentration of effective compounds in oil product used, and oily surface of wood to which dirt can be absorbed (Batish et al., 2008). On the other hand, essential oils are health-friendly (e.g. many of them are used in medicine, aromatherapy or cosmetics), and they cause only small problems in terms of liquidation of treated products after their service life.

Wood extracts are obtained from splinters, sawdust and bark of highly durable wood species, such as African cherry, black locust, oak, padouk, rosewood, sequoia, teak, and others – mainly tropical species. Wood extracts in aqueous, alcoholic or other liquid solutions contain bioactive compounds: polyphenolics (e.g. flavonoids, sterols, quinones), terpenoids (e.g. diterpenoids, triterpenoids, cedrol, agathadiol, epimanool, bornyl-acetate, cedrene), alkaloids and stilbenes. Such wood extracts are used for deep impregnation or surface treatment of less durable wood species with the aim to inhibit activity of various types of wood-damaging fungi (Gérardin et al., 2002; Castillo et al., 2004; Amusant et al., 2005; Sethy et al., 2005; Kazemi, 2007; Mburu et al., 2007). However, alteration of their chemical structure during extraction from durable woods, during treatment of nondurable woods or during exposure of treated wooden products can lead to loss of their original antifungal efficiency. Bio-efficient extracts are also obtained from the leaves of durable wood species; for example, acetone extracts from the leaves of *Dalberia sissoo* wood are efficient against wood-damaging fungi and bacteria (Kabir & Alam, 2007).

Chitosan is a 1,4-linked heterogeneous polymer of D-glucosamine (fraction of *N*-acetylations $F_A = 0$), which usually contains also *N*-acetyl-D-glucosamine units ($F_A \neq 0$). Chitosan is derived from crustacean shells, wing-cases of insects, and cellular walls of algae and fungi. It is partly water soluble under acidic conditions.

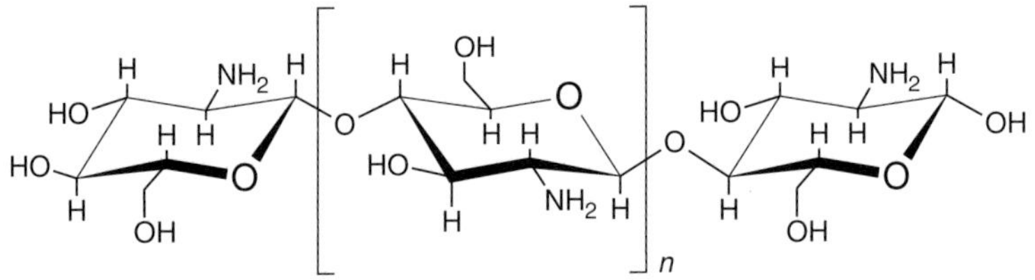

Chitosan (type $F_A = 0$, composed just of glucosamine units)

Chitosan can act fungistatically and at higher concentrations also as a fungicide (Eikenes et al., 2005). In 1–5% concentration it is efficient against some decay-causing fungi (e.g. the brown-rot *Coniophora puteana* and *Gloeophyllum trabeum* (Schmidt et al., 1995) and the white-rot *Trametes versicolor* (Maoz and Morrell, 2004)) and also against moulds (e.g. *Trichoderma harzianum* (Vesentini and Singh, 2006)). The efficacy of chitosan also depends on its molecular weight; for example, chitosan with a lower molecular weight has been found to be more effective against fungi *Trametes versicolor* and *Poria placenta*. The mechanism of its fungicide effect is not yet totally clear; however, there are several hypotheses according to which it attacks DNA and suppresses the synthesis of mRNA, or it impairs the membrane function of cells of fungi and thus also their permeability (Eikenes et al., 2005).

5.2.3 Insecticides

Insecticides for wood protection (Figure 5.3) are classified according to their chemical structure and origin to:

- inorganic
  - water soluble, able to be leached out from wood;
  - water soluble, fixable in wood;
- organic
  - creosote oils, on the basis of higher aromatic carbohydrates (see Section 5.2.2);
  - synthesized substances (e.g. pyrethroids, heterocycles, organophosphates, juvenile hormones, chitin-inhibitors), and also repellents and pheromones;
  - natural substances (e.g. pyrethrin).

Insecticides are classified according to their physiological effect on insects as follows:

- toxic
  - respiratory – that is, block the activity of respiratory system of insects;
  - protoplasmic – that is, cause the disintegration of live cells;
  - nervous – that is, block the activity of the central nervous system of insects;
- nontoxic
  - hormonal – that is, regulators of insect growth, development and behaviour;
  - repellents – that is, repulsing insects during their invasion to roundwood yards, and so on;
  - pheromones and other attractants – that is, attracting insects to a certain place, where they are killed.

Insecticides are classified also according to the method of their transport into insect body: contact (through the body surface), feeding (through digestive tract) and respiratory (through respiratory organs).

5.2.3.1 Toxic inorganic insecticides

Compounds of trivalent boron are known insecticides and fungicides effective against wood pests (see Section 5.2.2.1). They are used mainly in interiors and buildings against boring-beetles and termites; for example, H_3BO_3, $Na_2B_4O_7 \cdot 10H_2O$ and sodium octaborate (Morris et al., 2014). Their insecticide effect is protoplasmic. In exteriors they can be applied in the form of diffusion cartridges for the protection of wooden windows, bridges or utility poles against wood-damaging ants (Mankowski, 2007).

5.2.3.2 Toxic organic insecticides

In the past, chlorinated carbohydrates (DDT, hexachlorohexane), organophosphates and carbamates were used as efficient insecticides. Nowadays, they are replaced by pyrethroids, by some types of heterocycles based on pyridine and imidazole, as well as by QACs. They are applied similar to organic fungicides, in organic solvents or in water emulsions (Stirling & Temiz, 2014). Creosotes also have an insecticide effect, but their application is now very limited (see Section 5.2.2).

Pyrethroids have a similar chemical structure to natural pyrethrins, obtained by extraction from flowers of thermophilic plantation-grown chrysanthemums (*Chrysanthemum cinerariaefolium*, *Chrysanthemum coccineum*, etc.) growing in particular in Kenya, Tanzania and Brazil. To well-known synthetized pyrethroids belong permethrin, cypermethrin, cyfluthrin, deltamethrin, bifethrin, fenpropathrin and bioresmethrin. All these compounds are the dichloro- or dibromo-vinyl analogue esters of the [+]-*trans*-chrysanthemum acid present in the natural pyrethrin; that is, their structure is modified from the 3-phenoxy-benzyl-esters of the 2,2-dimethyl-3-(2,2-dihalogenvinyl)-cyclopropane-carboxylate-acid.

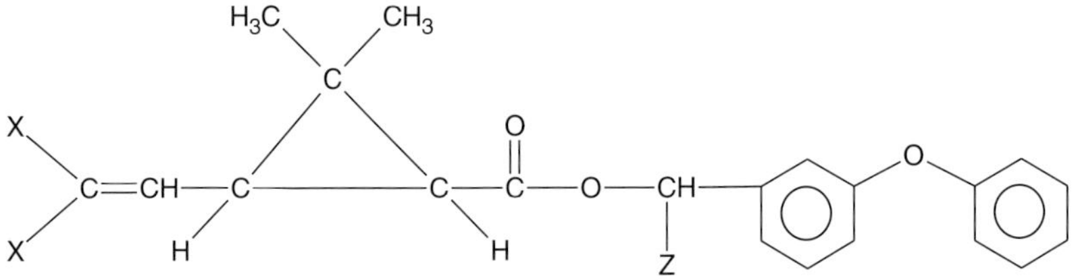

Pyrethroid

Pyrethroids block the central nervous system of insects. Their efficiencies are different; in particular, for deltamethrin : cypermethrin : cyfluthrin : permethrin against larvae of the house longhorn beetle (*Hylotrupes bajulus*) the efficiency is 25 : 7 : 7 : 1, and against larvae of the common furniture beetle (*Anobium punctatum*) it is 34 : 3 : 3 : 1.

Pyrethroids are only slightly toxic for warm-blooded animals. In the environment they are gradually degraded due to UV light, oxygen and water (Lloyd et al., 1998; Freeman et al., 2007); that is, the protection of wood through using them has to be regularly renewed, usually in 4- to 6-year cycles. Pyrethroids moderately corrode metals (e.g. steel, aluminium and copper). In practice, they are often applied in combined biocide preservatives (insecticide+fungicide); for example, in combination with triazoles, such as tebuconazole or propiconazole. Pyrethroids are effective against boring-beetles in use class 3 (Hunt et al., 2005), as well as against termites (Morris et al., 2014). They are used also for the protection of wooden composites; for example, they are added to the adhesive producing bio-stable plywood (Creffield & Scown, 2006).

Derivatives of 2-imidazolidinimine are highly efficient against wood-damaging beetles and termites – the main compound is 1-[(6-chloro-3-pyridinyl)-methyl]-4,5-dihydro-*N*-nitro-2-imidazolidinimine (imidacloprid). It dissolves in medium polar organic solvents. However, it can also be applied in water (0.06% aqueous solution at 20 °C). Imidacloprid acts on the central nervous system of insects. It is medium or slightly toxic against warm-blooded animals, with an $LD_{50\ (orally\ rats)}$ of 450–550 mg/kg.

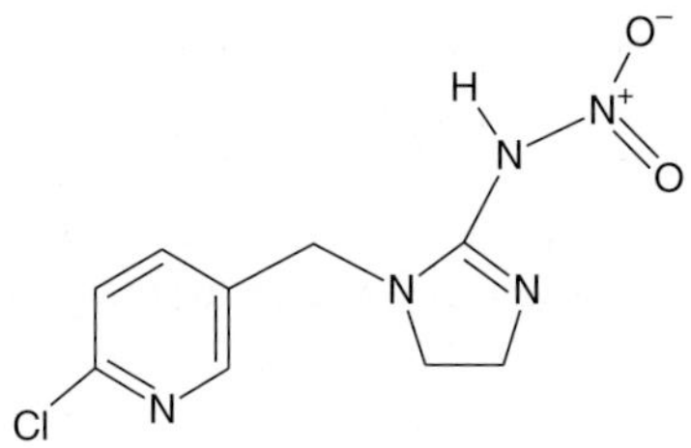

Imidacloprid

5.2.3.3 Nontoxic hormonal insecticides

Hormonal insecticides, known as insect growth regulators, influence certain development stages of insects, impairing the synchronizability of their hormones (Figure 5.4).

Analogues of insect hormones – the juvenile, anti-juvenile and skin-sloughing hormones – cause killing or sexual sterilization of insects. Their hormonal effects lie in:

- the introduction of one of the hormones (juvenile, skin-sloughing) to the insect body;
- the suppression of production of the juvenile hormones after the introduction of the anti-juvenile hormones to the insect body.

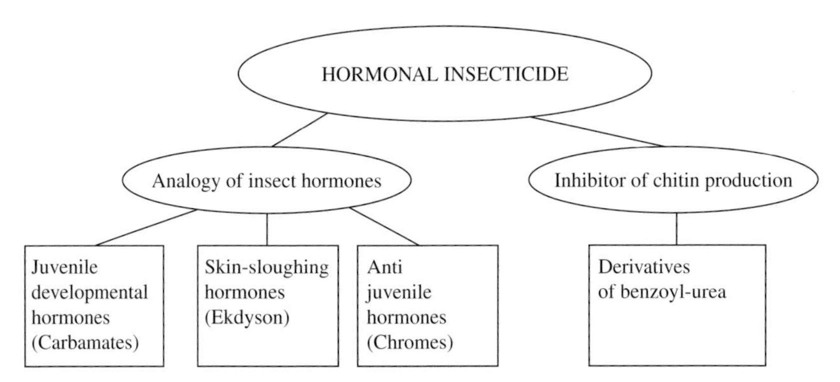

Figure 5.4 Classification of hormonal insecticides – insect growth regulators

Source: R., L. (2008) *Ochrana Dreva* (*Wood Protection*), Handbook, TU Zvolen, Slovakia, 453 p. Reproduced by permission of TU Zvolen

Fenoxycarb (ethyl-*N*-[2-(4-phenoxy-phenoxy)-ethyl]-carbamate) is a well-known representative of the juvenile hormones (Stirling & Temiz, 2014). Fenoxycarb protects wood against the house longhorn beetle (*Hylotrupes bajulus*), furniture beetles (Anobiidae), powder post beetles (Lyctidae) and other species of wood-damaging beetles; however, it is ineffective against termites. It is applied in a concentration of ∼0.04% (Pallaske, 1999). Its required retention is ∼1.0 g/m^3 in impregnated wood and ∼0.05 g/m^2 in surfaces of treated wood. It is not toxic to humans and other warm-blooded organisms (having an $LD_{50\ (orally\ rats)}$ ≥10000 mg/kg) and it does not corrode metals, so it is always a highly prospective insecticide for the protection of wooden structures.

O, O, NH–C(=O)–O–CH_2–CH_3

Fenoxycarb

Tebufenozid, halofenozid and azadirachtin are representatives of the skin-sloughing hormones.

Inhibitors of chitin production slow down biosynthesis of chitin and the production of cuticle (i.e. the external skeleton of insects). They are efficient against wood-damaging insects even in very small doses: 0.5–3 g/m^3 in impregnated wood or 0.02–0.05 g/m^2 on wood surfaces treated by spraying or immersion. Chitin production inhibitors are practically nontoxic for warm-blooded animals, and their ecotoxicity is negligible. The best known ones are the derivatives of benzoyl-urea; for example, flufenoxuron (Pallaske, 1999).

Flufenoxuron, *N*-[4-(2-chlor-α,α,α-trifluor-*p*-tolyloxy)-2-fluor-phenyl]-*N′*-(2,6-difluor-benzoyl)-urea, is virtually nontoxic to humans, with an

$LD_{50\ (orally\ rats)}$ ≥5000 mg/kg. It is stable in wood and highly resistant against weather impacts, which is contributed to by its minimal solubility in water and negligible volatility.

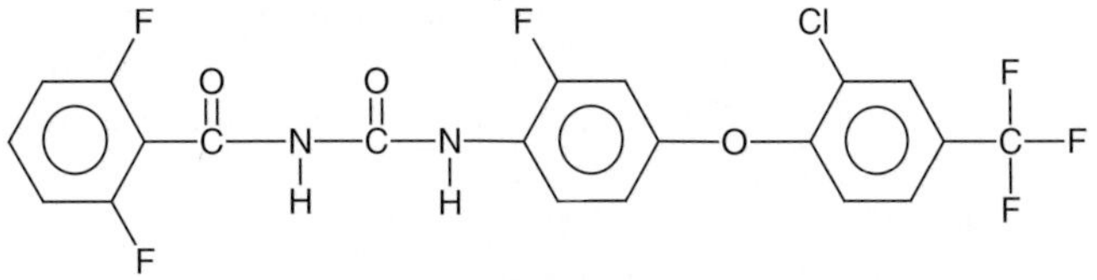

Flufenoxuron

5.2.3.4 Repellents and pheromones

Repellents are ecologically acceptable substances (e.g. some essential oils) that selectively repel some species of insects by smell or colour.

Pheromones are volatile substances produced by insects (or industrially synthesized) that affect the behaviour of individuals of the same species even in extremely small quantities and to distance of several kilometres. Pheromones are divided into several groups, from which the best known are:

- aggregation – ensures the accumulation of the individuals of both sexes of insects at a suitable source of food;
- sexual – regulates the contact of both insect sexes;
- dispersion – motivates the behaviour of insects in time and space.

Pheromones are used mainly on round-wood yards and in forests. For example, the synthesized pheromone *cis*-verbenol, which is produced naturally by female spruce bark beetles (*Ips typographus*), is applied in evaporators, in attracting trees, and so on. The effect of the pheromone on the selected insect species is combined with their subsequent mechanical or chemical killing.

5.2.4 Fire retardants

Fire retardants reduce the flammability of wood and other flammable materials (Horrocks & Price, 2001).

Fire retardants of wooden materials suppress the thermal decomposition and burning of wood by the following physical and chemical effects.

- They create a mass insulation layer, which restricts the ingress of oxygen to the wood and the directionally opposite transfer of the flammable gases from the wood.
- They create a thermal insulation layer, which prevent the ingress of heat from the external thermic sources to the wood; that is, either from the synthetic intumescent paints after their foaming and carbonization, or from

the wood itself, which in the presence of catalytic substances is converted to charcoal.

- They dilute flammable gases with inflammable gases, which leak from the wood during its thermal decomposition.
- They reduce the concentration of oxygen in the zone of active pyrolysis, whereas they supresses the course of exothermic thermo-oxidation reactions directly in the wood and also in the flammable gases escaping above its surfaces.
- They activate endothermic reactions, in particular dehydration reactions, and at the same time support the generation of a thermal insulation layer of charcoal, and in contrast suppress the generation of flammable gases.
- They prevent the total oxidation of charcoal to carbon dioxide, thus preventing the exothermic glow of charcoal, which is a potential source of other fires.

5.2.4.1 Physical effects of fire retardants

Restriction of mass transfer by creation of mass-nonpermeable layers (of oxygen to the wood and of flammable gases, such as methane and alcohols, from the wood:

- on the outside surfaces of wood (e.g. by meltable water-glasses, based on sodium and potassium silicates, or also by intumescent paints);
- inside the wood on the S_3 layers of its cell walls (e.g. by meltable borates, such as H_3BO_3 and $Na_2B_4O_7 \cdot 10H_2O$).

Restriction of heat transfer by intumescent thermal-insulation solid foams created from intumescent paints or foils on wood surfaces at temperatures above 150 °C (Figure 5.5). Intumescent foams have a microporous structure, which is a strong, nonflammable, and has a thermal-insulation effect (Horrocks & Price, 2001; Daniliuc et al., 2012). These foams protect the surfaces of wood for a rather long time (sometimes up to 30 min) against fire or radiant heat impact. Their expansion factor – that is, the relation of the thickness of the solid foam to the initial thickness of the paint – is ~50 : 1. In practice, at least 5 mm thickness of foam is desirable, which can be achieved by 300–500 g of paint per square metre of wood.

Intumescent paints and foils contain the following components, which at high temperatures create nonflammable gases and solid foam:

Macromolecular substance Foaming agent Catalyst and other additives	$\xrightarrow{t(°C)}$	(a) Nonflammable gases (b) Condensed phase in the form of solid foam

Dilution of flammable gases released from wood with inflammable gases H_2O, NH_3 or CO_2 with the aim to inhibit exothermic oxidation reactions in molecules of the flammable gases. The probability of collisions and reactions of molecules

Figure 5.5 The fire-stop effect of intumescent paint: the layer of solid carbonized foam (a) above the surface of undamaged wood (b)

Source: R., L. (2008) *Ochrana Dreva* (*Wood Protection*), Handbook, TU Zvolen, Slovakia, 453 p. Reproduced by permission of TU Zvolen

of flammable gases with molecules of oxygen is reduced in the presence of inflammable gases, whereas the propagation of exothermic reactions is also reduced. However, the inflammable gases, such as water vapour, carbon dioxide or ammonia may not be released prematurely, but at the moment of the thermal decomposition of wood connected with the generation of flammable gases. The sources of inflammable gases are, for example:

- hydrated aluminium sulphate – $Al_2(SO_4)_3 \cdot 18\ H_2O \rightarrow H_2O$
- ammonium dihydrogen phosphate or ammonium sulphate – $NH_4H_2PO_4$ or $(NH_4)_2SO_4 \rightarrow NH_3$
- potassium carbonate or sodium carbonate – K_2CO_3 or $Na_2CO_3 \rightarrow CO_2$.

5.2.4.2 Chemical effects of fire retardants

Oxidizable substances creating chemical bonds with oxygen decrease the concentration of oxygen molecules in the zone of active pyrolysis of the wood and block the burning process. Halides – for example, chlorides (Cl^-) and bromides (Br^-) – are mainly used for these purposes, as they are oxidized by oxygen to higher oxidation levels:

$$Cl^- + 0.5O_2 \rightarrow ClO^- + 0.5O_2 \rightarrow ClO_2 + 0.5O_2 \rightarrow ClO_3^- + 0.5O_2 \rightarrow ClO_4^- \quad (5.3)$$

Catalytic dehydration of polysaccharides and carbonization of wood – molecules of water are created from the polysaccharides during their endothermic dehydration, while the polysaccharides are also carbonized and so less are transformed to flammable gases. The dehydration of polysaccharides is catalysed by substances based on phosphoric acid; for example, $(NH_4)_2HPO_4$, $NH_4H_2PO_4$, K_3PO_4, or tri-arylphosphates. A similar dehydration effect on wood components is attributable also to salts of other strong inorganic acids (e.g. sulphuric acid and hydrochloric acid) or to boric compounds (e.g. H_3BO_3 and $Na_2B_4O_7 \cdot 10H_2O$).

5.2.5 Protective coatings against weather impacts

Coatings to protect wood used in exteriors from weather impacts are divided into two groups:

- film-forming and glazing – based on alkyd, polyacrylate, polyurethane or other macromolecules;
- low molecular weight, capable of penetration deep into the wood.

5.2.5.1 Film-forming and glazing-latex coatings

Film-forming coatings form a continuous nonporous layer on the surface of the wood.

Glazing-latex coatings partially penetrate the surface layers of wood and create porous layers; however, after several paint applications they can also create vapour-impermeable film layers.

Film-forming and glazing-latex coatings have a higher viscosity, their macromolecules are very large, and therefore these systems usually remain only on the outside wood surfaces, without penetrating the cell walls of wood. They preserve wood surfaces from UV radiation, mechanical wear and leaching out of extractive and depolymerized lignin–polysaccharide substances. In total, they improve the function and aesthetic value of wooden products. The transparent types of these coatings are convenient mainly for darker wood species in order to prevent the loss of their original colour; that is, they prevent their discoloration.

Various types of film-forming and glazing-latex coatings absorb and scatter UV radiation well, prevent transfer of moisture between wood and ambient air, thereby suppressing the creation of humidity stresses in wood and its consequent cracking. This is due to the creation of phase interfaces in the 'wood–coating' system, but mainly due to the impact of antioxidant-type additives, UV absorbents and pigments as reflectors of UV radiation (Morris & McFarling, 2006; Evans and Chowdhury, 2010).

Transparent coatings (varnishes) should contain more UV-effective substances than pigmented ones do. The UV additives used are (1) either heterocycles (e.g. hydroxyphenyl-*s*-triazines (Ozgenc et al., 2012; Forsthuber et al., 2013), hydroxyphenyl-benzotriazoles (Forsthuber et al., 2013) and imidized nanoparticles (Samyn et al., 2014)), (2) or polyvalent metal complexes. The UV blocking effect is also provided by mineral ZnO and TiO_2 particles (Cristea et al., 2010; Forsthuber et al., 2013). 'Hindered amine light stabilizers' (Schaller et al., 2009; Šomšák et al., 2015), antioxidants such as bark extracts (Saha et al., 2011) and lignin stabilizers such as succinic anhydride (Teacà et al., 2013) belong to the most progressive colour-stabilizing additives in the transparent or lightly pigmented coating systems for the finishing of outdoor wood products. However, in practice, and also in field and accelerated weathering tests, the durability of transparent coatings for protection of wood surfaces is lower than of pigmented ones (Reinprecht & Pánek, 2015).

The durability of transparent and pigmented coating systems present on wood surfaces is characterized by the state when their specific properties become technically dysfunctional or aesthetically unacceptable for the end user or from a technical point of view (Grüll et al., 2011). The most important properties of wood coatings are (1) colour stability and (2) functional stability, measured in terms of their hydrophobic properties, liquid water and vapour diffusivity, elasticity and adhesion strength.

Other types of coatings used nowadays for wood protection in exteriors are mainly coatings based on alkyds modified with drying oils and natural resins; and then there are the paints based on polyacrylates, acrylate copolymers, polyurethanes and their combinations (Petrič et al., 2003; Jacobsen & Evans, 2006; Reinprecht et al., 2011; Daniliuc et al., 2012; Petrič, 2013). Some of these coatings can partially penetrate into the wood as well. However, in the case of an unsuitable application (e.g. when a suitable input moisture of wood is not secured), these coatings may be damaged at the 'wood–coating' interface and lose their protective function (Figure 5.6).

Figure 5.6 Coatings on new spruce shingles were already damaged after 2 years due to a high wood moisture before painting

Source: R., L. (2008) *Ochrana Dreva* (*Wood Protection*), Handbook, TU Zvolen, Slovakia, 453 p. Reproduced by permission of TU Zvolen

5.2.5.2 Low molecular weight penetration systems

The main task of penetration systems used against weather impacts is making the wood surfaces hydrophobic and stabilizing them towards photo-oxidation processes, creation of cracks, and activity of staining fungi and moulds. These liquid systems comprise water-repellent waxes, vegetable oils, UV-effective substances and other stabilizers against weather factors, stains and/or pigments, resins (up to 10–20% maximum), fungicide against moulds and staining fungi, and solvents. They can easily flow through lumens, or can also penetrate to the cell walls of wood. For penetration to the cell walls of wood, all components of the penetration systems should be polar and their dimensions should be less than 10 nm (or even 1 nm) to 80 nm, with a molecular weight less than 1000–3000 Da (Williams, 2010). Cell walls treated with low molecular weight systems absorb less water and swell less than cell walls of natural wood. Waxes slow down water absorption and photodegradation of wood and prolong its durability (Lesar et al., 2011; Brischke & Melcher, 2015). For example, wood treated with modified *N*-methylol-waxes is suitable for class 3 use by EN 335 (Minh et al., 2007). Low molecular weight systems can be applied to both smooth and rough surfaces of wood by painting, spraying, dipping or vacuum-pressure technologies. Their durability on saw-textured or naturally-weathered wood is longer than on planed wood. On surfaces of new timbers there should be two layers applied. Reapplication of low molecular weight penetration systems on wood surfaces should be performed after 1–3 years, depending on the wood species and exposure (Williams, 2010).

Transparent and lightly coloured 'water repellents' finishes – in the past contained an organic solvent (e.g. turpentine or mineral spirit). However, as a result of ecological aspects and regulations concerning VOCs, only waterborne emulsion systems or paraffin oil formulations are commonly used. These finishes do not change the natural colour of wood, but sometimes the treated wood is brighter and has a golden tan (Williams & Feist, 1999).

Oil-based and water-based semi-transparent stains contain oil resins or oil–alkyd resins with a certain amount of pigments. They are applied in oils or water emulsions, however, without forming films on the wood surfaces. Their service life is longer than of water repellents.

5.2.6 Evaluation of new preservatives

New fungicides, insecticides, fire retardants and other active substances of preservatives, and also new commercial preservatives recommended for market, have to undergo numerous screening and standardization tests before being accepted. These tests are related to their:

- intended efficiency (fungicidal, fire retardation, weather resistance, etc.);
- treatability (solubility in various solvents, suitability for water dispersions, viscosity, penetration to wood, etc.);

- stability in wood (evaporation, leaching, etc.);
- corrosiveness for wood, but also for glues, metals, glasses, plastics and other materials potentially connected in/with wooden products (structure damage and strength loss);
- health risk to people and the environment (LD_{50}, etc.).

For example, in Europe the efficiency of new wood preservatives with intended application against fungal decay is now evaluated by several standardized laboratory and field tests; for example, the EN 113, EN 839, EN 330 and EN 252. For illustration, the basic principles of two tests for evaluation of new fungicides are shown:

- EN 113: Laboratory tests of decay resistance, using sap-pine and beech wood samples 50 × 25 × 15 mm^3 ($L \times R \times T$ – longitudinal × radial × tangential) preserved by vacuum technology with graded concentrations of new preservative, are performed on the basis of mass losses of samples after their 16 weeks' exposure in selected species of decay-causing fungi (*Coniophora puteana*, *Trametes versicolor*, *Poria placenta*, *Gloeophyllum trabeum*, or also other species). The minimal concentration of the new preservative needed is extrapolated to achieve a 3% mass loss of preserved wood.
- EN 252: Field test in contact with ground, using sap-pine or other wood stakes 500 × 50 × 25 mm^3 ($L \times R \times T$) preserved by vacuum-pressure impregnation with graded concentrations of new preservative, are performed on the basis of their 5-year resistance to attack by various types of decay-causing fungi and other organisms acting in unsterilized ground (250 mm of the stake's length is in the ground) and above ground. The minimal concentration of the new preservative needed is extrapolated from the degrees of damage to wooden stakes treated with the researched fungicide, to stakes treated with commercial established preservative and also unpreserved stakes.

5.3 Technologies of chemical protection of wood

The technologies of chemical protection of wooden products are decided on the basis of their exposure; for example, by defined classes of use according to EN 335. The total quality of wood protection is determined by the type of preservative (Section 5.2) and by its distribution and fixation in wood (Sections 5.3.3 and 5.3.4).

The *distribution of the preservative in the wood* is quantified by its retention R (kg/m^3 or g/m^2) and depth of penetration l (mm). Retention can also be quantified by other parameters; for example, by the WPG and pore-filling-ratio (PFR) (Section 6.1).

Table 5.2 Classes of wood impregnability (in accordance with EN 350-2)

Wood impregnability class		Species for particular wood zone		
		Sap-wood	**Mature wood**	**Heartwood (false heartwood)**
1	Easy	Acacia, alder, beech, maple, oak, poplar, walnut	Alder, beech, maple	
1–2		Birch	Birch	
2	Medium	Ash, fir, chestnut, larch		Ash
3	Difficult	Spruce, teak	Fir	Poplar, walnut
4	Extremely difficult		Spruce	Acacia, oak, chestnut, larch, teak, (beech)

All these parameters characterizing quality of wood impregnation are affected by several factors, formulated by Kurjatko and Reinprecht (1993), Hansmann et al. (2002), Tondi et al. (2013) and others:

- wood structure, which characterizes the active areas and coefficients of permeability of wood in its individual anatomical directions, and then also its impregnability (Sections 1.1 and 5.3.1; Table 5.2);
- wood moisture, which influences the opening/closing state of pits in coniferous wood (Section 1.1), and also the applicability of pressure and diffusion driving forces in transport of the preservative in wood (Section 5.3.3);
- application properties of the preservative – for example, viscosity, polarity and surface tension, which are significant in its transport in wood (Section 5.3.2);
- pressure and diffusion driving forces in transport of the preservative in wood (Section 5.3.3).

5.3.1 Improvement of permeability and impregnability of wood

Several operations should be carried out before the chemical protection of wooden products:

- debarking;
- adjustments to geometric shape and surface cleaning;
- improvement to their impregnability, if they are made from refractory species;
- optimization of moisture;
- marking;
- use of mechanical fittings or paints against creation of frontal and other cracks.

The improvement of permeability and impregnability of wooden products from refractory species, which are poorly impregnable (Table 5.2), can be achieved by their mechanical, physical, chemical or biological pretreatment (Morrell & Morris, 2002; Reinprecht, 2008; Lehringer et al., 2009).

Mechanical pretreatment of wood by cutting and puncturing is carried out on mechanized lines. Special knives or needles fixed in rollers are pressed to a predetermined depth of wood, usually 8–19 mm from peripheral areas of lumber and timber. In practice, the cutting and punching methods are often used for pretreatment of beams, fences and other squared assortments of spruce wood and other poorly impregnable species.

Larger holes in the wood, with the aim to improve the penetration process, are made with drill bits (e.g. into sleepers and utility poles). Railway sleepers are drilled in central parts from below in order to increase their frontal–axial areas and improve the penetration process, in particular at the centre of their length from the bottom loading area; for example, in the Czech Republic or Slovakia by eight bores with a diameter of 16 mm to a depth of 60–130 mm. Utility poles from spruce, fir or Douglas fir are drilled in a machine with several drill bits having a diameter of 3 mm to the depth of 30 $\pm$ 5 mm, usually only in their bottom part; for example, 400 mm above and 500 mm under the terrain.

Physical pretreatments of wood are based upon the local impairment of the wood structure and increasing its frontal–axial areas, using:

- laser perforations;
- hard regimes of drying and steaming, when microcracks are created in cell walls of all wood species, or the pit membranes in cell walls of tracheids in coniferous species are destroyed.

For intensification of the mass-transport processes in wooden products directly during their treatment, one can also use some physical forces (e.g. ultrasonic or pulse waves).

Chemical pretreatments of wood are based on the changes in chemical composition of the pits in cell walls of tracheids in coniferous species, using:

- extraction of terpenes or other accompanying substances by organic solvents, or by hot water solution of ammonium oxalate;
- chemical modification of –OH groups in pits (e.g. by acetylation), whereas the closing of more hydrophobic pits in tracheids on drying of wood will be worsened;
- degradation of pectins and other components present in the membrane (margo) of pits, either by aggressive chemicals (e.g. $NaClO_2$) or by steaming and boiling with the production of acetic acid, which has an aggressive effect on components in pits.

Biological pretreatments of wood are based either on a nonsterile immersion of round debarked timbers in natural dams for 1–2 months (Singh et al., 1998), or their placing in an environment infected by suitable types of microorganisms

(Figure 3.1). This can involve use of bacteria such as *Bacillus subtilis*, *Bacillus lichteniformis* and *Pseudomonans* sp. (Pánek & Reinprecht, 2011; Yildiz et al., 2012), microscopic fungi such as *Trichoderma gliocladium* and *Trichoderma viride* (Messner et al., 2003; Pánek et al., 2013), staining fungi, or white-rot fungi such as *Physisporinus vitreus* (Mai et al., 2004; Lehringer et al., 2010). It is possible to improve mainly the permeability of the sap-zone of coniferous species with a minimal change to their mechanical properties. The most suitable method for this is to use such bacteria and fungi that produce just some specific enzymes of the hydrolase type (e.g. xylanase, arabinase, amylase, pectinase) necessary for the attack of the structural components of bordered pits (e.g. pectins in their membranes), while they are unable to produce a broader complex of enzymes necessary for the damage of polysaccharides and lignin in the cell walls of wood.

5.3.2 Application properties of preservatives

The *application properties of the preservative* are predetermined by the treatment technology and also by the final quality of chemical protection of the wooden product and its service life. The most important are the following application properties of preservatives (Reinprecht, 2008).

- Sufficient storage life (e.g. without creation of sediments).
- Fast and uniform penetration into wood.
- Good applicability on wood surfaces.
- High physical stability in treated wood:
  - low leachability by water;
  - low volatility.
- High chemical stability in treated wood (e.g. without creation of less-effective agents from active substances):
  - in the presence of air;
  - upon the effect of sun (UV and other radiation);
  - in the presence of metals in technological devices (e.g. in autoclaves) and in the wooden products (e.g. in contact with nails and screws);
  - in contact with the wood (e.g. in the presence of tannins);
  - in contact with adhesives or paints used in surface treatment of preserved wood.
- Maintaining or improving the physical and mechanical properties of wood:
  - no corrosion of lignin–polysaccharide matrix of wood;
  - no increase in hygroscopicity and water absorptivity of wood;
  - no increase in electrical conductivity of wood (e.g. important for railway sleepers);
  - no impairment of wood aesthetics.
- Harmony with other substances present in wooden product:
  - no reduction of the strength of glued joints;
  - no impairment of the quality of surface treatments.

- A wide sphere of applicability, and both price and market availability:
  - applicability into air-dried wood, into wet wood or into wood with various moisture contents;
  - applicability within a wide range of temperatures;
  - applicability by various technologies, using various types of spraying, autoclave and other devices;
  - applicability approved by legal regulations in the majority of the world's countries, and thus also the possibility to export the chemically treated wood.

The application properties of preservatives depend on their physical–chemical and chemical properties, among which the following are the most important:

- single- or multi-component preservative system;
- physical state (i.e. solid, liquid, or gas);
- solubility in water and organic solvents;
- density;
- viscosity;
- diffusion coefficient;
- surface tension and wettability of wood surfaces;
- pH value;
- dissociation constant;
- polarity;
- spatial dimension of molecules or macromolecules;
- thermal points of phase transformations (melting point, boiling point);
- thermal stability (destruction point);
- chemical stability (e.g. reactivity with wood, adhesives, paints, oxygen).

5.3.3 Flow and diffusion transport of preservatives in wood

5.3.3.1 Pressure and diffusion driving forces

The preservatives are transported into the wood due to pressure and diffusion driving forces (Table 5.3):

- pressure forces (i.e. flow gradient $\mathrm{d}p/\mathrm{d}l$)
  - capillary pressure
  - hydrostatic pressure
  - dynamic pressure
  - mechanical pressure
  - vacuum in wood, produced in hot–cold or vacuum technologies
  - autoclave pressure, produced in vacuum-pressure technologies
- diffusion forces (i.e. gradient of diffusion $\mathrm{d}c/\mathrm{d}l$).

Table 5.3 Basic driving forces in the technologies of chemical protection of wood

Protection technology			
Principle	**Variations**	**Initial wood moisture**	**Driving force of transport**
Application of solution	1. Painting	Over-dried	Capillary pressure
	2. Spraying	Over-dried	Capillary pressure
	3. Immersion	Over-dried	Capillary pressure
	4. Immersion with storage	Raw (dry surfaces)	Capillary pressure + diffusion
	5. Panel method	Over-dried	Capillary pressure
	6. Panel method	Raw	Diffusion
	7. Injecting	Over-dried	Capillary pressure
Application of pastes	1. Painting with storage	Raw	Diffusion
	2. Bandages	Raw	Diffusion
	3. Injecting	Raw	Diffusion
Placing in the tank	1. Dipping	Over-dried	Capillary pressure
	2. Dipping	Raw	Diffusion
	3. Dipping after vacuum	Over-dried	External pressure
	4. Dipping with preheating	Over-dried	External pressure
Vacuum-pressure impregnation	1. Method of full saturation of cells (Bethell)	Over-dried	External pressure
	2. Method of partial saturation of cells (Lowry)	Over-dried	External pressure
	3. Method of empty cells (Rüping)	Over-dried	External pressure
Pressure–diffusion impregnation	Pulse methods with aqueous solutions	Raw (dry surfaces)	External pressure + diffusion
Combined drying and pressure impregnation	Cyclic methods with creosotes	Raw	External pressure

Source: R., L. (2008) *Ochrana Dreva* (*Wood Protection*), Handbook, TU Zvolen, Slovakia, 453 p. Reproduced by permission of TU Zvolen.

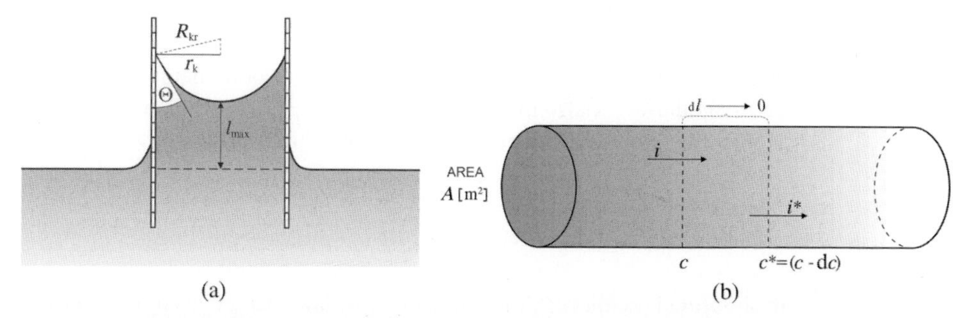

Figure 5.7 The simplest transport of preservatives into model cylindrical capillaries of wood: (a) due to a capillary pressure in an empty capillary; (b) due to a concentration gradient in a capillary filled with water or other solvent (Reinprecht 2008)

Source: R., L. (2008) *Ochrana Dreva* (*Wood Protection*), Handbook, TU Zvolen, Slovakia, 453 p. Reproduced by permission of TU Zvolen

All these driving forces usually decrease with prolonged time τ of the technology of wood protection; that is, $dp/dl = f(\tau)$ or $dc/dl = f(\tau)$. Therefore, usually only nonstationary flow or diffusion transport acts in the real chemical protection of wood (Siau, 1984; Kurjatko & Reinprecht, 1993).

Capillary pressure p_C is applied in all simple technologies, as is painting, spraying, immersion, dipping, bandages, panel impregnation or injecting through pre-bored channels (Figure 5.7a).

Hydrostatic pressure is applied together with capillary pressure in the technologies of immersion and dipping of wood, particularly in the directed flow of protective agent through the faces of green logs (Boucheri's method and its modifications by Gewecky).

Dynamic pressure is applied together with capillary pressure in the technology of wood spraying.

Local mechanical pressure is applied when liquid of the preservative is pressed into the capillaries of wood using a brush or a spatula, and mainly when injecting the wood.

Vacuum in wood (p_{hc}) *created by hot–cold* ($T_1 > T_2$) *technology*, and so a pressure gradient dp is generated:

$$dp\ (\text{Pa}) = p_{\text{atmospheric}} - p_{\text{hc}} = p_{\text{atmospheric}} - (p_{\text{w2}} + p_{\text{A2}}) \tag{5.4}$$

where $p_{\text{atmospheric}}$ is the atmospheric pressure consisting of the partial pressure of water vapour and the partial pressure of air $(p_{\text{w1}} + p_{\text{A1}})$ in heated wood at temperature T_1; p_{w2} is the partial pressure of water vapour in cooled wood at T_2, so $p_{\text{w2}} < p_{\text{w1}}$; p_{A2} is the partial pressure of air in cooled wood at T_2, so $p_{\text{A2}} < p_{\text{A1}}$.

Autoclave pressure is applied in autoclaves by various vacuum-pressure technologies.

5.3.3.2 Flow and diffusion transports in wood

Flow of liquid preservatives is dominant into air-dry wood; that is, in its lumens (macro-capillaries) or also in its cell walls (micro-capillaries). It takes place due to the effect of pressure driving forces dp/dl, which are produced

- naturally (e.g. capillary pressure) or
- intentionally (e.g. hydrostatic pressure or autoclave pressure).

In simplified conditions, the following flow laws can be used for modelling of wood preservation technologies.

Stationary viscous flow of a liquid preservative in wood does not exist, but it can be suitably idealized as the flow in model of cylindrical capillaries with defined length L under the following conditions:

- flow of liquid is viscous and linear ($V' \approx \mathrm{d}p$);
- liquid is homogeneous and incompressible;
- systems of cylindrical capillaries have the same cross-section – that is, a constant active area A_a in the model for flow along the entire length;
- there are no interactions between the liquid and surface of cylindrical capillaries (e.g. cell walls of vessels).

Under these conditions the flow rate of liquid V' can be expressed by the Poiseuille equation:

$$V'\ (\mathrm{m^3/s}) = \frac{V}{\tau} = \frac{nA\pi r^4_{(\text{C-mean})}}{8\eta L}\mathrm{d}p = \frac{AK\,\mathrm{d}p}{\eta L} \tag{5.5}$$

where V ($\mathrm{m^3}$) is the volume of liquid, τ (s) is time, A ($\mathrm{m^2}$) is the total area of the wood model perpendicular to the direction of flow, n ($1/\mathrm{m^2}$) is the number of capillaries per area unit A, $r_{(\text{C-mean})}$ (m) is the mean radius of capillaries in the direction of flow ($r_{(\text{C-mean})} = (\sum(n_j r^4_{Cj})/\sum n_j)^{1/4}$), η (Pa s) is the dynamic viscosity of the liquid, p (Pa) is the pressure, L (m) is the total length of capillaries in the wood model in the direction of liquid flow, dp/L (Pa/m) is the pressure driving force (gradient) of flow, and K ($\mathrm{m^2}$) is the coefficient of specific permeability of the capillaries in the wood model ($K = (n\pi r^4_{(\text{C-mean})})/8$).

The intensity of viscous flow of liquid i_{flow} in such an idealized model of wood is defined under the idealized stationary conditions by Darcy's law:

$$i_{\text{flow}}\ (\mathrm{m/s}) = \left(\frac{V}{\tau A}\right) = -\left(\frac{K}{\eta}\right)\left(\frac{\mathrm{d}p}{L}\right) = -k\left(\frac{\mathrm{d}p}{L}\right) \tag{5.6}$$

where $k = K/\eta$ ($\mathrm{m^2}$/(Pas) is the coefficient of real permeability of wood for a particular liquid.

The *nonstationary viscous flow of a liquid preservative* in wood – for the simplest nonstationary case ($dl = f(\tau)$; dp and A_a are constants) – can be analysed using the model of parallel uniform cylindrical capillaries having a constant diameter r and a constant active area A_a. The depth of liquid penetration l in such a capillary model is directly proportional to the square root of pressure difference dp and the time of preservation process τ, and inversely proportional to the square root of the dynamic viscosity of the liquid η:

$$l\,(\mathrm{m}) = r\sqrt{\frac{\mathrm{d}p\,\tau}{4\eta}} \tag{5.7}$$

However, in practice, the penetration of liquid preservatives to the wood is the most intense in the first phases of the technological process and then it slows down, which may be aptly characterized by an exponential increase of the individual parameters of wood impregnation (Figure 5.8).

Diffusion of water-soluble preservatives takes place on the basis of concentration driving forces – gradients dc/dl. It is dominant in wet wood containing free and bound water. Diffusion is very important also after the treatment process; that is, during long-term transport of preservatives or other substances to cell walls filled with bound water (Jakes et al., 2013). For example, Petrič et al. (2000), using transmission electron microscopy with X-ray microanalysis, found that the highest amount of copper and zinc wood preservatives in cell walls of spruce wood was located in the cell corners and middle lamellae having the highest portion of lignin. Diffusion is applied also upon the penetration of toxic or nontoxic gases in the chemical sterilization of wood (see Section 7.4).

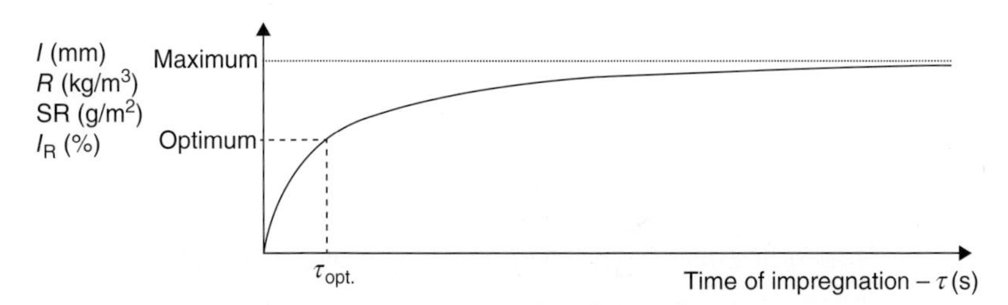

Figure 5.8 Nonlinear (usually exponential) increase of the depth of penetration *l*, the retention *R*, the surface retention SR, and l_R (equal to PFR – see Section 6.1) of the preservative into the wood during the impregnation process, and specification of the optimal impregnation time τ_{opt}

Source: R., L. (2008) *Ochrana Dreva* (*Wood Protection*), Handbook, TU Zvolen, Slovakia, 453 p. Reproduced by permission of TU Zvolen

Intensity of diffusion transport of individual molecules of preservative $i_{diffusion}$ in the capillaries of wood with a known length L is defined under stationary conditions by Fick's first law (Figure 5.7b):

$$i_{\text{diffusion}}\ (\text{mol}/(\text{m}^2\text{s})) = \left(\frac{N}{\tau A}\right) = -D\left(\frac{\text{d}c}{L}\right) \tag{5.8}$$

where N (mol) is the amount of molecules of an active substance present in the preservative, τ (s) is time, A (m^2) is the total area of wood perpendicular to the direction of diffusion, c (mol/m^3) is the concentration of an active substance of the preservative, L (m) is the total length of capillaries in the direction of diffusion, dc/L (mol/m^4) is the concentration driving force (gradient) of diffusion, and D (m^2/s) is the diffusion coefficient of an active substance of the preservative in the fluid medium used (e.g. in free water filling the lumens of wood cells).

The intensity of diffusion increases with an increase in temperature. At a higher temperature a higher concentration c (or c_{max}) of an active substance of the preservative in the fluid medium (e.g. in water) is achieved, and the mobility of its molecules, defined by the coefficient of diffusion D, also increases. Reinprecht and Makovíny (1990) documented this knowledge for the diffusion of $CuSO_4 \cdot 5H_2O$ in a wet beech wood (*Fagus sylvatica*) at temperatures of 18, 25 and 50 °C. Diffusion of H_3BO_3 and other active water-soluble substances of various commercial preservatives is higher in a wetter wood, and this knowledge is successfully utilized in practice.

5.3.4 Fixation of preservatives in wood

The fixation of preservative active substances in the wood is important in particular for wooden products and structures used in exteriors. The fixation of copper, boron or other preservatives in wood may be provided by their chemical reactions – grafting onto wood components, creation of stable complexes, adsorption interactions and so on, usually in which –OH or other functional groups of wood participate as well (Table 5.4).

The fixation processes should be performed only after the time when transport processes of the preservative into the wood structure are completed, with the aim not to affect its penetration and retention.

5.3.5 Non-autoclave technologies of chemical protection of wood

Wooden elements of log cabins, ceilings, trusses and other constructions exposed under roofs (by EN 335 in use classes 1 and 2, or also 3) can be preserved by painting and other non-autoclave technologies.

5.3.5.1 Painting, spraying, immersion and dipping

Painting and other common non-autoclave technologies are performed at atmospheric pressure. They are based especially upon the capillary forces (in

Table 5.4 Basic methods of fixation of preservative active substances in wood

- Mechanical latching of the water-insoluble and low-volatile preservative into the lumens of wood cells (creosote oils, heterocycles, etc.)
- Adsorption of the preservative active substance on the polysaccharide and lignin components of wood
- Creation of complexes between the preservative active substance and the wood components, in particular with guaiacyl components of lignin (fixation of copper Cu^{2+}, chromium $Cr_2O_7^{2+}$, CrO_3, etc.)
- Ion-exchange reactions of the preservative active substance with carbonyl groups of hemicelluloses and lignin (fixation of QACs, etc.)
- Oxidation–reduction reactions of the preservative active substance or fixative excipient with –OH or other functional groups of wood (e.g. fixative chromium $Cr^{6+} \rightarrow Cr^{3+}$ creates insoluble complex compounds into which active biocides are incorporated, such as Cu^{2+}, F^-, and others from copper/chromate/borate, copper/flour and other salts)
- Change in the pH-value of wood after the technology of protection and subsequent crystallization of the active substance from its aqueous solution in wood (fixation of Cu-HDO fungicide, etc.)

flow) or concentration gradient (in diffusion). However, some of them also depend to a lesser degree on hydrostatic forces (immersion and dipping), local mechanical forces (painting) or dynamic forces (spraying). Immersion lasts from several seconds to 15 min, while dipping can take several hours or days. When dipping, the additional driving force of shock waves can also be applied (Jagadeesh et al., 2005).

All these technologies are realized discontinuously or continuously. For example, discontinuous immersion and dipping technologies are usually performed in stationary tanks of various shapes and dimensions (Figure 5.9a). Discontinuous spraying is often used for maintenance treatment of log cabins and other wooden constructions (Figure 5.9b), while continuous spraying is applied for treatment of new windows and other wooden products directly in factories.

5.3.5.2 Panel impregnation, bandages, cartridges and injecting

Panel impregnation is a special technology for the protection of compact vertical panel structures (e.g., log cabins and shingle facing) lasting from several days to weeks (Figure 5.10). It is applied predominantly when reconstructing damaged wooden structures without their disassembly.

Bandages are diffusion methods for the additional protection of those local zones of wood in which optimum moisture conditions for the activity of biological pests were created or from which the preservative has already been partially washed out or evaporated. The principle of the bandages lies in the primary wrapping of the surface of moist wood with a porous cloth soaked in a polar protective agent and its subsequent wrapping with a vapour-impermeable foil.

(a)

(b)

Figure 5.9 Technology of wood treatment by immersion or dipping in a stationary dip-tank (a) and by discontinuous spraying (b)

Sources: (a) Reproduced from prospect. (b) R., L. (2008) *Ochrana Dreva* (*Wood Protection*), Handbook, TU Zvolen, Slovakia, 453 p. Reproduced by permission of TU Zvolen

For example, utility poles in the transition zones of ground and air can be additionally protected by bandages (Figure 5.11).

Cartridges filled with paste, powder or crystals of water-soluble biocides are inserted into the pre-bored holes in the wood. They are used for prevention, particularly for curative protection on risky places; for example, in L-joints of frames and wings of windows (Figure 5.12). The cartridges are usually of a

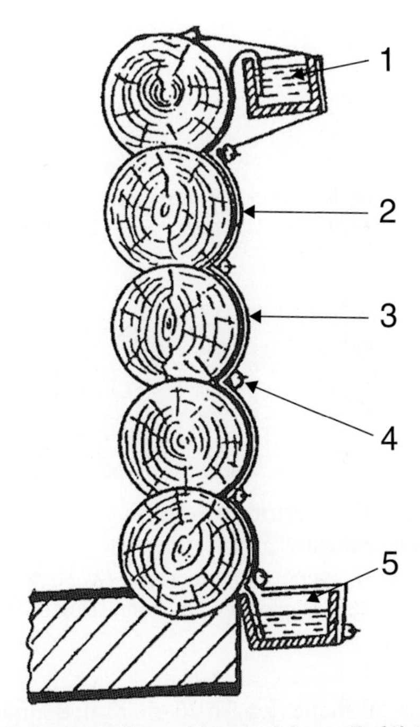

Figure 5.10 Panel impregnation of log-cabin wall: (1) top tank of preservative; (2) porous cloth; (3) vapour-impermeable foil (e.g. polythene, poly(vinyl chloride)); (4) pressing bar; (5) lower tank of preservative

Source: R., L. (2008) *Ochrana Dreva* (*Wood Protection*), Handbook, TU Zvolen, Slovakia, 453 p. Reproduced by permission of TU Zvolen

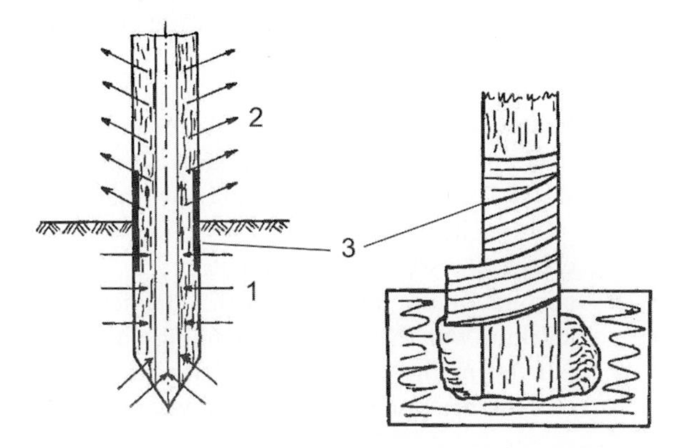

Figure 5.11 Bandage for local chemical protection of a wood column: (1) penetration of water; (2) evaporation of water; (3) bandaging

Source: R., L. (2008) *Ochrana Dreva* (*Wood Protection*), Handbook, TU Zvolen, Slovakia, 453 p. Reproduced by permission of TU Zvolen

Figure 5.12 Cartridge with biocide inserted into a risky L-joint of a wooden window

Source: R., L. (2008) *Ochrana Dreva* (*Wood Protection*), Handbook, TU Zvolen, Slovakia, 453 p. Reproduced by permission of TU Zvolen

cylindrical shape (length 20–50 mm, diameter 5–15 mm) made of paper, cardboard or some other porous material that after soaking with water facilitates the diffusion of biocide from them to the surrounding wetted zones of wood.

Injecting is a local protection of selected zones of wood with aqueous or organic solutions of the preservative. It can be performed either into prebored holes in wood or into natural holes in wood created before by wood-damaging insects. When applying aqueous solutions into air-dried wood it will be necessary to pay attention to local swelling of the wood and the subsequent possibility of it cracking. Injecting can also be undertaken using other processes:

- hollow screws of diameter 4–10 mm and length 30–100 mm having small openings along their side surfaces are inserted into drilled holes in the wood; liquid preservative can be subsequently fed through these openings in the hollow screws from a reservoir via a pressure hose by applying an external pressure of 0.1–1.5 MPa;
- boring of holes into wood with a diameter of 5–25 mm and length 100–250 mm, with their subsequent filling with liquid or powdered preservative;
- puncturing the wood with steel needles to a depth of 10–50 mm and concurrent injection of the preservative.

5.3.6 Autoclave technologies of chemical protection of wood

Railway sleepers, utility poles (for telecommunications, electric distribution, etc.), palisades, stakes and rods for grapevine support, mining timber, roof and facing shingles, pergolas, children's playgrounds, gazeboes, garden furniture and other assortments manufactured in great numbers for exterior use (by EN 335 in use classes 3, 4 and 5) should be chemically protected using autoclave technologies – vacuum-pressure for air-dried wood and pressure-diffusion for wet wood.

5.3.6.1 Vacuum–pressure impregnations of the air-dried wood

Impregnation of air-dried wood can be performed by more vacuum-pressure technologies which differ from each other by regulation of the vacuum (V), atmospheric pressure (A) and increased pressure (P) stages. By their suitable regulation in an autoclave (Figure 5.13) the required depth of penetration and retention of the liquid preservative (and then also its active substances) into wood shall be secured (Bergman, 2003).

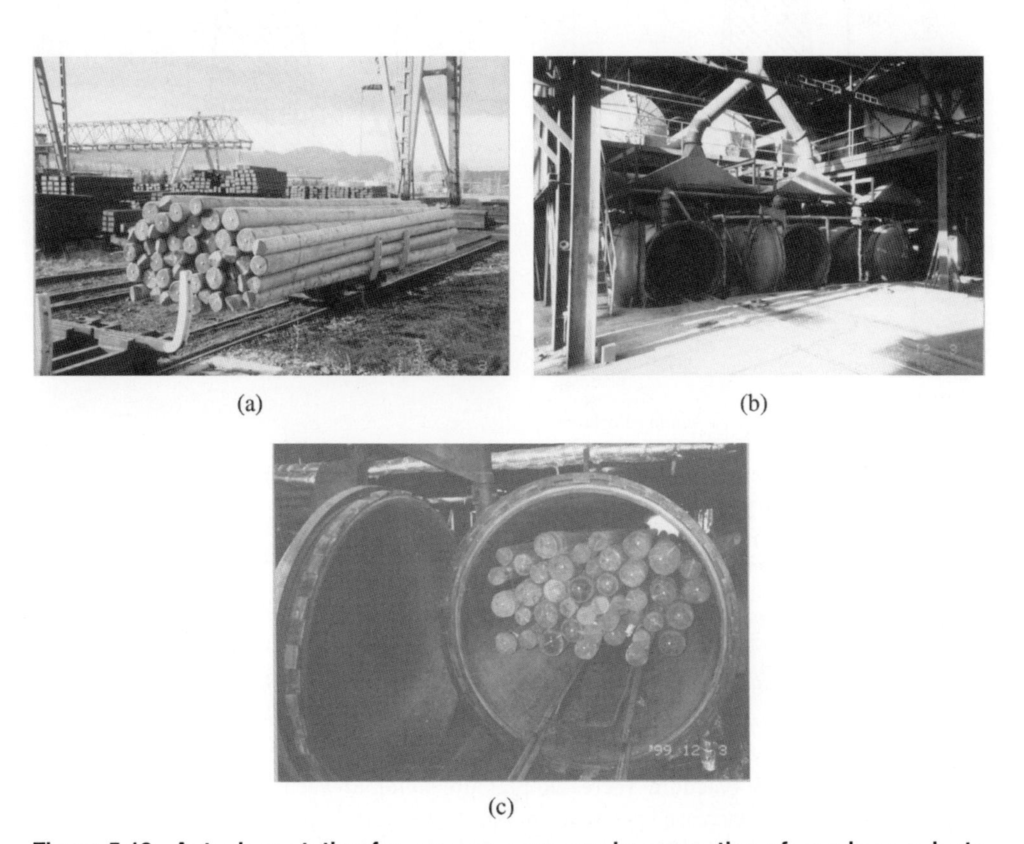

(a) (b) (c)

Figure 5.13 Autoclave station for vacuum–pressure impregnation of wooden products

Source: R., L. (2008) *Ochrana Dreva* (*Wood Protection*), Handbook, TU Zvolen, Slovakia, 453 p. Reproduced by permission of TU Zvolen

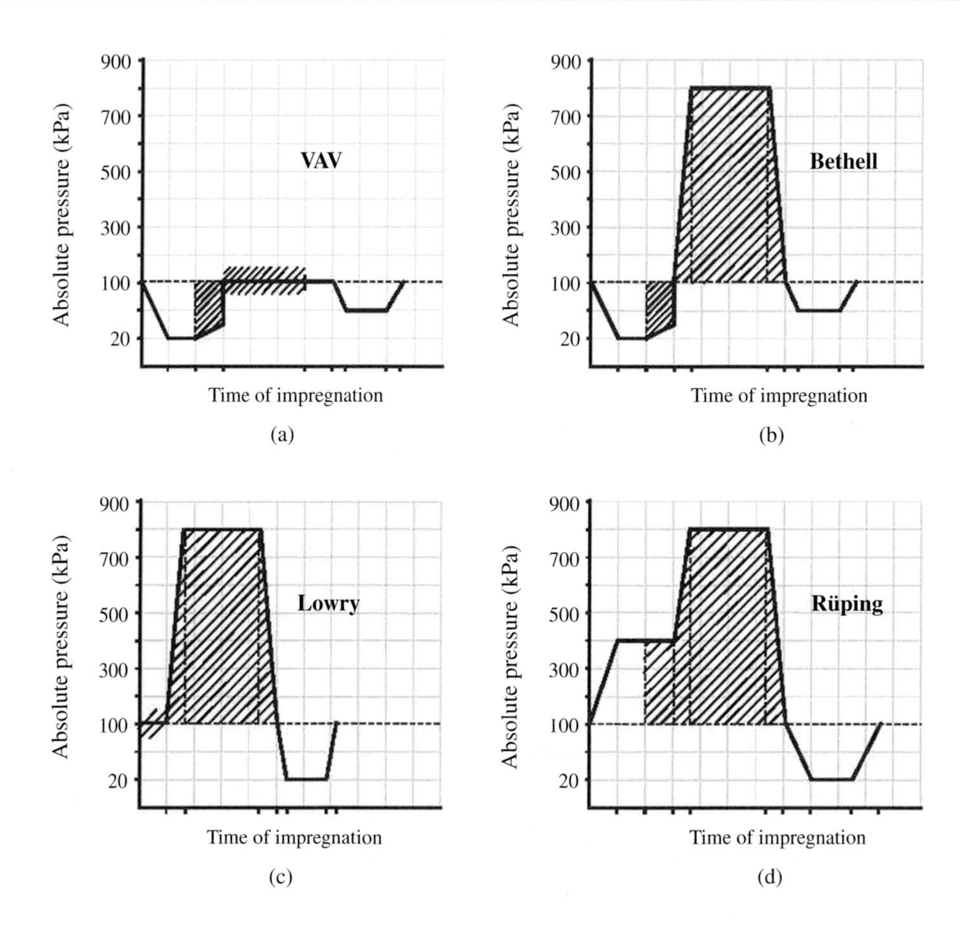

Figure 5.14 Basic technologies for impregnation of air-dried wood: (a) VAV; (b) VPV – Bethell; (c) PV – Lowry; (d) PPV – Rüping

Source: R., L. (2008) *Ochrana Dreva* (*Wood Protection*), Handbook, TU Zvolen, Slovakia, 453 p. Reproduced by permission of TU Zvolen

The best-known vacuum-pressure technologies for impregnation of the air-dried wood that are applied in practice today are the following (Figure 5.14):

- VAV (vacuum–atmospheric pressure on liquid–vacuum);
- VPV (vacuum–increased pressure on liquid–vacuum), Bethell;
- PV (increased pressure on liquid – vacuum), Lowry;
- PPV (increased pressure on wood–more increased pressure on liquid–vacuum), Rüping;
- combined technologies (e.g. Ruetger's technology, technology of a gradual increase of pressure, technology of a double Rüping impregnation).

The *VPV technology (method of complete saturation of wood cells) proposed by Bethell* is convenient mainly for water-soluble preservatives. A vacuum of 20 kPa is first created in the autoclave above the wood products, usually lasting 30–60 min (Figure 5.14b). Then the preservative is introduced to the autoclave, the vacuum is removed and an increased pressure of 800–900 kPa applied to the level of the liquid preservative over a period of ~1–6 h. After the termination of increased pressure and draining the liquid preservative from the autoclave, a final vacuum of 60–20 kPa is applied for 30–60 min. VPV technology is advantageously used for the impregnation of wood with aqueous solutions of preservative, when its retention may be regulated also by means of its concentration in the aqueous solution. VPV technology is suitable for the treatment of utility poles, palisades or shingles using, for example, Cu-HDO fungicide in aqueous solution. Creosotes can also be applied by this technology, but today only for products for which it is difficult and extremely difficult to impregnate wood species (e.g. spruce and fir) subject to outdoor exposures (e.g. poles).

The *PPV technology (method of empty wood cells) as proposed by Rüping* was only for creosotes. First, the wood products are subjected to an increased pressure of 250–400 kPa (Figure 5.14d) maintained for 15–100 min, with the aim to increase pressure of air in the lumens of wood cells. Creosote oil is poured into the autoclave only afterwards, while the pressure acting on the level of creosote oil is increased as well, initially to 350–500 kPa (0.5–2 h) and then to 800–900 kPa (1–8 h). The time of this increased pressure acting on the level of creosote oil is longer than with the Bethell technology – when the same depth of penetration should be achieved – since there are smaller pressure driving forces dp/dl created in the Rüping technology. Evacuation of the creosote oil from the autoclave and creation of a vacuum is the last stage of the PPV technology, similar to the cases of VAV, VPV and PV technologies. However, the quantity of creosote evacuation from the lumens of wood cells is substantially greater in this PPV technology (Figure 5.15), which is confirmed by both practice and theoretical calculations (Reinprecht, 1995). Creosote runs out from the impregnated wood due to the compressed air accumulated during impregnation, particularly in less conductive elements of wood (e.g. in the libriform of broadleaved species) or in the central zones of a wooden product. In the last stage of impregnation, 30–60% of creosote oil returns from the wood to the autoclave. Creosote introduced to the wood remains more or less just on the surface of cell walls from the side of lumens on the S_3 layer (creosote, being a nonpolar substance, cannot get into the cell walls); that is, it does not fill the entire space of the lumens (Figure 5.15). For example, the retention of creosote into beech sleepers, under the condition of a total impregnation of their volume, is around half in the '2 × PPV' technology (double Rüping technology) when R is 160–175 kg/m^3, compared with the 'VPV' technology (Bethell) when R is ~350 kg/m^3.

5.3.6.2 Pressure–diffusion impregnations of wet wood

Impregnation of the wet wood can be performed with several technologies:

- primary drying and subsequent vacuum–pressure impregnation;

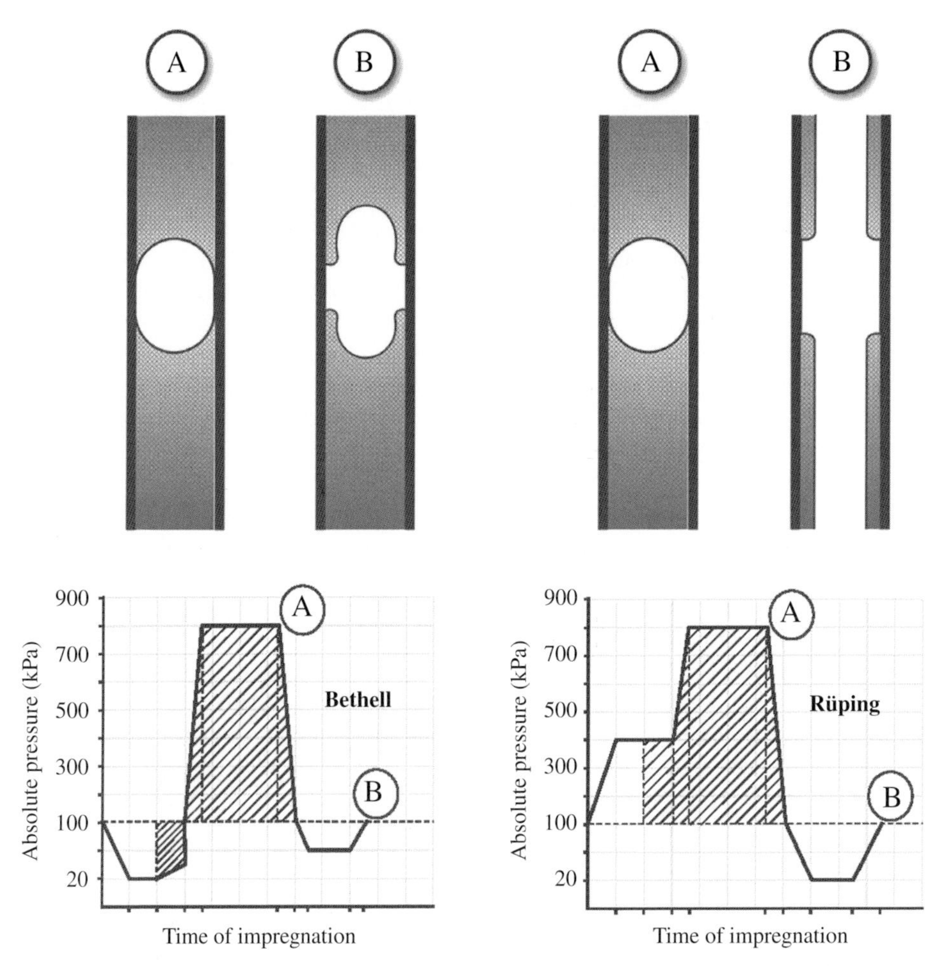

Figure 5.15 Scheme of filling of lumens of cells of wood in the VPV (Bethell) and PPV (Rüping) technologies – that is, prior to evacuation of creosote from the autoclave (stage 'A') and after evacuation of creosote from the autoclave (stage 'B')

Source: R., L. (2008) *Ochrana Dreva* (*Wood Protection*), Handbook, TU Zvolen, Slovakia, 453 p. Reproduced by permission of TU Zvolen

- cyclic vacuum–pressure impregnations with vacuum drying phases;
- pulse pressure–diffusion impregnation.

Pulse pressure–diffusion impregnation technology was proposed for impregnation of wet utility poles prepared from refractory, fresh coniferous wood as soon as possible after cutting. Bordered pits in tracheids of spruce, fir and other refractory species have yet to be opened, which is the state in fresh wood above the fibre saturation point (~30%). The wet, freshly debarked wood poles are

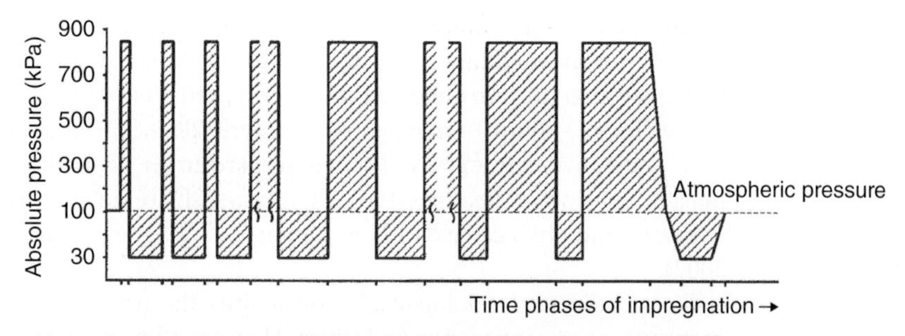

Figure 5.16 Pulse pressure–diffusion technology with changing of increased pressure and vacuum stages, used for impregnation of wet wood with water-soluble preservatives

Source: R., L. (2008) *Ochrana Dreva* (*Wood Protection*), Handbook, TU Zvolen, Slovakia, 453 p. Reproduced by permission of TU Zvolen

exposed to short (~5 min) pulses in the autoclave, consisting of increased pressure created on the liquid preservative and then of vacuum or atmospheric pressure (Peek, 1987). This means the stages of increased pressure (gradually from 1 to 4 min) are cyclically altered with stages of vacuum or atmospheric pressure (gradually from 4 to 1 min). Pulse technologies consist of a minimum of 120 cycles (Figure 5.16). Generally, they are suitable only for the protection of wet poles or other round products from fresh coniferous wood using water-soluble and fixable biocides (e.g. Cu-HDO).

5.3.7 Nanotechnologies and nano-compounds for chemical protection of wood

Nanotechnologies are well known in various parts of industry, medicine, and so on. They are also promising candidates for wood preservation. The basic characteristics of elemental metals, metal oxides and some other preservatives can be totally different when in the state of nano-metals and other nano-compounds. This means that nano-systems potentially attractive as wood preservatives may perform in an unusual manner and may also present unknown and unpredictable risks for people and the environment.

Nano-preservatives are created similarly to other nano-compounds; that is, by altering particulate size either in the liquid phase (e.g. metal oxide salts, colloidal metals), gas phase (e.g. flame synthesis, plasma-based vapour-phase synthesis) or solid phase (e.g. high-energy ball milling) via chemical reaction, heating or refluxing (Clausen, 2007).

Nano-preservatives have several characteristics (e.g. size, type, charge, concentration range, dispersion stability) that are important during their application into wood and which can improve their efficiency and stability in protected wood. Complete penetration and uniform distribution of nano-compounds in wood and in its cells walls can be expected if their particle size is smaller than the pore diameter in the bordered pits (<150 nm) or even in the cell walls (<1–80 nm) (Freeman & McIntyre, 2008; Kartal et al., 2009). However, also convenient for wood protection are larger nano-silver particles having a dimension in the range 100–200 nm, which are not able to penetrate into cell walls and are localized only in lumens of vessels or other cells (Ellis et al., 2007).

Biologically active nanoparticles are mainly the nanosized and micronized preparations of copper, zinc and silver. They are efficient against fungi and termites, as was summarized in the work of Terziev et al. (2016). The antifungal efficiency of nano-metals depends primarily on their chemical composition. For example, copper, zinc and other heavy metals are more effective against white-rot fungi than against brown-rot fungi (Green & Clausen, 2005). In this regard, Reinprecht et al. (2015) found that lime wood preserved with a 1% water dispersion of zinc oxide nanoparticles increased its resistance against the white-rot fungus *Trametes versicolor* about ~56%, and about ~40% against the brown-rot fungus *Coniophora puteana*. Similar results have been achieved by Kartal et al. (2009), Clausen et al. (2010), Akhtari and Arefkhani (2013), Mantanis et al. (2014) and Rezazadeh et al. (2014).

The biological mechanism of silver uncharged nanoparticles (Ag) is in their oxidation in a wet wood with the creation of stable silver ions (Ag^{+}) having antimicrobial properties. For wood protection the size of silver nanoparticles is important as well because, for the same mass, smaller particles have a higher surface area available, meaning a quicker oxidation of silver atoms and a more rapid biocide effect.

For solid wood with small amounts of trace elements (Mn, Fe, Co, Ni, Cu, Zn, Mo, Pb, etc.), which have a positive or negative physiological influence on the mycelium growth and enzymatic activity of fungi, the additional treatment with nanoparticles of silver, copper, zinc, aluminium or other metals (nano-biocides) can be effective only if their concentration is sufficient; that is, greater than the lower limit of the physiological toxicity range (Wažny & Kundzewicz, 2008).

The dispersion stability of nano-metals should be high. Together with their low viscosity, this will allows for greatly improved preservative penetration, non-leachable or hydrophobic treatment, and stable finishes for above-ground applications (Clausen, 2007).

Nanoparticles active against fire and weathering are mainly nano-B_2O_3, nano-TiO_2, nano-Fe_2O_3 and nano-SiO_2.

Nano-packing of preservatives relates to the nano-carrier delivery systems and to the use of nanotubules, mainly with the aim to control the kinetics of preservatives release in wood exposed in wet conditions. Laks and Heiden (2004) designed a nano-carrier system based on 100 nm beads embedded with biocides, for their slow, controlled release to the surrounding area (e.g. to wooden joints or surfaces at their increased humidity). The kinetics of the biocide release depend on the pore size in the nanotubules and on the biocide solubility and diffusion coefficient. This system could be convenient mainly for protection of wooden composites and wood surfaces. Nanotubules may be surface coated or capillary loaded in technology of wood preservation, or may be added to a mixture of wood particles and other components in production of wooden composites. However, in wooden composites they had to be compatible with resins, and the resin should not be a barrier to release the biocide. Proper nano-packing of preservatives could lead to increased applications for biocides that are otherwise unsuitable for exteriors (Clausen, 2007).

Generally, prospective uses of nanotechnologies for wood preservation should preferentially be aimed at increasing:

- The photostability of wood; for example, with nanoparticles of titanium dioxide (Forsthuber et al., 2013), zinc oxide (Clausen et al., 2010; Cristea et al., 2010; Salla et al., 2012), iron oxide (Fe_2O_3), or others added to surface coatings in the function of UV absorbers and diffusers of UV light.
- The fire resistance of wood; for example, with preservatives containing nanoparticles of silica (Wang et al., 2010) and titanium dioxide (Wang et al., 2007).
- The biological resistance of wood against various pests; for example, with nanoparticles of silver (Akhtari & Arefkhanil 2013; Moya et al., 2014), zinc oxide (Clausen et al., 2010; Akhtari & Arefkhani, 2013; Stanković et al., 2013; Mantanis et al., 2014), zinc borate (Lykidis et al., 2013; Mantanis et al., 2014), copper (Akhtari & Arefkhani, 2013) and copper oxide (Mantanis et al., 2014), or titanium dioxide (Marzbani & Mohammadnia-Afrouzi, 2014).

Nano-compounds also have the potential to be used in mixtures with polyacrylates for the conservation of wooden artefacts (Trăistaru et al., 2012; Reinprecht et al., 2015).

5.3.8 Quality control of chemically protected wood

The *quality of chemically protected wood* is assessed on the basis of meeting the following requirements: (1) use of the proper technological process of wood treatment (i.e. adherence to technological regulations, such as the concentration of the preservative and the pressure, temperature and time of the treatment); (2) achieving the desired penetration, retention, fixation and effectiveness of the preservative in the treated wood; and (3) ensuring the proper and safe use of the preserved wood in practice (i.e. in terms of the long-term stability of the preservative in the wood under particular exposure and, at the same time, in terms of current and future requirements for the safety and environmental friendliness of the treated wood).

Wooden products treated with preservatives shall be awarded a certificate if they are preserved in compliance with the given technology and if they meet quality requirements. The certificate is issued in written form and also in the form of a seared or fixed mark fitted to each product. For example, in the USA, in accordance with the American Lumber Standard Committee, the mark includes the name of the chemically treated product, the year of treatment, the estimated year of the end of its lifetime, the name of the preservative, the third-party inspection agency, AWPA use category, retention of the preservative in wood, and treatment company (Lebow, 2010).

5.3.8.1 Inspection of the penetration and retention of preservatives

The *requirements for penetration and retention of preservatives in wood* are defined in the appropriate standards valid for the particular territory of the world. For example, for European countries, EN 351-1 defines nine penetration classes (P1–P9) and five retention classes (R1–R5). The particular penetration and retention class must be achieved in the process of surface treatment or deep impregnation of the wooden product, depending upon the level of impregnability of the given wood species (EN 350-2), and also upon the class of use of the wooden product (EN 335). The penetration l (mm) and retention R (kg/m^3 or g/m^2) of preservatives into treated wood are evaluated using direct and indirect inspections of the technological process.

Direct inspection of the penetration and retention of the preservative into wood is carried out by analysing small samples taken from a sample part of chemically treated wooden product; that is, from utility poles, sleepers, beams, and so on. The number of sample parts can be stated in accordance with ISO 2859-1 where, for example, in the most frequently used level II inspection it is necessary to randomly select eight sample parts from a set of 26–50 parts. If the treatment technology is adhered to, a maximum of two sample parts may show faults in the retention and/or penetration of the preservative.

In accordance with EN 351-1, the maximum tolerance for penetration is 10% for products made of easily impregnable wood species (class 1 impregnability in accordance with EN 350-2 – Table 5.2) and 25% for products of wood species that are medium to very difficultly to impregnate (classes 2 to 4 impregnability). The penetration into wood in a radial and possibly also in a tangential direction can be evaluated from boreholes taken from poles (round wood) and beams or from cuts taken from poles (EN 351-2). Penetration of the preservative into wood in an axial direction is evaluated using transverse and longitudinal cuts (EN 351-2). Differentiation of the retentions (retentions are usually lower in a greater depth of treated wood) can be stated via analysing thin layers of wood gradually removed using, for example, a Forstner drill bit or by analysing thin cuts taken with a microtome and having thickness no more than 0.2 mm (EN 212). Qualitative proof or analysis of the depth penetration of colourless preservatives into wood is carried out using colouring indication agents applied to a sample part, or using X-ray, γ-ray, Fourier transform infrared (FTIR) spectroscopy, mass spectrometry, chromatography and other diagnostic methods convenient for detection of damaged wood as well (see Section 7.3).

Analysis of the retention of preservatives in wood is carried out either using classic chemical methods or special physical–chemical methods (chromatography, near-infrared (NIR) spectroscopy, FTIR spectroscopy, etc.) using extracts from an analysed sample of wood or from wood shavings taken from the sample.

Indirect inspection of the retention of the preservative into wood can only be carried out if a reliable dependence between its retention and a measurable parameter of the wood treatment process was stated in advance. For example, this measurable parameter is a change in the weight of the collection of wood (e.g. a trolley with railway sleepers) before and after impregnation, from which the consumption of a liquid preservative during impregnation in litres or kilograms per cubic metre of wooden product can be computed.

5.3.8.2 Inspection of the fixation and effectiveness of preservatives

Along with the retention and penetration of a preservative into wood, the final effectiveness and stability of the chemical protection of wood is also a very important. These qualitative parameters depend upon the type and required effectiveness of the preservative in particular application conditions (e.g. a potential decrease of a biocidal effectiveness of the preservative due to the influence of extractive substances in the particular type of treated wood) and, at the same time, upon its fixation in the protected wood connected with its long-term stability in various climatic conditions.

Requirements for the fixation of a preservative in preserved wood are determined by its planned exposure.

Preservatives for exterior use should be stable against a complex of weather factors acting in a region of the treated wood exposure. Leaching, evaporation and other effects of weather factors are evaluated in field and accelerated tests. For example, in Europe, accelerated ageing tests of protected wood are evaluated either for individual weather factors – leaching with water (e.g. by ENV 1250-2), evaporation caused by wind (e.g. by ENV 1250-1), evaporation caused by temperature >40 °C (e.g. by EN 73) – or the a combination of weather factors (solar radiation, temperature changes, effect of rainfall – e.g. by accelerated ageing in accordance with EN 927-6). However, field exposures remain an important tool for evaluating new test methods (Lebow, 2014).

Preservatives applied in interiors should not be in any way volatile, so that they do not threaten human health and simultaneously do not lower their effectiveness.

Fixation of biocides and other preservatives in wood is currently evaluated using modern physical–chemical methods or physical analytical methods such as NIR spectroscopy, FTIR spectroscopy, gas and liquid chromatography, X-rays, γ-rays, scanning electron microscope. Analysis may be carried out with liquid leaches, evaporated gasses or weathered components of wood alone that have undergone previous leaching or another model ageing load.

Efficacy of preservatives must be predefined with the aim of a suitably, environmentally friendly and economically increased resistance of treated wood against damaging factors.

Biocides should sufficiently increase wood resistance to wood-damaging pests. For developing and commercial types of biocides, this requirement is evaluated using screening tests and then tests on wood samples in accordance with valid standards. In Europe, more standardized tests are used for evaluation of the activity of pests in chemically protected wood, with separate tests for wood-decaying fungi in accordance with EN 113 and EN 252, staining fungi in accordance with EN 152, woodworm beetles in accordance with EN 49-2, powder-post beetles in accordance with EN 20-1 and EN 20-2, house longhorn beetle in accordance with EN 46-1 and EN 46-2, termites in accordance with EN 117 and EN 118, or marine borers in accordance with EN 275. Analogical anti-pest-activity tests can also be performed with similarly preserved wooden samples in a prior aged model (e.g. in accordance with EN 73, EN 84 and/or EN 927-6), or with samples taken from chemically treated wooden products after a defined exposure time.

Fire retardants should increase resistance of the treated wood against various thermal loads, modelling conditions of its combustion and real fire. In Europe, solid woods, wooden composites and other construction materials are today classified in accordance with EN 13501-1 into seven classes with regard to their fire behaviour: A1, A2, B, C, D, E and F (Table 2.2). For this classification they are tested and evaluated on the basis of more fire-resistant tests (EN 1363-1), mentioned in Section 2.2.2.3; that is, applying first EN ISO 1182 (burns–does not burn, at 750 °C), then EN ISO 1716 (combustion heat in a pure oxygen atmosphere in a bomb calorimeter), EN 13823 (single burning item test, evaluating the fire growth rate, the lateral flame spread, total heat release and smoke production) and EN ISO 11925-2 (flame spread). The efficiency of fire retardants is recommended to be performed not only on laboratory-prepared samples, but (similar to testing biocides) also on samples taken from preserved wooden elements, or to undertake fire tests with completed wooden constructions.

The aesthetic and functional sides of anti-weather coatings in demanding exterior conditions are also evaluated by valid artificial and practical tests. For anti-weather coatings, the most important are their colour changes, thinning, stability against cracks, peeling and crumbling as a result of loss of elasticity and decrease of adhesion to wood.

5.4 Chemical protection of wooden composites

5.4.1 Wooden composites and their susceptibility to damage

Wooden composites are wooden materials consisting of wood in various degrees of disintegration (lamellae, veneers, chips, fibres, etc.) and suitable additives (adhesives, waxes, preservatives, etc.). The most well-known types of wooden composites include glued lamella wood (glulam), block board, plywood, LVL, parallel strand lumber (PSL), laminated strand lumber (LSL), PB, OSB, wafer board (WB), fibreboard (FB) of soft, medium or hard type, wood–cement particleboard (CCB), and in recent years also wood–plastic composite (WPC).

In various buildings, wood and wooden composites are also combined with metals, plastic and other materials. For example, panels for wooden buildings apart from wood prisms and wooden composite boards (OSB, CCB or PB), also contain inorganic materials (mineral wool, plaster in plasterboard, metal film, etc.) and organic materials (textiles, polyurethane and other plastics in the role of insulating foams, coating films, etc.).

Properties of the wooden composites, including their resistance to abiotic and biological damage, are determined by:

- the type of wood, and mode and level of its disintegration;
- the type and amount of additives used;
- the mechanism for joining disintegrated parts of wood into composite (e.g. by creating adhesive joints in plywood or PB; or by creating hydrogen bonds as well as covalent bonds between wood particles in FB or cardboard);
- the technology of composite production (e.g. a pressing diagram).

The properties of a wooden composite (C) depend upon the properties, proportion and location of its components: (A) wood, (B1) adhesive, (B2) water repellent, (B3) biocide and/or fire retardant, (B4) other specific agents; however, they also depend significantly upon the synergetic relationship between these components (Figures 1.2 and 5.17). Wooden composites for specific purposes should have better selected properties in comparison with properties of solid wood (A) due to presence and location of additives of known properties (B = B1, B2, B3, B4), which means C > A + B.

The *susceptibility of a wooden composite to damage* is given by the type and proportion of disintegrated wooden components prepared from wood species

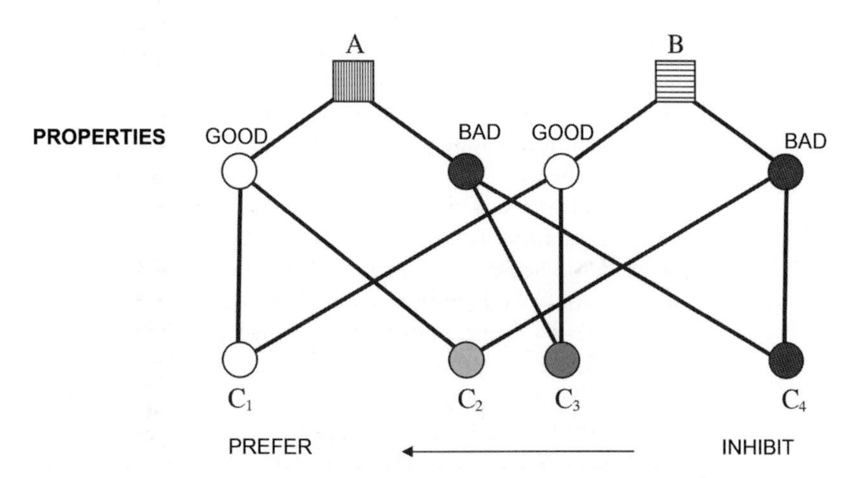

Figure 5.17 The positive and negative synergy when transforming initial good and bad properties of the basic components (A and B) of a wooden composite into its final properties (C_1 (preferred), C_2, C_3, C_4)

more or less prone to damage (Chapters 1–3) and the type and proportion of various inorganic and organic additives:

- Inorganic additives (e.g. plaster and cement) usually better resist damage than solid wood. Examples are the highly biologically and fire-resistant CCB and wood–concrete blocks, as well as plasterboards.
- Organic additives of natural type (e.g. starch and casein) are similarly or more susceptible to damage than wood. Synthetic organic additives can have a higher resistance than wood; for example, amino-resin adhesives increase the fire resistance of wooden composites, whilst phenolic adhesives increase their resistance to brown-rot fungi.

Weather and dimensional stability of wooden composites is mainly important in the building industry, housing and transport. For a humid environment, we can only use wooden composites with an adhesive resistant to water and aggressive atmospheric, chemical and biological degrading factors. Good examples are the lamella woods glued with polyurethane, plywood glued with phenolic resins, or CCBs. For WPCs, used, for example, as decking, siding, fences, windows and door frames, their stability against UV light is important (Hyvärinen et al., 2013).

Biological resistance of conventional wooden composites (plywood, PB, OSB, FB, etc.) against algae, bacteria, fungal decay, moulds and termites is poorer comparing to WPCs, mainly those containing chlorine atoms; for example, the PVC-based WPC (Xu et al., 2013). This is mainly thanks to a relatively small amount of synthetic or inorganic additives in such materials in relation to the amount of wood. However, some resins are also biodegradable. For example, Gusse et al. (2006) found that wooden composites bonded with phenolic resins can be well attacked by white-rot fungi due to degradation of this resin whose structure is similar to lignin.

On the other hand, wooden composites, apart from glued lamella wood (glulam), are well resistant to wood-boring beetles since small wooden particles are coated with synthetic adhesive, cement or another joining agent that does not form an interesting food source for larvae of woodworm and other wood-damaging beetles.

Flammability and fire resistance of common wooden composites, such as glued lamella wood, plywood, PB or fibre board, are comparable to solid wood of the same density. However, CCB is much less flammable. Organic adhesives in wooden composites undergo thermal decomposition and they do not usually have a significant effect upon the thermal characteristics of composites or the accumulation of heat in them. A comparison of composites joined using amino urea-formaldehyde and melamine-formaldehyde resins, phenolic phenol-formaldehyde resins and epoxy resins shows that the least flammable are composites joined using amino resins and the most flammable are composites joined using epoxies. The cause is the creation of inflammable gases (nitrogen, ammonia) from the amino resins during their thermal decomposition. Generally, on burning, wooden composites produce more smoke than solid wood, which is mainly typical of composites with a proportion of epoxy. Metal films (aluminium or steel with a thickness of ≥0.5 mm) on the surface of wooden composites significantly suppress their combustion and the spread of flames.

5.4.2 Principles and technologies of chemical protection of wooden composites

5.4.2.1 Principles of chemical protection of wooden composites

The *modes of chemical protection of wooden composites* are determined by their type and exposure. For example, wooden composites with decreased flammability (e.g. for using in buildings) have to be treated with fire retardants. Unlike solid wood, several types of composites cannot be exposed to rainwater or to an environment with a markedly changing relative humidity of air.

Several types of wooden composites can be totally damaged due to increased humidity, mainly as a result of (1) irreversible structural changes in adhesives (e.g. hydrolysis of amino resins) and (2) cyclic changes in the dimensions of wooden particles due to sorption and desorption processes.

This means that the application of biocides into non-water-resistant composites is totally pointless. Although these composites would become resistant to fungi and other pests, in wet conditions they would quickly lose their proposed functionality. So, only water-resistant wooden composites are suitable for a wet environment. At the same time they must be resistant to biological pests, which is achieved by (1) using particles of highly durable wood species, (2) using particles of acetylated or thermally modified wood, (3) applying water-stable biocides or/and (4) applying suitable lamination or coating layers.

Types of preservatives for the chemical protection of wooden composites are usually the same as those commonly used for solid wood preservation. In practice, mainly fungicides and fire retardants are applied (Reinprecht et al., 1986; Jeihooni et al., 1994; Smart & Wall, 2006), or in tropical and subtropical countries the application of insecticides against termites is also important (Becker, 1972). From biocides, nano-metals, copper compounds, triazoles, pyrethroids and others are used (Table 5.5). In practice, the issues of the effects of preservative upon the strength characteristics of wooden composites, upon their water resistance and exposure stability are very important (Kirkpatrick & Barnes, 2006; Taghiyari & Farajpour 2013; Reinprecht & Kmeťová 2014; see Section 5.4.2.4).

The preservatives in the wooden composite must be sufficiently stable during its production. For example, during the pressing process, at an increased temperature of 100–170 °C, they must neither decompose into less effective components nor become more toxic to humans. A constant problem is the disturbance of adhesive joints in the composite, since the preservatives block the surface of the wood to molecules of adhesive and some preservatives also decrease the reactivity of the adhesive (Gardner et al., 2003).

Variability in the technologies for the chemical protection of wooden composites is greater than for treatment of solid wood. Together with the traditional technologies suitable for wooden composites treatment (e.g. protection of wood in

Table 5.5 Some preservatives suitable for protection of wooden composites

Wooden composite	Preservatives
Glued lamella wood (glulam)	ACQ compounds CuNs IPBC Creosotes
Plywood	ACQ compounds Boroglycols Triazoles (propiconazole, tebuconazole, etc.) Pyrethroids (bifenthrin, deltamethrin, permethrin, etc.) Imidacloprid
Oriented strand board (OSB)	Zinc borates CuNs IPBC Pyrethroids
Particleboard (PB)	Fire retardants Pyrethroids
Fibreboard (FB)	Fire retardants Zinc borates and H_3BO_3
Wood–plastic composites (WPC)	Zinc borates

Sources: Gardner et al. (2003), Smith and Wu (2005), Creffield and Scown (2006), Kirkpatrick and Barnes (2006) and Reinprecht (2008).

the form of disintegrated particles), new technologies are also currently being developed. For example, organic biocides for protecting ready-made composites are applied in the supercritical carbon dioxide (SC-CO_2) gaseous carrier medium. According to Muin et al. (2001), with this technology the mechanical properties of ready-made plywood, PBs and FBs change minimally, and only with OSBs is change more apparent. Treatment of wooden composites using SC-CO_2 technology is implemented, for example, using IPBC-silafluofen biocide at a temperature of 35 °C and a pressure of 7.85 MPa, with significant improvement in their resistance to wood-decaying fungi and termites (Tsonuda & Muin, 2003).

Technologies for the chemical protection of wooden composites are incorporated into two basic groups:

- in-process treatment (IPT);
- post-manufacture treatment (PMT).

They can be combined; for example, when high fire resistance of a composite is required, it is suitable to introduce inorganic fire retardants into the composite during its production (IPT) and then treat its surface using an intumescent coating (PMT).

5.4.2.2 In-process treatment of wooden composites

Preservatives for IPT technology should meet the following criteria (Kirkpatrick & Barnes, 2006):

- thermal stability during the production of a composite, mainly during pressing;
- no negative interaction with an adhesive and no effect upon glued joints;
- minimum volatility and leachability with water;
- no negative effect upon the strength of the composite;
- no influence upon the surface finishing of the composite;
- negligible environmental impact;
- low price without a market impact upon the finances of the production.

IPT of wooden composites can be implemented in the form of:

- protection of disintegrated wooden particles – for example, lamellae, veneers, chips or fibres;
- protection of adhesives, waxes, minerals and other additives, as well as foils or coatings for the final surface treatment of the composite;
- usage of special adhesives, additives or foils that simultaneously act as a biocide or fire retardant;
- protection of all components of the composite during its production – for example, when applying an adhesive, when layering woodchips or during pressing of PBs.

Protection of disintegrated particles of wood (Figure 5.18) is carried out using similar technology to the chemical protection of solid wood. The most

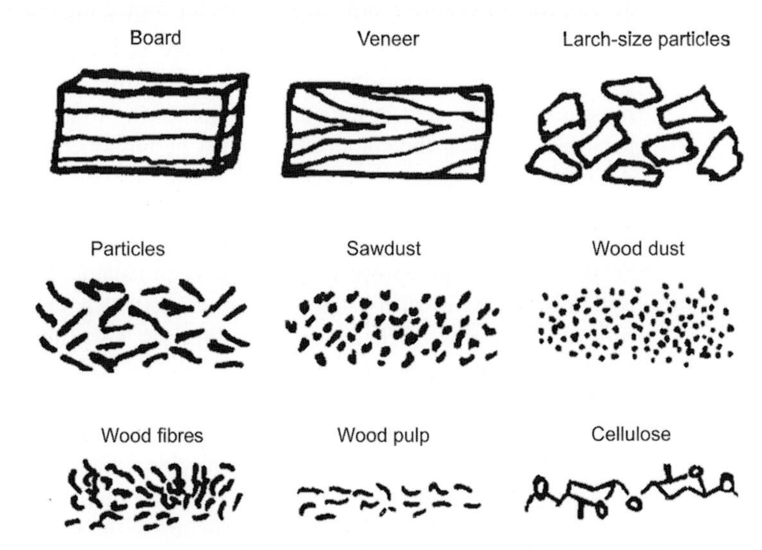

Figure 5.18 Wood particles in various degrees of disintegration prepared for chemical protection

Table 5.6 The advantages and disadvantages of the IPT of disintegrated wood in production of chemically protected wooden composites

Advantages
- The possibility to select the level of protection of individual wood particles (lamellae, wood chips, veneers, etc.) designed for the various layers of the wooden composite, and therefore the possibility to regulate the retention of preservatives through the composite's cross-section (in practice, a higher level of protection is usually selected for the surface layers of a composite)
- The possibility for perfect (deep) protection of wood particles

Disadvantages
- Increased costs related to the additional drying of disintegrated and chemically treated wood particles to achieve their technological moisture content w before the production of a composite; that is, a veneer to $w \approx 5$–7% or wood chips to $w \approx 4$–6%
- Cannot be used for wood fibre composites produced using wet technology since there is the danger of leakage of the preservatives into waste water (ecological problem) with a simultaneous decrease in its content in the composite (efficacy problem)

advantageous method is spraying or dipping. Some vacuum or vacuum-pressure technology can also be used for lamellae of greater dimensions and for veneers made of poorly impregnable wood species. The protection of disintegrated wood has some advantages and disadvantages (Table 5.6).

The required retention of a preservative by wood particles can be regulated via the concentration of the solution (only water solutions or water emulsions are usually used), via the spray dose, or the time and hydro-module of dipping.

Protection of adhesive, hydrophobic agent or other additives requires considering the issue of their interaction with the preservative. For example, it should not negatively affect gluing quality of the adhesive, nor its other properties such as viscosity, surface tension or pH value. The preservative must homogenize well in the adhesive and, owing to its smaller amount and local placement in the composite, it must be highly effective (Figure 5.19a). An effective fire retardant is tri-arylphosphate $[R-(C_6H_4)-O]_3P{=}O$ (where R is an alkyl, hydroxyl or other functional group), which takes part in a cross-linking reaction with the phenolic adhesive during hot pressing to create a solid organophosphate–phenolic resin (Mahút et al., 1985). Another option is the use of such catalysts in adhesives which, apart from their basic function in curing of thermosetting adhesives, also fulfil a fungicidal (e.g. ZnO), insecticide or fire-retardant function.

Special adhesives with a protecting effect are prepared by their manufacturers. They contain functional groups with a fungicidal, fire retardant or other protecting effect. In the past, an example of a biocide adhesive was acrylate polymer with integrated organotin biocide (tri-*n*-butyl-tin-methacrylate); currently, triazoles are integrated into the polymer structure of the adhesive. An insufficient fire resistance of epoxy and phenolic adhesives can be improved by integration of fire-resistant organic substances with a proportion of phosphorus or nitrogen atoms directly in their polymer chain. Particularly in the production of fire-resistant epoxy adhesives, instead of 2,2-bis-(*p*-hydroxyphenyl)propane $HO{-}(C_6H_4){-}C(CH_3)_2{-}(C_6H_4){-}OH$, a suitable organic compound is used that contains a phosphorus atom and a minimum of two hydroxyl or two amino groups. An example is bis-(*m*-hydroxyphenyl)-ethyl-phosphine oxide with two hydroxyl groups, which reacts with two molecules of epichlorohydrin $CH_2{=}(O){=}CH{-}CH_2{-}Cl$ with the creation of a basic linear molecule of epoxy resin having a higher fire resistance.

H_2C—(O)—$CH{-}CH_2{-}O{-}C_6H_4{-}P(=O)(CH_2{-}CH_3){-}C_6H_4{-}O{-}CH_2{-}HC$—(O)—$CH_2$

Fire-resistant molecule of epoxy

Addition of preservatives to wood components during composite production is suitable for preparing PB, OSB, FB and some other agglomerated boards. Preservative is usually added during:

- homogenization of wooden particles with additives in a mixing drum;
- layering of chips, strands or fibres;
- pressing.

Preservatives are added as concentrates (liquids, solids), solutions, dispersions or in the form of special films which at pressing are located as inner layers or surface layers of the composite.

The homogenization of wooden particles with preservatives in a mixing drum (Figure 5.19c), before or during application of adhesive, is useful in the

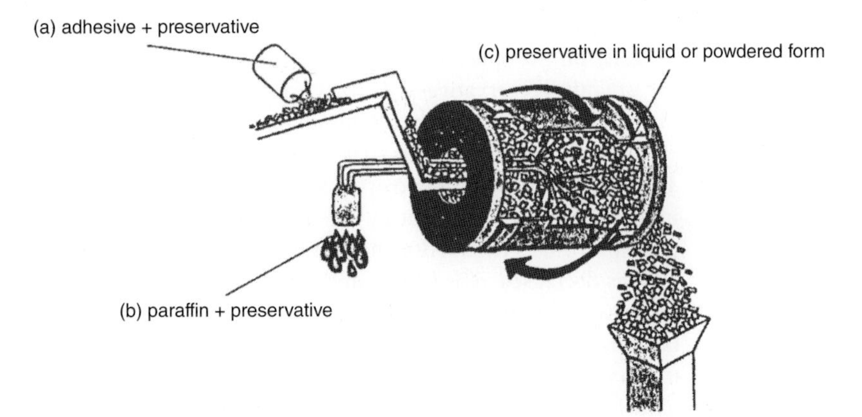

Figure 5.19 Application of preservatives to a wooden composite during its production: initially to the adhesive (a) and another additive (b), or to the mixing drum (c)

production of PB and OSB. Preservatives are applied in powdered form or as a highly concentrated solution (biocides: H_3BO_3, zinc borate, nano-zinc oxide, etc.; fire retardants: ammonium dihydrogen phosphate, etc.). The application of preservatives in powdered form brings a problem with their uneven distribution in the wooden composite. The reason is their insufficient homogenization between wooden particles in the mixing drum, as well as the fall of heavier powdered particles of the preservative to the lower layers of the carpet of composite board during layering. In comparison with the method of direct protection of wooden particles, this disadvantage must be compensated by adding ~30–60% more of the preservative to the composite. However, this method is beneficial since it does not require additional drying of wooden particles.

Pressing of protective films or embedding mouldable preservatives into the surface layers of a wooden composite are methods suitable for specific situations. For example, protective films are pressed to plywood designed for exteriors. The method of embedding preservatives requires optimization of the pressing process (pressure, temperature and time) and of the initial humidity of the surface layers of the composite.

5.4.2.3 Post-manufacture treatment of wooden composites

PMT of wooden composites can be carried out using pressure-free technology (e.g. coatings, spraying or dipping), as well as vacuum–pressure technology. Today, with the prospect of SC-CO_2 technology, the preservative is introduced to the wooden composite in the gaseous phase using SC-CO_2 as a carrier (Morrell et al., 2005), or also steaming technology using gaseous tri-methyl borate (Murphy, 1994). The SC-CO_2 and steaming technologies rank among the best. Using these the mechanical properties of wooden composites worsen only negligibly since they do not cause any marked changes in the structure of composites.

The *technology of integrating preservatives into wooden composites in SC-CO_2* is based on the specific properties of CO_2, which acts as a carrying medium for organic preservatives. SC-CO_2 can behave like a liquid and dilute various organic substances, but it simultaneously behaves like a gas, able to penetrate easily into the porous structure of the wooden composite or solid wood. Technology for treating air-dried wooden materials using an 'organic biocide in SC-CO_2' is carried out at a temperature of 30–40 °C and an increased pressure of 6–9 MPa. After finishing the treatment the pressure in the autoclave is decreased. The CO_2 evaporates from the wooden material in the form of a gas whilst the preservative remains in the wooden material. An advantage of this technology is the safety of the CO_2 carrier of the preservative in terms of health, its low cost and flame resistance. SC-CO_2 technology can be used for treatment of various types of wooden composites (Muin et al., 2001, Tsonuda & Muin, 2003), but also of solid wood (Acda et al., 2001). However, the preserved wooden composite, similar to composites preserved with other technologies, should not be exposed to water and polar or aggressive liquids since these could disturb glued joints and also cause irreversible changes in its shape.

5.4.2.4 Effect of preservatives on properties of wooden composites

The properties of wooden composites change specifically under the influence of preservatives. The change depends upon the creation of interactions in the system 'wooden substance–preservative–adhesive–other additives'.

The *strength properties* of wooden composites often worsen under the influence of chemical substances used as preservatives for IPT technologies. This is mainly caused by the influence of inorganic substances, typically borates and ammonium sulphate, but also organic biocides (Sean et al., 1989; Reinprecht & Perlác, 1995), which negatively affect:

- the curing process of adhesive (e.g. salts of acids accelerate curing of amino and phenolic adhesives);
- the interphase interface between wood and adhesive, when the preservative first creates a continuous film on the surface of the wooden particles (e.g. meltable H_3BO_3) and then, in the pressing stage, this film is transformed into a mouldable glaze that forms a separating interlayer between the preserved wood and the adhesive.

A negative effect of preservatives upon the strength properties of wooden composites is most marked when technology for the initial treatment of disintegrated wooden particles is used. This is caused by the creation of continuous separating interlayers on the surface of the wooden particles. On the other hand, the chemical protection of finished wooden composites using gaseous trimethyl borate or SC-CO_2 technology retains their strength or decreases it only slightly (Hashim et al., 1994; Jones et al., 2001; Barnes et al., 2002).

Water resistance is mainly important for wooden composites designed for exteriors, which should also be sufficiently bio-resistant. It is pointless adding biocides (mainly fungicides) to those composites in which the wooden particles are joined using an adhesive that is not sufficiently water resistant (e.g. urea-formaldehyde resin). On the other hand, it is convenient to add preservatives into composites bonded with melamine-formaldehyde-, phenol-formaldehyde-, polyurethane- or epoxy-based adhesives, which contribute towards increasing their water resistance.

Ecotoxicological viewpoints related to protected wooden composites are important not only during their production (issues of VOCs, disintegration of biocides and fire retardants during pressing at increased temperature, etc.) and use, but also for their disposal at the end of their service life (issues of recycling).

The *price* of chemically treated wooden composites in relation to their increased durability and applicability must also be reasonable. This means that the chemical protection of composites (similar to the protection of products made of solid wood) should be specifically selected depending upon their exposure and the risk of damage by biological pests and abiotic factors.

References

Acda, M. N., Morrell, J. J. & Levien, K. L. (2001) Supercritical fluid impregnation of selected wood species with tebuconazole. *Wood Science and Technology* 35, 127–136.

Akhtari, M. & Arefkhani, M. (2013) Study of microscopy properties of wood impregnated with nanoparticles during exposed to white-rot fungus. *Agriculture Science Developments* 2(11), 116–119.

Amusant, N., Moretti, C., Richard, B., et al. (2005) Chemical compounds from *Eperua falcata* and *Eperua grandiflora* heartwood and their biological activities against wood destroying fungi. In *IRG/WP 05*, Bangalore, India. IRG/WP 05-30373.

Amusant. N., Thévenon. M-F., Leménager, N. & Wozniak, E. (2009). Potential of antifungal and antitermitic activity of several essential oils. In *IRG/WP 09*, Beijing, China. IRG/WP 09-30515.

Asamoah, A. & Antwi-Boasiako, C. (2007) Treatment of selected lesser used timber species against subterranean termites using heartwood extracts from teak and dahoma. In *IRG/WP 07*, Jackson, MS, USA. IRG/WP 07-30434.

Aschacher, G. & Gründlinger, R. (2000) Ecotoxicological risks of anti-sapstain preservatives washed off from treated timber. In *IRG/WP 00*, Kona, HI, USA. IRG/WP 00-50144.

Ash, M. & Ash, I. (2004) *Handbook of Preservatives*. Synapse Information Resources Inc., New York.

Bakhsous, B., Dumarcay, S., Gelhaye, E. & Gérardin, P. (2006) Investigation of new wood preservation formulations based on synergies between antioxidant, 2-HPNO and propiconazole. In *IRG/WP 06*, Tromsø, Norway. IRG/WP 06-30401.

Barnes, H. M., Murphy, R. J. & Via, B. K. (2002) Bending and tensile properties of vapour boron-treated composites. In *IRG/WP 02*, Cardiff, Wales, UK. IRG/WP 02-40228.

Batish, D. R., Singh, H. P., Kohli, R. K. & Kaur, S. (2008) Eucalyptus essential oil as a natural pesticide. *Forest Ecology and Management* 256(12), 2166–2174.

Becker, G. (1972) Protection of wood particleboard against termites. *Wood Science and Technology* 6, 239–248.

Bergman, Ö. (2003) Influence of different treatment parameters on penetration, retention and bleeding of creosote. In *IRG/WP 03*, Brisbane, Australia. IRG/WP 03-40255.

Bota, P., Baines, E., Mead, A. & Watkinson, S. C. (2010) Antifungal and wood preservative efficacy of IPBC is enhanced by α-aminoisobutyric acid. In *IRG/WP 10*, Biarritz, France. IRG/WP 10-30544.

Brischke, C. & Melcher, E. (2015) Performance of wax-impregnated timber out of ground contact: results from long-term field testing. *Wood Science and Technology* 49, 189–204.

Budija, F., Humar, M., Kričej, B. & Petrič, M. (2008) Possibilities of use of propolis for wood finishing. *Wood Research* 53(2), 91–101.

Cassens, D. L. & Eslyn, W. E. (1981) Fungicides to prevent sapstain and mold on hardwood lumber. *Forest Products Journal* 31(9), 39–42.

Castillo, A., Cabrera, Y., Preston, A. & Morris, R. (2004) Evaluation of teak sawdust *Tectona grandis* L. Fil as a potential source to obtain a natural wood preservative in Colombia. In *IRG/WP 04*, Ljubljana, Slovenia. IRG/WP 04-30356.

Chittenden, C. & Singh, T. (2011) Antifungal activity of essential oils against wood degrading fungi and their applications as wood preservatives. *International Wood Products Journal* 2(1), 44–48.

Clausen, C. A. (2007) Nanotechnology: implications for wood preservation industry. In *IRG/WP 07*, Jackson, MS, USA. IRG/WP 07-30415.

Clausen, C. A., Yang, V. W., Arango, R. A. & Green III, F. (2010) Feasibility of nanozinc oxide as a wood preservative. In *Proceedings of the 105th Annual Meeting of the American Wood Protection Association*. American Wood Protection Association, Birmingham, AL, pp. 255–260.

Crawford, D. M., Lebow, P. K. & DeGroot, R. C. (2000) Evaluation of new creosote formulations after extended exposures in fungal cellar tests and field plot tests. In *IRG/WP 00*, Kona, HI, USA. IRG/WP 00-30228.

Creffield, J. W. & Scown, D. K. (2006) Laboratory and field evaluation of imidacloprid and cypermethin as glueline treatments for softwood plywood. In *IRG/WP 06*, Tromsø, Norway. IRG/WP 06-30405.

Cristea, M. V., Riedl, B. & Blanchet, P. (2010) Enhancing the performance of exterior waterborne coatings for wood by inorganic nanosized UV absorbers. *Progress in Organic Coatings* 69(4), 432–441.

Daniliuc, A., Deppe, B., Deppe, O., et al. (2012) New trends in wood coatings and fire retardants. *European Coating Journal* 7(8), 20–25.

Dhyani, S. & Tripathi, S. (2006) Protection of hard and softwood through Neem leaves extracts and oil – a direction towards development of eco-friendly wood preservatives. In *IRG/WP 06*, Tromsø, Norway. IRG/WP 06-30394.

Eaton, R. A. & Hale, M. D. C. (1993) *Wood – Decay, Pests and Protection*. Chapman & Hall, London.

Eikenes, M., Alfredsen, G., Larnoy, E., et al. (2005) Chitosan for wood protection – state of the art. In *IRG/WP 05*, Bangalore, India. IRG/WP 05-30378.

Ellis, J. R., Jayachandran, K. & Nicholas, D. (2007) Silver – the next generation wood preservative. In *IRG/WP 07*, Jackson, MS, USA. IRG/WP 07-30419.

Evans, P. D. & Chowdhury, M. J. (2010) Photostabilization of wood with higher molecular weight UV absorbers. In *IRG/WP 10*, Biarritz, France. IRG/WP 10-30524.

Evans, P. D., Schmalzl, K. J., Forsyth, C. M., et al. (2008) Formation and structure of metal azole complexes. In *IRG/WP 08*, Istanbul, Turkey. IRG/WP 08-30469.

Forsthuber, B., Schaller, C. & Grüll, G. (2013) Evaluation of the photo stabilizing efficiency of clear coatings comprising organic UV absorbers and mineral UV screeners on wood surfaces. *Wood Science and Technology* 47(2), 281–297.

Freeman, M. H. & McIntyre, C. R. (2008) Comprehensive review of copper-based wood preservatives. *Forest Products Journal* 58(11), 6–27.

Freeman, M. H., Obanda, D. N. & Shupe, T. F. (2007) Permethrin – a critical review of an effective wood preservative insecticide. In *IRG/WP 07*, Jackson, MS, USA IRG/WP 07-30413.

Gardner, D. J., Tascioglu, C. & Walinder, M. E. (2003) Wood composite protection. In *Wood Deterioration and Preservation: Advances in Our Changing World*, Goodell, B., Nicholas, D. D. & Schultz, T. P (eds). ACS Symposium Series, vol. 845. American Chemical Society, Washington, DC, pp. 399–419.

Gérardin, P., Baya, M., Delbarre, N., et al. (2002) Evaluation of tropolone as a wood preservative – activity and mode of action. In *IRG/WP 02*, Cardiff, Wales, UK. IRG/WP 02-30282.

Göttsche, R. & Marx, H.-N. (1989) Kupfer-HDO – ein vielseitiger Wirkstoff im Holzschutz. *Holz als Roh- und Werkstoff* 47, 509–513.

Green III, F. & Arango, R. A. (2007) Wood protection by commercial silver formulations against eastern subterranean termites. In *IRG/WP 07*, Jackson, MS, USA. IRG/WP 07-30422.

Green III, F. & Clausen, C. A. (2005) Copper tolerance of brown rot fungi: oxalic acid production in southern pine treated with arsenic-free preservatives. *International Biodeterioration & Biodegradation* 56, 75–79.

Grüll, G., Truskaller, M., Podgorski, L., et al. (2011) Maintenance procedures and definition of limit states for exterior wood coatings. *European Journal of Wood and Wood Products* 69(3), 443–450.

Gusse, A. C., Miller, P. D. & Volk, T. I. (2006) White-rot fungi demonstrate first biodegradation of phenolic resin. *Environmental Science and Technology* 40, 4196–4199.

Hansmann, C. H., Gindl, W., Wimmer, R. & Teischinger, A. (2002) Permeability of wood – a review. *Wood Research* 47(4), 1–16.

Härtner, H. & Barth, V. (1996) Effectiveness and synergistic effects between copper and polymer betaine. In *IRG/WP 96*, Guadeloupe. IRG/WP 96-30097.

Hashim, R., Murphy, R. J., Dickinson, D. J. & Dinwoodie, J. M. (1994) The mechanical properties of boards treated with vapour boron. *Forest Products Journal* 44(10), 73–79.

Hastrup, A. Ch. S., Green III, F., Clausen, C. & Jensen, B. (2005) *Serpula lacrymans* – the dry rot fungus tolerance towards copper-based wood preservatives. In *IRG/WP 05*, Bangalore, India. IRG/WP 05-10555.

Hegarty, B., Yu, B. & Leightley, L. (1997) The suitability of isothiazolone microemulsions as long-term wood preservatives. In *IRG/WP 97*, Whistler, BC, Canada. IRG/WP 97-30150.

Horrocks A.R. & Price, D. (2001) *Fire Retardant Materials*. Woodhead/CRC Press, Abington Hall, UK.

Humar, M., Petric, M. & Pohleven, F. (2001) Changes of the pH value of impregnated wood during exposure to wood-rotting fungi. *Holz als Roh- und Werkstoff* 59, 288–293.

Humphrey, D. G., Duggan, P. J., Tyndall, et al. (2002) New boron-based biocides for the protection of wood. In *IRG/WP 02*, Cardiff, Wales, UK. IRG/WP 02-30283.

Hunt, A. C., Humprey, D. G., Wearne, R. & Cookson, L. J. (2005) Performance of permethrin and bifentrin in framing timbers subjected to hazard class 3 exposure. In *IRG/WP 05*, Bangalore, India. IRG/WP 05-30383.

Hunt, G. M. & Garratt, G. A. (1967) *Wood Preservation*, 3rd edn. McGraw-Hill, New York.

Hyvärinen, M., Väntsi, O., Butylina, S. & Kärki, T. (2013) Ultraviolet light protection of wood–plastic composites: a review of the current situation. *Journal of Computational and Theoretical Nanoscience* 19(1), 320–324.

Jacobsen, B. & Evans, F. G. (2006) Surface and system treatments of wood for outdoor use. In *IRG/WP 06*, Tromsø, Norway. IRG/WP 06-30412.

Jagadeesh, G., Lal, R., Ravikumar, G. & Rao, K. S. (2005) Shockwaves in wood preservation. In *IRG/WP 05*, Bangalore, India. IRG/WP 05-40308.

Jakes, J. E., Plaza, N., Stone, D. S., et al. (2013) Mechanism of transport through wood cell wall polymers. *Journal of Forest Products and Industries* 2(6), 10–13.

Jebrane, M. (2015) Covalent fixation of boron in wood through transesterification reaction with boron-bearing vinyl ester. In *The Eighth European Conference on Wood Modification*, Helsinki, Finland, pp. 174–181.

Jeihooni, A., Krahmer, R. L. & Morrell, J. J. (1994) Properties and decay resistance of preservative-treated Douglas fir flake-board. *Forest Products Journal* 44(11), 23–24.

Jones, W. A., Barnes, H. M. & Murphy, R. J. (2001) Ancillary properties of vapour boron-treated composites. In *IRG/WP 01*, Nara, Japan. IRG/WP 01-40210.

Kabir, A. H. & Alam, M. F. (2007) Reduction of environmental toxicity through eco-friendly wood biopreservative. In *IRG/WP 07*, Jackson, MS, USA. IRG/WP 07-50243.

Kartal, S. N., Hwang, W-J., Shinoda, K. & Imamura, Y. (2004) Decay and termite resistance of wood treated with boron-containing quaternary ammonia compound, didecyl dimethyl ammonium tetrafluoroborate (DBF) incorporated with acryl-silicon type resin. In *IRG/WP 04*, Ljubljana, Slovenia. IRG/WP 04-03334.

Kartal, S. N., Hwang, W-J. & Imamura, Y. (2005) Preliminary evaluation of new quaternary ammonia compound didecyl dimethyl ammonium tetrafluoroborate for preventing fungal decay and termite attack. In *IRG/WP 05*, Bangalore, India. IRG/WP 05-30375.

Kartal, S. N., Brischke, C., Rapp, A. O. & Imamura, Y. (2006) Biocidal resistance of didecyl dimethyl ammonium tetrafluoroborate (DBF)-treated wood in soil-bed and basidiomycetes tests. In *IRG/WP 06*, Tromsø, Norway. IRG/WP 06-30393.

Kartal, S. N., Green III, F. & Clausen, C. A. (2009) Do the unique properties of nanometals affect leachability or efficacy against fungi and termites? *International Biodeterioration & Biodegradation* 63(4), 490–495.

Kazemi, S. M. (2007) Impregnation of beech, maple, alder and lime under different treatments of extractives and fungal attack. *Journal of Biological Sciences* 7, 1463–1467.

Kirkpatrick, J. W. & Barnes, H. M. (2006) Biocide treatments for wood composites – a review. In *IRG/WP 06*, Tromsø, Norway. IRG/WP 06-40323.

Krajewski, K. J. & Strzelczyk-Urbańska, A. A. (2009) European biocidal products market after implementation of Directive 98/8/EC – report from the Commission to the Council and the European Parliament. *Forestry and Wood Technology* 67, 156–167.

Kurjatko, S. & Reinprecht, L. (1993) *Transport of Substances in Wood.* Monograph 7/1993, TU Zvolen, Slovakia (in Slovak, English abstract).

Kües, U., Mai, C. & Militz, H. (2007) Biological wood protection against decay: microbial staining, fungal moulding and insect pests. In *Wood Production, Wood Technology, and Biotechnological Impacts*, Kües, U. (ed.). Universitätsverlag Göttingen, pp. 273–294.

Lahary, A. K., Hasan, M. R. & Chowdhury, M. A. J. H. (2006) Criteria for environmentally and socially sound and sustainable wood preservation industry. In *IRG/WP/06*, Tromsø, Norway. IRG/WP/06-50237.

Laks, P. & Heiden, P. A. (2004) Compositions and methods for wood preservation. *US Patent, 6,753,035.*

Lebow, S. T. (2010) Wood preservation. In *Wood Handbook: Wood as an Engineering Material.* Forest Product Laboratory, Madison, WI, chapter 15.

Lebow, S. T. (2014) Evaluating the leaching of biocides from preservative-treated wood products. In *Deterioration and Protection of Sustainable Biomaterials*, Schultz, T. P., Goodell, B. & Nicholas, D. D. (eds). ACS Symposium Series 1158, American Chemical Society, Washington, DC, chapter 14, pp. 239–254.

Lehringer, C. H., Richter, K., Schwarze, F. W. M. R. & Militz, H. (2009) A review on promising approaches for liquid permeability improvement in softwoods. *Wood and Fiber Science* 41(4), 373–385.

Lehringer, Ch., Hillebrand, K., Richter, K., et al. (2010) Anatomy of bioincised Norway spruce wood. *International Biodeterioration & Biodegradation* 64, 346–355.

Leithoff, H. & Blancquaert, P. (2006) The future of wood protection in the light of the BPD – Biocidal Products Directive 98/8/EC of 16.02.1998. In *COST Action E37 – Sustainability through New Technologies for Enhanced Wood Durability*, Poznań, Poland.

Leithoff, H., Blancquaer, P., van der Flaas, M. & Valcke, A. (2008) Wood protection in Europe: developments expected up to 2010. In *Development of Commercial Wood Preservatives*, Schultz, T. P., Militz, H., Freeman, M. H., et al. (eds). ACS Symposium Series, vol. 982. American Chemical Society, Washington, DC, pp. 564–581.

Lekounougou, S., Ondo, J. P., Jacquot J. P., et al. (2007) Effects of caffeine on growth of wood-decaying fungi. In *IRG/WP 07*, Jackson, MS, USA. IRG/WP 07-30427.

Lesar, B., Pavlic, M., Petric, M., et al. (2011) Wax treatment of wood slows photodegradation. *Polymer Degradation and Stability* 96, 1271–1278.

Li, S., Freitag, C. & Morrell, J. J. (2007) Preventing fungal attack of freshly sawn lumber using cinnamon extracts. In *IRG/WP 07*, Jackson, MS, USA. IRG/WP 07-30432.

Lin, L. & Furuno, T. (2001) Biological resistance of wood–metaborate composites using the borax solution system. In *IRG/WP 01*, Nara, Japan. IRG/WP 01-30259.

Lloyd, J. D. (1998) Borates and their biological application. In *IRG/WP 98*, Maastricht, Netherlands. IRG/WP 98-30178.

Lloyd, J. D., Schoeman, M. W. & Brownsill, F. (1998) Losses of pyrethroids from treated wood due to photodegradation. In *IRG/WP 98*, Maastricht, Netherlands. IRG/WP 98-30177.

Lloyd, J. D., Fogel. J. L. & Vizel, A. (2001) The use of zirconium as an inert fixative for borates in preservation. In *IRG/WP 01*, Nara, Japan. IRG/WP 01-30256.

Lykidis, G., Mantanis, G., Adamopoulos, S., et al. (2013) Effects of nano-sized zinc oxide and zinc borate impregnation on brown-rot resistance of black pine (*Pinus nigra* L.) wood. *Wood Material Science and Engineering* 8(4), 242–244.

Lyon, F., Thevénon, M-F., Pizzi, A. & Gril, J. (2009) Resistance to decay fungi of ammonium borate oleate treated wood. In *IRG/WP 09*, Beijing, China. IRG/WP 09-30505.

Mahút, J., Osvald, A., Reinprecht, L. & Krakovský, A. (1985) Fire retardant treatment of phenoplasts with triarylphosphate. In *Flammability of Polymers*, SAV Bratislava, Smolenice, Czechoslovakia, pp. 44–53.

Mai, C. & Militz, H. (2007) Wood preservation. In *Wood Production, Wood Technology, and Biotechnological Impacts*, Kües, U. (ed.). Universitätsverlag Göttingen, pp. 259–271.

Mai, C., Kües, U. & Militz, H. (2004) Biotechnology in the wood industry. *Applied Microbiology and Biotechnology* 63, 477–494.

Mankowski, M. (2007) Potential for controlling carpenter ants in utility poles with borates. In *IRG/WP 07*, Jackson, MS, USA. IRG/WP 07-10623.

Mantanis, G., Terzi, E., Kartal, S. N. & Papadopoulos, A. N. (2014) Evaluation of mould, decay and termite resistance of pine wood treated with zinc- and copper-based nanocompounds. *International Biodeterioration & Biodegradation* 90, 140–144.

Maoz, M. & Morrell, J. J. (2004) Ability of chitosans to limit wood decay under laboratory conditions. In *IRG/WP 04*, Ljubljana, Slovenia. IRG/WP 04–30339.

Marzbani, P. & Mohammadnia-Afrouzi, Y. (2014) Investigation on leaching and decay resistance of wood treated with nano-titanium dioxide. *Advances in Environmental Biology* 8(10), 974–978.

Matsunaga, H., Kiguchi, M. & Evans, P. (2007) Micro-distribution of metals in wood treated with a nano-copper wood preservative. In *IRG/WP 07*, Jackson, MS, USA. IRG/WP 07–40360.

Mazela, B., Bartkowiak, M. & Ratajczak, I. (2007) Animal protein impact on fungicidal properties of treatment formulations. *Wood Research* 52(1), 13–22.

Mburu, F., Dumarcay, S., Thévenon, M. F. & Gérardin, P. (2007) On the reason of *Prunus africana* natural durability. In *IRG/WP 07*, Jackson, MS, USA. IRG/WP 07-10611.

McIntyre, C. R. (2010) Comparison of micronized copper particle sizes. In *IRG/WP 10*, Biarritz, France. IRG/WP 10-30538.

Messner, K., Bruce, A. & Bongers, H. P. M. (2003) Treatability of refractory wood species after fungal pre-treatment. In *The Second European Conference on Wood Modification*, Göttingen, Germany, pp. 389–401.

Micales-Glaeser., Lloyd, J. D. & Woods, T. L. (2004) Efficacy of didecyl dimethyl ammonium chloride (DDAC), disodium octaborate tetrahydrate (DOT), and chlorothalonil (CTL) against common mold fungi. In *IRG/WP 04*, Ljubljana, Slovenia. IRG/WP 04-30338.

Minh, N. H., Militz, H. & Mai, C. (2007) Protection of wood for above ground application through modification with a fatty acid modified *N*-methylol/paraffin formulation. In *IRG/WP 07*, Jackson, MS, USA. IRG/WP 07-40378.

Mohareb, A., Thévenon, M-F., Wozniak, E. & Gérardin, P. (2010) Effects of monoglycerides on leachability and efficacy of boron wood preservatives against decay and termites. *International Biodeterioration & Biodegradation* 64, 135–138.

Mohareb, A. S. O., Badawy, M. E. I. & Abdelgaleil, S. A. M. (2013) Antifungal activity of essential oils isolated from Egyptian plants against wood decay fungi. *Journal of Wood Science* 59(6), 499–505.

Morrell, J. J. & Morris, P. I. (2002) Methods for improving preservative penetration into wood – a review. In *IRG/WP 02*, Cardiff, Wales, UK. IRG/WP 02-40227.

Morrell, J. J., Acda, M. N. & Zahora, A. R. (2005) Performance of oriented strandboard, medium density fiberboard, plywood, and particleboard treated with tebuconazole in supercritical carbon dioxide. In *IRG/WP 05*, Bangalore, India. IRG/WP 05-30364.

Morris, P. I. & McFarling, S. (2006) Enhancing the performance of transparent coatings by UV protective pre-treatments. In *IRG/WP 06*, Tromsø, Norway. IRG/WP 06-30399.

Morris, P. I., Grace, J. K. Yoshimura, T. & Tsunoda, K. (2014) An international termite field test of wood treated with insecticides in a buffered amine oxide carrier. *Forest Products Journal* 64(5–6), 156–160.

Moya, R., Berrocal, A., Rodriguez-Zuñiga, A., et al. (2014) Effect of silver nanoparticles on white-rot wood decay and some physical properties of three tropical wood species. *Wood and Fiber Science* 46(4), 527–538.

Muin, M., Adachi, A. & Tsunoda, K. (2001) Applicability of supercritical carbon dioxide to the preservative treatment of wood-based composites. In *IRG/WP 01*, Nara, Japan. IRG/WP 01-40199.

Murphy, R. J. (1994) Vapour phase treatments for wood products. In *Wood Preservation in the 90s and Beyond*. Proceedings No. 7308. Forest Product Society, Madison, WI, pp. 83–88.

Nicholas, D. D., Williams, A. D., Preston, A. F. & Zhang, S. (1991) Distribution and permanency of DDAC in southern pine sapwood treated by the full-cell process. *Forest Products Journal* 41(1), 41–45.

Ozgenc, O., Hiziroglu, S. & Yildiz, U. C. (2012) Weathering properties of wood species treated with different coating applications. *BioResources* 7(4), 4875–4888.

Pallaske, M. (1999) Insect growth regulators – modes of action and mode of action-dependent peculiarities in the evaluation of the efficacy for their use in wood preservation. In *Reconstruction and Conservation of Historical Wood*, 2nd International Symposium, TU Zvolen, Slovakia, pp. 239–248.

Pallaske, M. (2004) Chemical wood protection: improvement of biocides in use. In. *COST Action E22 – Environmental Optimisation of Wood Protection*, Estoril, Portugal.

Pánek, M., & Reinprecht, L. (2011) *Bacillus subtilis* for improving spruce wood impregnability. *BioResources* 6(3), 2912–2931.
Pánek, M., Reinprecht, L. & Mamoňová, M. (2013) *Trichoderma viride* for improving spruce wood impregnability. *BioResources* 8(2), 1731–1746.
Pánek, M., Reinprecht, L. & Hulla, M. (2014) Ten essential oils for beech wood protection – efficacy against wood-destroying fungi and moulds, and effect on wood discoloration. *BioResources* 9(3), 5588–5603.
Pankras, S., Cooper, P., Ung, T. & Awoyemi, L. (2009) Copper to quat ratio in alkaline copper quat (ACQ) wood preservative: effects on fixation and leaching of preservative components. In *IRG/WP 09*, Beijing, China. IRG/WP 09-30496.
Papadopoulos, C. J., Carson, C. F., Hammer, K. A. & Riley, T. V. (2006) Susceptibility of *Pseudomonas* to *Melaleuca alternifolia* (tea-tree) oil and components. *Journal of Antimicrobial Chemotherapy* 58(2), 449–451.
Peek, R.-D. (1987) Application of oscillating pressure to improve treatability of refractory species. In *IRG/WP 87*, Honey Harbour, Ontario, Canada. IRG/WP 87-3449.
Petrič, M. (2013) Surface modification of wood: a critical review. *Reviews of Adhesion and Adhesives* 1(2), 216–247.
Petrič, M., Murphy, R. J. & Morris, I. (2000) Microdistribution of some copper and zinc containing waterborne and organic solvent preservatives in spruce wood cell walls. *Holzforschung* 54(1), 23–26.
Petrič, M., Pavlič, M., Kričej, B., et al. (2003) Blue stain testing of alkyd and acrylic stains. In *IRG/WP 03*, Brisbane, Australia. IRG/WP 03-20273.
Pizzi, A. & Baecker, A. (1996) A boron preservative fixation mechanism. *Holzforschung* 50(6), 507–510.
Ray, M., Dickinson, D. J. & Archer, K. (2010) A comparison of the performance of related copper based preservatives against soft rot. In *IRG/WP 10*, Biarritz, France. IRG/WP 10-30540.
Reinprecht, L. (1995) Suggestion of the outflow model for specific stages of wood drying and impregnating. In *Vacuum Drying of Wood '95, International Conference*, High Tatras. TU Zvolen, Slovakia, pp. 119–130.
Reinprecht, L. (1996) *TCMTB and Organotin Fungicides for Wood Preservation – Efficacy, Ageing, and Applicability*. Monograph 10/96/A, TU Zvolen.
Reinprecht, L. (2007) Selected laboratory tests of boron efficacy against wood-damaging fungi. *ProLigno* 3(3), 27–37.
Reinprecht, L. (2008) *Wood Protection*. Handbook. TU Zvolen, Slovakia (in Slovak).
Reinprecht, L. (2010) Fungicides for wood protection – world viewpoint and evaluation/testing in Slovakia. In *Fungicides*, Carisse, O. (ed.). InTech, Rijeka, Croatia, chapter 5, pp. 95–122.
Reinprecht, L. (2013) *Wood Protection*, Handbook. TU Zvolen, Slovakia.
Reinprecht, L. & Kmeťová, L. (2014) Fungal resistance and physical-mechanical properties of beech plywood having durable veneers or fungicides in surfaces. *European Journal of Wood and Wood Products* 72(4), 433–443.
Reinprecht, L. & Makovíny, I. (1990) Diffusion of inorganic salt $CuSO_4 \cdot 5H_2O$ in wood structure. In *Latest Achievements in Research of Wood Structure and Physics*, IUFRO Conference. TU Zvolen, Slovakia, pp. 203–214 (in Russian, English abstract).
Reinprecht, L. & Pánek, M. (2015) Effects of wood roughness, light pigments, and water repellents on the color stability of painted spruce subjected to natural and accelerated weathering. *BioResources* 10(4), 7203–7219.

Reinprecht, L. & Perlác, J. (1995) Properties of particleboards protected by TBTN and TCMTB fungicides. In *Wood Modification '95, 10th Symposium*, Poznaň, Poland, pp. 255–264.

Reinprecht, L., Baculák, J. & Pánek, M. (2011) Natural and accelerated ageing of pains for wooden windows. *Acta Facultatis Xylologiae Zvolen* 53(1), 21–31 (in Slovak, English abstract).

Reinprecht, L., Mahút, J. & Osvald, A. (1986) Effect of preservatives on the anti-mould resistance of plywood. *Drevo (Wood)* 41(10), 301–303 (in Slovak, English abstract).

Reinprecht, L., Vidholdová, Z. & Kožienka, M. (2015) Decay inhibition of lime wood with zinc oxide nanoparticles in combination with acrylic resin. *Acta Facultatis Xylologiae Zvolen* 57(1), 43–52 (in Slovak, English abstract).

Rezazadeh, A., Farahani, M. R. M., Afrouzi, Y. M. & Khalaji, A. A. D. (2014) Investigation on rot resistance of *Populus deltoids* wood treated with nano-zinc oxide. *World of Science Journal* 2(2), 19–28.

Richardson, B. A. (1993) *Wood Preservation*. E & FN Spon, London.

Saha, S., Kocaefe, D., Boluk, Y. & Pichette, A. (2011) Enhancing exterior durability of jack pine by photo-stabilization of acrylic polyurethane coating using bark extract. Part 1: effect of UV on color change and ATR-FT-IR analysis. *Progress in Organic Coatings* 70(4), 376–382.

Salla, J., Pandey, K. K. & Srinivas, K. (2012) Improvement of UV resistance of wood surfaces using ZnO nanoparticles. *Polymer Degradation and Stability* 97, 592–596.

Samyn, P., Stassens, D., Paredes, A. & Becker, G. (2014) Performance of organic nanoparticle coatings for hydrophobization of hardwood surfaces. *Journal of Coatings Technology Research* 11(3), 461–471.

Schaller, Ch., Rogez, D. & Braig, A. (2009) Hindered amine light stabilizers in pigmented coatings. *Journal of Coatings Technology Research* 6(1), 81–88.

Schilling, S. J. & Inda, J. J. (2010) Toward an assessment of copper bioavailability in treated wood. In *IRG/WP 10*, Biarritz, France. IRG/WP 10-20445.

Schmidt, O., Müller, J. & Moreth, U. (1995) Potentielle schutzwirkung von Chitosan gegen holzpilze. *Holz-Zentralblatt* 121(150), 2503.

Sean, T., Brunette, G. & Côte, F. (1989) Protection of oriented strandboard with borate. *Forest Products Journal*, 39(5), 47–51.

Sethy, A. K., Nagaveni, H. C., Mohan, S. & Chandrashekar, K. T. (2005) Fungal decay resistance of rubber wood treated with heartwood extract of rosewood. In *IRG/WP 05*, Bangalore, India. IRG/WP 05-30367.

Siau, J. F. (1984) *Transport Processes in Wood*. Springer Verlag, Berlin.

Singh, A. P., Gallager, S. S., Schmitt, U., et al. (1998) Ponding of radiata pine – the effect of ponding on coating penetration into wood. In *IRG/WP 98*, Maastricht, Netherlands. IRG/WP 98-10249.

Singh, T., Chittenden, C. & Vesentini, D. (2006) In vitro antifungal activity of chilli against wood degrading fungi. In *IRG/WP 06*, Tromsø, Norway. IRG/WP 06-10572.

Smart, R. & Wall, W. (2006) Copper borate for protection of engineered wood composites. In *IRG/WP 06*, Tromsø, Norway. IRG/WP 06-40334.

Smith, W. R. & Wu, Q. (2005) Durability improvement for structural wood composites through chemical treatment. *Forest Products Journal* 55(2), 8–17.

Solár, R. & Reinprecht, L. (2004) Theoretical survey to thermal processing of by biocides impregnated wooden materials after their service life. In *COST Action E37 – Sustainability through New Technologies for Enhanced Wood Durability, Managing the Environmental Risk*, Reinbek, Germany. http://www.academia.edu/18754692/

Theoretical_survey_to_thermal_processing_of_bybiocides_impregnated_wooden_materials_after_their_service_life (accessed 11 May 2016).

Šomšák, M., Reinprecht, L. & Tiňo, R. (2015) Effect of plasma and UV-absorbers in transparent acrylic coatings on photostability of spruce wood in exterior. *Acta Facultatis Xylologiae Zvolen* 57(1), 63–73 (in Slovak, English abstract).

Stanković, A., Dimitrijević, S. & Uskoković, D. (2013) Influence of size and morphology on bacterial properties of ZnO powders hydrothermally synthesized using different surface stabilizing agents. *Colloids and Surfaces B: Biointerfaces* 102, 21–28.

Stirling, R. & Morris, P. I. (2010) Degradation of carbon-based preservatives by black-stain fungi. In *IRG/WP 10*, Biarritz, France. IRG/WP 10-30533.

Stirling, R. & Temiz, A. (2014) Fungicides and insecticides used in wood protection. In *Deterioration and Protection of Sustainable Biomaterials*, Schultz, T. P., Goodell, B. & Nicholas, D. D. (eds). ACS Symposium Series 1158, American Chemical Society, Washington, DC, pp. 185–201.

Stook, K., Tolaymat, T., Ward, M., et al. (2005) Relative leaching and aquatic toxicity of pressure-treated wood using batch leaching test. *Environmental Science and Technology* 39, 155–163.

Taghiyari, H. R. & Farajpour B. O. (2013) Effect of copper nanoparticles on permeability, physical, and mechanical properties of particleboard. *European Journal of Wood and Wood Products* 71(1), 69–77.

Teacà, C. A., Rosu, D., Bodîrlàu, R. & Rosu, L. (2013) Structural changes in wood under artificial UV light irradiation by FTIR spectroscopy and color measurements – a brief review. *BioResources* 8(1), 1478–1507.

Terziev, E., Kartal, S. N., Yilgor, N., et al. (2016) Role of various nano-particles in prevention of fungal decay, mold growth and termite attack in wood, and their effect on weathering properties and water repellency. *International Biodegradation & Biodeterioration* 107, 77–87.

Thevenon, M. F. & Pizzi, A. (2003) Polyborate ions influence on the durability of wood treated with non-toxic protein borate preservatives. *Holz als Roh- und Werkstoff* 61, 457–464.

Tondi, G., Wieland, S., Lemenager, N., et al. (2012) Efficacy of tannin in fixing boron in wood. *BioResources* 7(1), 1238–1252.

Tondi, G., Thevenon, M-F., Mies, B., et al. (2013) Impregnation of Scots pine and beech with tannin solutions: effect of viscosity and wood anatomy in wood infiltration. *Wood Science and Technology* 47(3), 615–626.

Toussaint-Dauvergne, E., Soulounganga, P., Gérardin, P. & Loubinoux, B. (2000) Glycerol/glyoxal: a new boron fixation system for wood preservation and dimensional stabilization. *Holzforschung* 54, 123–126.

Townsend, T. G. & Solo-Gabriele, M. (2006) *Environmental Impacts of Treated Wood.* CRC Press.

Trăistaru, A. A. T., Timar, C. M., Câmpean, M., et al. (2012) Paraloid B72 versus Paraloid B72 with nano-ZnO additive as consolidants for wooden artefacts. *Materiale Plastice* 49(4), 293–300.

Tsonuda, K. & Muin, M. (2003) Preservative treatment of wood-based composites with mixture formulation of IPBC-silafluofen using supercritical carbon dioxide as carrier gas. In *IRG/WP 03*, Brisbane, Australia. IRG/WP 03-40251.

Vesentini, D. & Singh, T. (2006) The effect of chitosan on the growth and physiology of two wood-inhabiting fungi. In *IRG/WP 06*, Tromsø, Norway. IRG/WP 06-10590.

Wallace, D. F. & Dickinson, D. J. (2006) The bacterial transformation of organic biocides: a common mechanism? In *IRG/WP 06*, Tromsø, Norway. IRG/WP 06-10585.

Wang, S. Y., Chen, P. F. & Chang, S. T. (2005) Antifungal activities of essential oils and their constituents from indigenous cinnamon (*Cinnamomum osmophloeum*) leaves against wood decay fungi. *Bioresource Technology* 96, 813–818.

Wang, Z., Han, E., Liu, F. & Ke, W. (2007) Thermal behavior of nano-TiO_2 in fire-resistant coating. *Journal of Materials Science and Technology* 23(4), 547–550.

Wang, Z., Han, E., Liu, F. & Ke, W. (2010) Fire and corrosion resistances of intumescent nano-coating – containing nano-SiO_2 in salt spray condition. *Journal of Materials Science and Technology* 26(1), 75–81.

Wasserbauer, R. (2000) *Biological Decomposition of Buildings.* ABF–ARCH, Prague (in Czech).

Ważny, J. & Kundzievicz, A. (2008) Conditions and possibilities of nanobiocides formulation for wood protection. *Drewno* (*Wood*) 51(180), 107–115.

Williams, R. S. (2010) Finishing of wood. In *Wood Handbook – Wood as an Engineering Material.* Forest Product Laboratory, Madison, WI, chapter 16.

Williams, R. S. & Feist, W. C. (1999) *Water repellents and water-repellent preservatives for wood.* General Technical Report FPL-GTR-109. USDA, Forest Products Laboratory, Madison, WI.

Williams, T. M. (2007) The mechanism of action of isothiazolone biocides. *PowerPlant Chemistry* 9(1), 14–22.

Wong, S. Y., Grant, I. R., Friedman, M., et al. (2008) Antibacterial activities of naturally occurring compounds against *Mycobacterium avium* ssp. *paratuberculosis. Applied and Environmental Microbiology* 73(19), 5986–5990.

Xu, K., Li. K., Yun, H., et al. (2013) A comparative study on the inhibitory ability of various wood-based composites against harmful biological species. *BioResources* 8(4), 5749–5760.

Xue, W., Kennepohl, P. & Ruddick, J. N. R. (2010) A comparison of the chemistry of alkaline copper and micronized copper treated wood. In *IRG/WP 10*, Biarritz, France. IRG/WP 10-30528.

Yang, V. W. & Clausen, C. A. (2006) Moldicidal properties of seven essential oils. In *IRG/WP 06*, Tromsø, Norway. IRG/WP 06-30404.

Yildiz, S., Canakci, S., Yildiz, U. C., et al. (2012) Improving of the impregnability of refractory spruce wood by *Bacillus lichteniformis* pretreatment. *BioResources* 7(1), 565–577.

Yamaguchi, H. (2001) Silicic acid–boric acid complexes as wood preservatives. In *IRG/WP 01*, Nara, Japan. IRG/WP 01-30273.

Zabielska-Matejuk, B. & Skrzypczak, A. (2006) Ionic liquids with organic and inorganic anions as highly active wood preservatives. In *IRG/WP 06*, Tromsø, Norway. IRG/WP 06-30411.

Zhang, Z., Jiang, W., Jian, W., et al. (2015) Residues and dissipation kinetics of triazole fungicides difenoconazole and propiconazole in wheat and soil in Chinese fields. *Food Chemistry* 168, 396–403.

Standards

EN 20-1 (1992) Wood preservatives – Determination of the protective effectiveness against *Lyctus brunneus* (Stephens) – Part 1: Application by surface treatment (Laboratory method).

EN 20-2 (1993) Wood preservatives – Determination of the protective effectiveness against *Lyctus brunneus* (Stephens) – Part 2: Application by impregnation (Laboratory method).

EN 46-1 (2009) Wood preservatives – Determination of the preventive action against recently hatched larvae of *Hylotrupes bajulus* (Linnaeus) – Part 1: Application by surface treatment (laboratory method).

EN 46-2 (2009) Wood preservatives – Determination of the preventive action against recently hatched larvae of *Hylotrupes bajulus* (Linnaeus) – Part 2: Ovicidal effect (laboratory method).

EN 49-2 (2015) Wood preservatives – Determination of the protective effectiveness against *Anobium punctatum* (De Geer) by egg-laying and larval survival – Part 2: Application by impregnation (Laboratory method).

EN 73 (2014) Wood preservatives – Accelerated ageing of treated wood prior to biological testing – Evaporative ageing procedure.

EN 84 (1997) Wood preservatives – Accelerated ageing of treated wood prior to biological testing – Leaching procedure.

EN 113 (1996) Wood preservatives – Test method for determining the protective effectiveness against wood destroying basidiomycetes – Determination of the toxic values.

EN 117 (2012) Wood preservatives – Determination of toxic values against *Reticulitermes* species (European termites) (laboratory method).

EN 118 (2013) Wood preservatives – Determination of preventive action against *Reticulitermes* species (European termites) (Laboratory method).

EN 152 (2011) Wood preservatives – Determination of the protective effectiveness of a preservative treatment against blue stain in wood in service – Laboratory method.

EN 212 (2003) Wood preservatives – General guidance on sampling and preparation for analysis of wood preservatives and treated timber.

EN 252 (2014) Field test method for determining the relative protective effectiveness of a wood preservative in ground contact.

EN 275 (1992) Wood preservatives – determination of the protective effectiveness against marine borers.

EN 330 (2014) Wood preservatives – Determination of the relative protective effectiveness of a wood preservative for use under a coating and exposed out-of-ground contact – Field test: L-joint method.

EN 335 (2013) Durability of wood and wood-based products – Use classes: definitions, application to solid wood and wood-based products.

EN 350-2 (1994) Durability of wood and wood-based products – Natural durability of solid wood – Part 2: Guide to natural durability and treatability of selected wood species of importance in Europe.

EN 351-1 (2007) Durability of wood and wood-based products – Preservative-treated solid wood – Part 1: Classification of preservative penetration and retention.

EN 351-2 (2007) Durability of wood and wood-based products – Preservative-treated solid wood – Part 2: Guidance on sampling for the analysis of preservative-treated wood.

EN 460 (1994) Durability of wood and wood-based products – Natural durability of solid wood – Guide to the durability requirements for wood to be used in hazard classes.

EN 599-1 (2014) Durability of wood and wood-based products – Efficacy of preventive wood preservatives as determined by biological tests – Specification according to use class.

EN 599-2 (2015) Durability of wood and wood-based products – Efficacy of preventive wood preservatives as determined by biological tests – Part 2: Labelling.

EN 839 (2014) Wood preservatives – Determination of the protective effectiveness against wood destroying basidiomycetes – Application by surface treatment.

EN 927-6 (2006) Paints and varnishes – Coating materials and coating systems for exterior wood – Part 6: Exposure of wood coatings to artificial weathering using fluorescent UV lamps and water.
ENV 1250-1 (1994) Wood preservatives – Methods for measuring losses of active ingredients and other preservative ingredients from treated timber – Part 1: Laboratory method for obtaining samples for analysis to measure losses by evaporation to air.
ENV 1250-2 (1994) Wood preservatives – Methods of measuring losses of active ingredients and other preservative ingredients from treated timber – Part 2: Laboratory method for obtaining samples for analysis to measure losses by leaching into water or synthetic sea water.
EN 13501-1+ A1 (2009) Fire classification of construction products and building elements – Part 1: Classification using data from reaction to fire tests.
EN 13823 (2010) Reaction to fire tests for building products – Building products excluding floorings exposed to the thermal attack by a single burning item.
EN 1363-1 (2012) Fire resistance tests – Part 1: General Requirements.
EN 16402 (2013) Paints and varnishes – Assessment of emissions of substances from coatings into indoor air – Sampling, conditioning and testing.
EN ISO 1182 (2010) Reaction to fire tests for products – Non-combustibility test.
EN ISO 1716 (2010) Reaction to fire tests for products – Determination of the gross heat of combustion (calorific value).
EN ISO 11925-2 (2010) Reaction to fire tests – Ignitability of products subjected to direct impingement of flame – Part 2: Single-flame source test.
ISO 2859-1 (1999) Sampling procedures for inspection by attributes – Part 1: Sampling schemes indexed by acceptance quality limit (AQL) for lot-by-lot inspection.

Directives

Biocidal Product Directive 98/8/EC.
Directive 2007/1451/ES.
The Paints Directive 2004/42/EC.

6 Modifying Protection of Wood

6.1 Methodology, ecology and effectiveness of wood modification

The modification of wood is targeted intervention into its structure in order to improve its properties, such as resistance to biological pests, fire, radiation, water, aggressive chemicals, dimensional changes or mechanical loads, as well as its aesthetics.

Wood modification has been developed over the last 50 years as a potential alternative to (1) chemical preservative treatments and (2) replacement of rare durable exotic woods, mainly for products proposed for damp conditions convenient for activity of pests. For example, studies (e.g. Ringman et al., 2014; Zelinka et al., 2015) of decay resistance of modified wood have shown that to increase it the dominant role should be a reduction in transport of enzymes and other catalysts of fungi in the 'cross-linked' cell walls of modified wood and also a lower equilibrium moisture content.

Suitably modified products of wood and wooden composites are increasingly being applied in housing, the building industry, transport, etc. By modifying less-durable and reasonably priced species of wood (e.g. beech, ash, pine, spruce), modified woods with properties similar to rare and durable exotic woods are obtained. Today, the total production of modified wood in Europe is estimated to be 300,000–400,000 m^3 per year (Militz, 2015). In practice, this is mainly thermally modified woods and some types of chemically modified woods.

Recently, it is generally the following factors that prevent wider use of modified woods:

- complicated production processes;
- high production costs;

- a decrease in the strength and an increase in the fragility of wood due to thermal and some chemical modifications;
- only well-impregnable wood species can be chemically modified;
- insufficient promotion between purchasers and potential users.

6.1.1 Methods of wood modification: mechanical, physical, chemical and biological

Individual structural levels of wood can be modified by various methods: (1) mechanical, (2) plasma, laser, thermal, hydrothermal and other physical, (3) chemical and (4) biological, as well as by their combination (Table 6.1).

Mechanical modification of wood is based on plasticizing wood and its subsequent compression. Lignin in the midlle lamella, primary layer and other layers of the cell walls of wood is thermoplastic with a glass transition temperature $T_g \cong 170$ °C. This relatively high temperature can be reduced in the presence of bound water (hydrothermal softening of wood) or in the presence of plasticizers and softeners (chemical softening of wood by ammonia, urea, dimethyl urea, dimethyl sulphoxide, etc.). During the first stage of the mechanical modification of wood lignin in the cell walls is softened – plasticized; in the second stage, the wood with a softened lignin is compressed in a metal mould (Rowell, 1999). The compressed wood has a higher density, hardness and compression strength compared with the untreated wood. The shape of the compressed wood is stabilized by its hydrothermal pretreatment or post-treatment (Laine et al., 2013). For example, Inoue et al. (2008) showed that springback of mechanically modified wood can be reduced by pre-steaming at 180–220 °C for 10–20 min.

In the past, several types of densified wood have been produced; for example, 'Staypak/Staybwood' in the USA, 'Lignostone' in Germany or 'Bukolis' and 'Lignamon' in Czechoslovakia. Mechanically densified wood is suitable mainly for interiors (since it resists fire well), as well as for special products exposed to

Table 6.1 Basic methods for modification of solid wood

Modification method	Change in the wood structure		
	Geometric	Anatomical and morphological	Molecular
Mechanical	+	+	–
Plasma, laser	(+)*	+*	+*
Thermal	–	(+)	+
Chemical			
filling lumens	–	+	–
blocking —OH groups	– (+)	(+)	+
Biological	– (+)	– (+)	+

+, a significant change; (+), an insignificant change; –, without an apparent change, *, changes in surfaces.

increased mechanical stresses in industry and buildings, or for musical instruments. However, its original hygroscopicity, dimensional stability and resistance against wood-damaging fungi do not apparently improve (Solár et al., 2005), and when exposed to exterior environments it should be treated with weather-resistant biocides and coatings.

Plasma and laser modifications of wood surfaces are based on intentional effects of energy fields on wood components present in its surfaces. Plasma is mixture of ionized molecules of gas. It is created either at high temperatures (10^6–10^8 K – high-thermal plasma) or in the presence of electrical fields at normal temperatures (3×10^2 – 2×10^4 K – cold and low-thermal plasma). However, in practice, only cold or non-thermal plasma discharges (atmospheric pressure, dielectric barrier magnetron, pulsed glow, corona, laser fusion, etc.) in an air, nitrogen or argon medium are usually used (Odrášková et al., 2008; Petrič, 2013). CO_2-laser irradiation has a defined power output; for example, the LCS 400 device has a maximum power output of 400 W (Kubovský & Kačík, 2013). Owing to the effect of these two energy fields, plasma and laser treatment of the wood surfaces (1) increases or decreases its hydrophobicity, (2) decreases roughness, (3) improves bonding capacity, (4) improves weather stability, and/or (5) improves healthy characteristics after sterilization killing of microorganisms. Studies (e.g. Podgorski et al., 2000; Odrášková et al., 2008; Wolkenhauer et al., 2008; Busnel et al., 2010; Šomšák et al., 2015) have shown that plasma treatment positively affects wood surface characteristics mainly by increasing the polar component of surface energy. Surfaces of wood treated with a CO_2-laser also have some additional benefits compared with untreated ones; for example, their colour is more stable on weathering (Kubovský et al., 2016) and they resist mould attacks better (Vidholdová et al., 2016, unpublished results).

Thermal modification of wood is based on thermal and hydrothermal modifications of wood at high temperatures from 160 to 260 °C. At high temperatures the wood polymers, mainly hemicelluloses, change their structure due to hydrolytic, dehydration and cross-linking reactions. Various hydrophobic water-insoluble substances, and also substances with a toxic or repellent effect to biological pests, are created in the wood. The thermal modification of wood is best implemented in an inert environment; that is, in a vacuum or nitrogen gas, in order to prevent increased degradation of the cellulose and a decrease in the wood's strength. Methods for the thermal modification of wood are described in more detail in Section 6.2.

Chemical modification of wood is based on treating the wood structure with such types of chemical substances that do not have an apparent biocidal, fire retardant or other protective effect but which create the required protective effect and improvement in selected properties directly in the wood. When used for surface modification, they remain on the surface of the wood or only partly penetrate its surface layers. When used for inner modification they are located deeper in the wood, either only in the lumens of its cells or also penetrating the cell walls where several of them can chemically react with polysaccharides and lignin. The principle of increasing the resistance of chemically modified wood to water, UV radiation and various biological pests is based on creating new chemical or only mechanical interactions between the modifying substance and the wood. Thermosets (e.g. phenolic resins, amino resins and furfuryl-alcohol resins) as well as thermoplastic polyacrylates are used preferentially for filling the lumens of wood cells. Only low molecular weight and polar thermosets or acrylates penetrate the cell walls, as well. Intentional reactions with —OH or other groups of the wood can create reactive monomers; for example, carboxylic acids and their anhydrides, aldehydes, lactones, isocyanates, nitriles, alkyl sulphates, halogen derivatives, epoxides and others (Rowell, 1984). Methods for the chemical modification of wood are described in more detail in Section 6.3.

Biological modification of wood is based on the principle of the antagonistic relationships between pests dangerous for wood and other biological organisms. Such organisms initially inhabit the wood without any apparent damage to it (e.g. some bacteria, moulds or sap-staining fungi) and prevent the activities of wood-decaying fungi or other more significant pests of wood. The biological modification of wood also includes its enzymatic modification; for example, decreasing the proportion of —OH phenolic groups in lignin using oxidases and other types of enzymes. Methods for the biological modification of wood are described in more detail in Section 6.4.

Specially modified wooden composites are made in a similar manner to the common wooden composites (glulam, plywood, FBs, etc.) from disintegrated wood particles, adhesives and various additives, including preservatives. Their service life depends upon specific modifications made to the disintegrated wood particles and/or the final composite. Known examples include acetylated veneers and veneers modified with phenolic resins, both used for the production of highly durable plywood (Bicke & Militz, 2014). Special more water- and bio-resistant FBs are made of A-Cell acetylated wood fibres.

6.1.2 Ecology of wood modification

Products made of modified wood should not burden the environment after their service life more than products made of solid wood do. However, several components of modified wood have high stability in ecosystems (e.g. furfuryl resins in Kebony wood (Lande et al., 2004)) or may affect the high-temperature burning process and other forms of disposing of products at the end of their usage (e.g. acetyl groups in Accoya wood). Antagonistic organisms used for the biological modification of wood (e.g. some bacteria, moulds, sap-staining fungi) can also be harmful to human health, mainly if they produce mycotoxins.

Nevertheless, the ecological and health side of products made of modified wood should certainly be better than products made of wood protected with fungicides, insecticides or other types of preservatives.

6.1.3 Effectiveness of wood modification

The effectiveness of wood modification is characterized by its durability increase and improvement of its selected properties (Equations 6.1, 6.2, 6.3

and 6.6). On chemical modification, the effectiveness is also related to the distribution and retention (weight-percent-gain, WPG; pore-filling ratio, PFR) of the modification substance in wood (Equations 6.4 and 6.5).

Durability increase (DI) defines the ratio of the durability of modified wood (D_{mod}) to unmodified wood (D) as a percentage:

$$\text{DI}\ (\%) = \frac{D_{mod}}{D} \times 100 \tag{6.1}$$

Anti-swelling efficiency (ASE) defines by what percentage the modified wood is more resistant to swelling in water ($\beta_{V\ mod}$) in relation to unmodified wood (β_V):

$$\text{ASE}\ (\%) = \frac{\beta_V - \beta_{V\ mod}}{\beta_V} \times 100 \tag{6.2}$$

Mechanical property increase (MPI) defines the ratio of the mechanical property of modified wood (MP_{mod}) to unmodified wood (MP) as a percentage:

$$\text{MPI}\ (\%) = \frac{\text{MP}_{mod}}{\text{MP}} \times 100 \tag{6.3}$$

WPG is the percentage amount of the dry mass of modifying substance ($m_{mod} = m_{0mod} - m_0$) retained by the solid absolutely dry wood (m_0), whose weight after modification adequately increased in an absolutely dry state (m_{0mod}):

$$\text{WPG}\ (\%) = \frac{m_{0\ mod} - m_0}{m_0} \times 100 \tag{6.4}$$

PFR is the percentage volume of the dry mass of modifying substance ($V_{\text{modifying substance}}$) retained by the solid wood with known volume of its pores before modification ($V_{\text{wood pores}}$); this parameter is important if lumens of wood cells are modified:

$$\text{PFR}\ (\%) = \frac{V_{\text{modification substance}}}{V_{\text{wood pores}}} \times 100 \tag{6.5}$$

Efficiency of chemical wood modification (ψ_{mod}) defines the ratio of the change in the evaluated property of chemically modified wood ($P_{\text{MW-Change}}$ in per cent) to the retention of the modifying substance by the wood (WPG in per cent; or PFR in per cent):

$$\Psi_{mod} = \frac{P_{\text{MW-Change}}}{\text{WPG}} \tag{6.6}$$

6.2 Thermally modified wood

Thermally modified wood has been produced for ~20 years, mainly in Finland and other countries of western Europe. It is known for its increased dimensional stability and durability – properties that predetermine its use in conditions of external exposure without ground contact. It is made from non-durable and less-durable wood species; for example, spruce, pine, beech, birch, aspen and ash (Militz, 2002; Esteves & Pereira, 2009; Gérardin, 2015).

6.2.1 Principles, methods and technology of thermal wood modification

6.2.1.1 Principles and methods of the thermal modification of wood

Thermal modification of wood is a process in which the molecular structure of wood changes due to the increased temperatures, usually from 160 to 260 °C, for a period of 15 min to 24 h. Larger changes occur mainly in hemicelluloses, where chemically aggressive acetic acid is generated, whilst smaller changes occur in lignin connected with cross-link reactions and in cellulose connected with its depolymerization and crystallinity increase (Yildiz et al., 2006; Gonzalez-Pena et al., 2009). Different types of polyaromatic compounds are also produced in thermo-treated wood (Lovaglio et al., 2015), but in conifers the amount of terpenoids decreases (Kačík et al., 2012).

Thermal modification of wood improves its resistance to water and biological pests. The types and scope of changes in the molecular structure of the wood (disturbing the hemicellulose molecules and their interaction with lignin), but also subsequent changes in the anatomical structure (thinning of cell walls) and geometric structure (reduced density and browning) of modified wood, depend upon the temperature level and time of treatment. However, these changes also depend upon the wood species used, its dimensions and initial humidity, as well as upon the environment in which the thermal modification takes place; that is, an oxidation environment (air with a proportion of oxygen) or a less aggressive inert environment (nitrogen, vegetable oil, etc.).

Wood already becomes more dimensionally stable under oven-drying at a temperature of ~110 °C. At temperatures above 160 °C, new ester and other chemical bonds are created in the lignin–saccharide matrix, thanks to which the thermally modified wood has less —OH groups and becomes a hydrophobic material. Hemicelluloses and lignin are partially decomposed with the creation of water, carbon dioxide, formic acid, acetic acid, furaldehyde and other substances (Section 2.4.2). They may enter into subsequent condensation reactions (i.e. between each other and also with wood polymers), by which the thermally modified wood gains high firmness but also fragility. Generally, during the production of thermally modified wood, more hydrophobic, more bio-resistant, and more stiffening substances are created in its structure (Reinprecht & Vidholdová, 2008a; Esteves & Pereira, 2009).

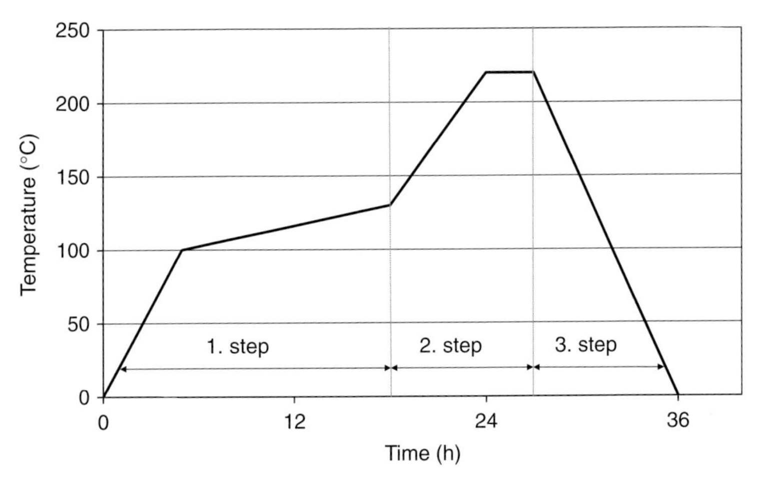

Figure 6.1 Scheme of the ThermoWood production

Source: Mayes, D. and Oksanen, O. (2002) *ThermoWood® Handbook*. Stora Enso Timber, Finnforest, Finland, 52 p

6.2.1.2 Technologies of the thermal modification of wood

Thermally modified wood is produced using several technologies; in Europe, this is mainly via the ThermoWood process, Plato process, OHT process and the rectification process.

1. The ThermoWood process (VTT, Stora, Finnforest, Finland) consists of three stages (Figure 6.1):
   - *Temperature increase and drying.* In this first stage, the temperature in the oven quickly increases to 100 °C and then slowly to ~130 °C. Hot air or water vapour media are used in the drying. The wood is dried to zero moisture content.
   - *Thermal modification.* In this second stage, the temperature increases to 185 to 215–230 °C for 2–3 h, depending on the classification of the ThermoWood® product; that is, Thermo-S or Thermo-D (Table 6.2). Similarly, but in vacuum, Silvapro® wood is produced at temperatures between 220 and 230 °C.
   - *Cooling and conditioning.* In this third stage, the thermally treated wood is gradually cooled and, at 80–90 °C, it starts to be wetted in order to achieve its final humidity of ~4–7% (Jones et al., 2006).
2. The Plato process (Plato: Providing Lasting Advanced Timber Option, The Netherlands) has four stages, implemented at atmospheric as well as increased pressure:
   - *Hydrothermolysis.* This first stage is hydrothermal modification of fresh or air-dried wood using water vapour or hot air at 150–190 °C and increased pressure of 0.6–1 MPa for a period of 4–5 h.

Table 6.2 Examples of the use of the ThermoWood®

Wood type	Thermo-S	Thermo-D
Coniferous (e.g. pine, spruce)	• In interiors ○ furniture ○ cladding ○ windows and doors ○ flooring ○ saunas	• In interiors and exteriors ○ cladding ○ tiling ○ saunas and bathroom furniture ○ flooring ○ garden furniture ○ children's playgrounds ○ noise barriers
Broadleaves (e.g. birch, aspen)	• In interiors ○ furniture ○ flooring ○ saunas	• In interiors as Thermo-S (however, ThermoWood treated with a higher temperature is darker)

Temperature used in production of ThermoWood®: Thermo-S, maximum 190 ± 3 °C (coniferous) or 185 ± 3 °C (broadleaves); Thermo-D, maximum 212 ± 3 °C (coniferous) or 200 ± 3 °C (broadleaves)

- *Drying.* This second stage is decreasing the humidity of the wood in the oven to ~8–10%, which usually takes 3–5 days.
- *Curing.* This third stage is about stabilizing the wood at 150–190 °C and atmospheric pressure of 0.1 MPa for a period of 12–16 h. The access of air, and therefore also oxygen, to the wood is limited. The cured wood has a humidity of less than 1%.
- *Conditioning.* This fourth stage is repeated humidification of the wood in the oven chamber to a value of 4–6%, which is usually achieved within 3 days (Militz & Tjeerdsma, 2001).

3. In the *OHT process (OHT: Oil Heat Treatment, Menz Holz, Germany)* the thermal modification of wood is carried out in autoclaves in a medium of hot vegetable oil at a temperature of 200–220 °C. With this technology, only a small amount of oxygen is present in the wood, which limits undesired thermal oxidation processes. The wood is firstly impregnated with hot oil. The internal zones of the wood should be heated to ~180–200 °C for 2–4 h. For example, the total duration of the OHT process for 4 m long poles of 0.1 m diameter is 18 h, including initial heating and final cooling. The most frequently used vegetable oils are linseed and rapeseed oils, which allow fast heating and an even transfer of heat to the internal zones of the modified wood. When selecting vegetable oils, the issue of their potential additional polymerization due to oxygen in the air is also important, in order to optimize their stability in thermally modified wood. For example, triglycerides of unsaturated linseed oil's fatty acids can partially penetrate to the cell walls of wood and after polymerization they anchor securely within the wood (Rosenqvist, 2000). Vegetable oils modified with maleic acid anhydride can

cross-link in the wood and become highly stable. However, in this situation, it is already a combination of the thermal and chemical modification of wood with a synergetic effect to improve its bio-resistance (Tjeerdsma et al., 2005).

4. The principle of the *rectification process (NOW: New Option Wood, France)* is based on gradually increasing the temperature of air-dried wood, with an initial humidity of ~12%, up to 210–260 °C (Armines, 1986). Heating is carried out in an inert nitrogen environment with a requirement for not exceeding 2% oxygen. At such high temperatures, the wood undergoes slight pyrolysis, but the thermal oxidation reaction without access of oxygen still does not take place to a great extent. The original firmness of the wood therefore decreases only slightly and the wood gains good hydrophobic and biocidal properties as a result of the creation of various condensed and less-polar substances in the cell walls.

The various technologies of wood thermal modification are protected by patents; for example, EP0018446 (1982), EP0612595 (1994), EP0623433 (1994), EP0622163 (1994), EP0759137 (1995) and US5678324 (1997). And other thermal processes are continually being investigated (e.g. Royal or Stellac), using various heating media, including other types of vegetable oils and oil resins (Wang & Cooper, 2005; Spear et al., 2006; Reinprecht & Vidholdová, 2008b). Technologies for the thermal modification of wood are optimized taking into account the applied species of coniferous or broadleaved woods, their intended usage and practical knowledge. More than 300,000 m^3 of thermally modified wood is currently produced annually in Europe, mainly in Finland, Germany, France and the Netherlands.

6.2.1.3 Classification of thermally modified woods

Several types of thermally modified woods are produced nowadays. For example, in the Netherlands, under the name Plato®Wood, Plato®Hout or Platonium®; in Finland, under the name ThermoWood® (Finnforest), in the two classes Thermo-D and Thermo-S. Classes of ThermoWood® differ from each other in the technological parameters of their production, with subsequent recommendations for their use (Table 6.2).

6.2.2 Durability and other properties of thermally modified wood

The *biological resistance of thermally modified wood* is sufficient against more pests. Wood species with low resistance against fungal decay (e.g. with durability classes 4 and 5 in accordance with EN 350-1) after thermal modification become durable to very durable (durability classes 2 to 1). An example is pine sapwood, which, after modification using the ThermoWood process at 205 °C, becomes durable comparable to white cedar or yew woods (Jämsä & Viitaniemi, 2001), or at temperatures over 210 °C it becomes very durable comparable to teak wood (Figure 6.2). The Plato process also significantly improves the resistance of less durable wood species against fungal decay (Figure 6.3). The

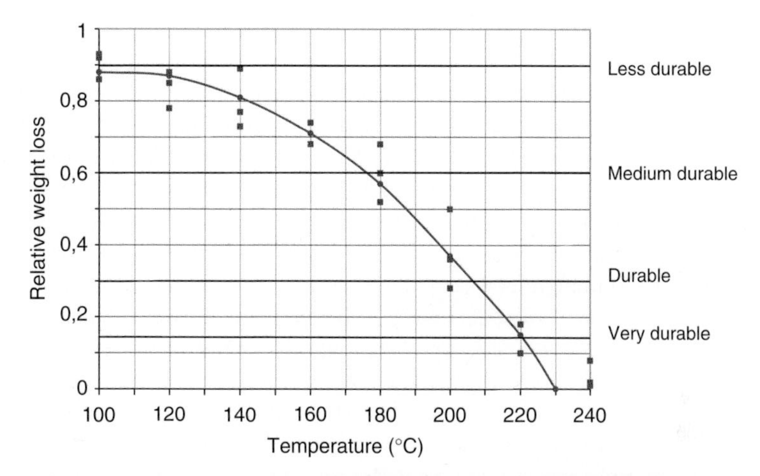

Figure 6.2 An increase in the durability of pine sapwood thermally modified at 100–240 °C – that is, a reduction in its relative loss of weight when undergoing fungal decay in comparison with untreated wood

Source: Mayes, D. and Oksanen, O. (2002) *ThermoWood® Handbook*. Stora Enso Timber, Finnforest, Finland, 52 p

durability of OHT woods increases with the temperature of the vegetable oil; for example, pine wood treated with oils heated over 200 °C showed a fungal decay of only 1–2% mass loss after 16 weeks (Rapp & Sailer, 2001). However, the OHT process does not significantly increase the resistance of pine sapwood or beech wood to moulds (Reinprecht & Vidholdová, 2008b).

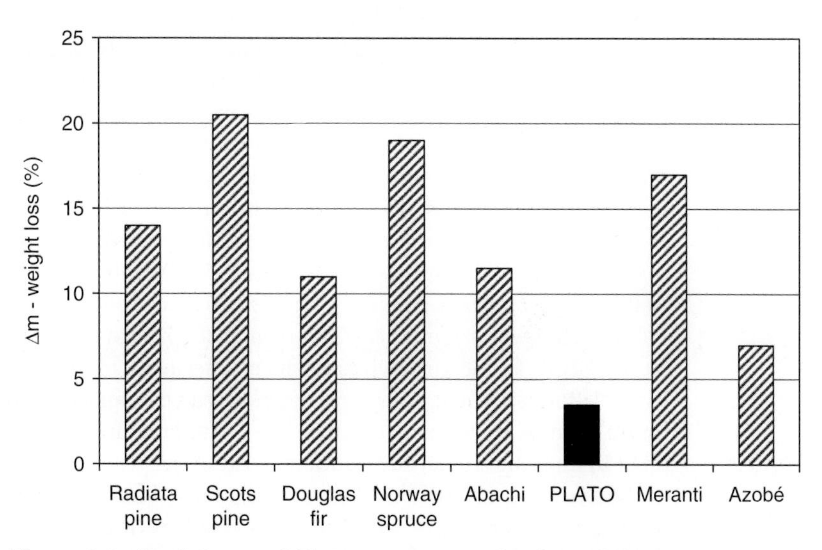

Figure 6.3 Resistance of Plato spruce-wood to fungal decay compared with untreated woods

Source: Anonymous (2002) *The Plato Technology – A Novel Wood Upgrading Technology*. Platowood B.V., Arnhem, The Netherlands, 14 p

It is generally true that thermally modified wood has better biological resistance mainly against brown-rot and white-rot fungi, whilst its bio-resistance increases with the intensity of the thermal process (Reinprecht, 1999; Militz & Tjeerdsma, 2001; Tjeerdsma et al., 2005; Mburu et al., 2006; Boonstra et al., 2007; Esteves & Pereira, 2009). An improvement in the resistance of thermally modified wood to moulds, sap-staining fungi and soft-rot fungi is, however, not always explicit (Kamdem et al., 1999; Kartal, 2007). Wood after thermal modifications is usually sufficiently resistant to wood-boring beetles (Jämsä & Viitaniemi 2001), but its resistance to termites can be lower than of untreated wood (Doi et al., 2004). Willems et al. (2013) explained the durability of thermally modified wood by an increase in the intrinsic chemical resistance to bio-oxidation, using a semi-empirical thermodynamic correlation with the O/C-ratio decrease in thermal degradation of hemicelluloses.

Thermally modified woods are materials suitable for interiors as well as exteriors without contact with ground in class 3 use (EN 335). Their suitability for classes 4 and 5 use, however, is still being verified via long-term tests in contact with the ground and fresh and salt water (Scheiding et al., 2005). According to Welzbacher and Rapp (2005), practical experience and laboratory and field tests to date have shown that the resistance of thermally modified woods from non-durable species against white-rot and soft-rot fungi (class 5 by EN 350-1, poor durability) is the best if they are produced using the OHT process (class 2, very good durability), slightly worse using the rectification process (class 3, medium durability), whilst the Plato process is usually less suitable for products permanently exposed in wet conditions (class 4, little durability).

The *hygroscopicity* of thermally modified wood is lower compared with untreated wood. This is directly reflected as well in its increased dimensional stability and biological resistance (Wang & Cooper, 2005; Awoyemi, 2006). By recent views, hygroscopicity and fungal durability of thermally modified woods are argued to become controlled by the physical state of polymers in the cell walls of wood rather than by changes in their chemical composition; that is, glassy polymers of wood may assist its dimensional and fungal stability (Willems, 2015). In a normal interior climate at a temperature of 20 °C and relative air humidity of 60–65%, the equilibrium moisture content of solid wood is ~10–12.5%, whilst after thermal modification by rectification or ThermoWood processes it is markedly lower, being only 3–5% (Vernois, 2001). Its dimensional stability is a very good, when swelling and shrinkage only reach 50% of the value of untreated wood (Jämsä & Viitaniemi, 2001; Živković et al., 2008; Zanuncio et al., 2014). In the given circumstances, thermally modified wood is mainly suitable for use where, along with its increased biological resistance, the minimum changes to its shape during exposure are required; for example, for parquet in interiors and facade cladding in exteriors.

Weather resistance of thermally modified wood in rainy exposures is not the best since its surfaces of a darker brown shade turn to grey relatively quickly. It is therefore beneficial to apply surface treatment (e.g. using plant oils or pigmented coatings). Water absorption of wood treated below a temperature of 200 °C can even increase due to formation of larger pores, microcracks and opening of bordered pits (Metsä-Kortelainen et al., 2006).

Flammability and fire resistance of thermally modified wood is usually comparable to untreated wood. However, thanks to the presence of residual plant oils and waxes, wood prepared using the OHT process has partially decreased resistance to the spread of flames (Wang & Cooper, 2007).

Strength, firmness and hardness of thermal-treated wood change depending upon the technological conditions of its production and also upon the wood species (Bengtsson et al., 2002; Esteves et al., 2005; Esteves & Pereira, 2009). The strength of heat-treated wood usually decreases, because at high temperatures its components, mainly the hemicelluloses, are disturbed (Section 2.4.2). Thermally modified wood is noticeably more fragile and prone to cracking (Scheiding et al., 2015), having lower bending and tensile strength, usually about 10–30% (Figure 6.4a). On the other hand, its firmness and usually also its surface hardness remain unchanged or even increase slightly (Figure 6.4b). It is generally recommended that, owing to decreased strength, all types of thermally modified woods should not be used for the load-bearing elements in wooden structures.

The *colour* of thermally modified wood turns from yellow to brown (Feher et al., 2014), and is similar to several tropical wood species (e.g. teak or cedar). A higher temperature and longer heating time causes a darker colour to develop. Thermal-treated woods also obtain a typical caramel odour. In terms of weathering resistance, it has been reported that the short-term colour stability of thermally modified wood exposed to artificial UV radiation is better than in untreated wood (Ayadi et al., 2003). However, in practice, architects and costumers expected a better colour stability of products from thermally modified woods – in which they gradually obtain a grey shade similar to native woods.

The *technical properties* of thermal-treated woods change specifically. They have increased abrasiveness due to a decrease in density and disturbance of the hemicelluloses. For example, staircase steps suffer wear and tear more easily than ones from untreated wood (Brischke & Rapp, 2004). Thermally modified wood is relatively well painted with weather-resistant alkyd coatings and well glued with polyurethane and phenol-formaldehyde (PF) adhesives. However, its joints are clearly weaker when using PVAc water-dispersed adhesive, which penetrates less into its hydrophobic surface layers. It is easier to remove screws from thermally modified wood, whilst resistance to the extraction of nails does not significantly change (Bengtsson et al., 2003). An issue is the increased corrosion of nails and other metal fasteners, probably due to the presence of a residue of formic and acetic acids created during the thermal modification processes (Jermer & Andersson, 2005).

The *acoustic properties* of thermally modified wood can be optimized in such a way that it can be used for producing various musical instruments, such as guitars (Zauer et al., 2015). At relatively lower temperatures, it is possible to prepare a special type of thermally modified wood whose structure and properties are identical to wood stored naturally for a long time, but with a slightly higher Young's modulus of elasticity. Its acoustic properties are often identical to wood in old, historically valuable musical instruments, and this fact can also be used in their restoration (Pfriem et al., 2005). An important factor is also increased

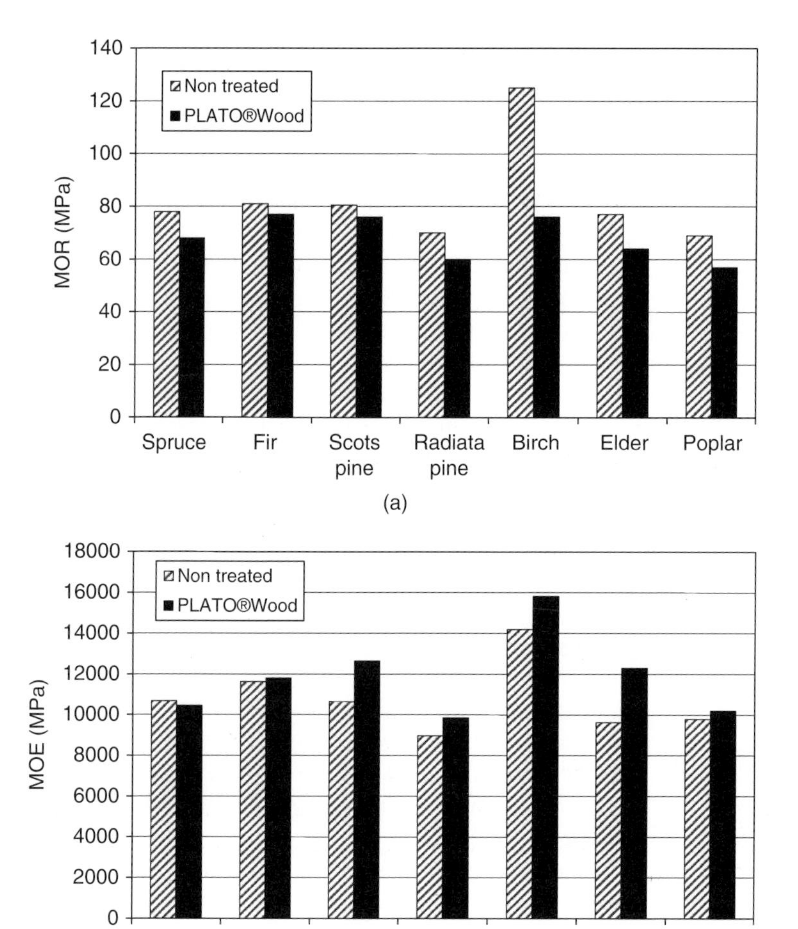

Figure 6.4 Bending strength MOR (a) and stiffness MOE (b) of untreated woods and of the same woods after thermal modification by the Plato process

Source: Anonymous (2002) *The Plato Technology – A Novel Wood Upgrading Technology*. Platowood B.V., Arnhem, The Netherlands, 14 p

dimensional stability of thermally treated wood, mainly in those musical instruments that are used in markedly changing climatic conditions.

6.2.3 Applications of thermally modified wood

Thermally modified wood is an ideal material for interior products; for example, parquet, tiles, panels, kitchen furniture and sauna walls or floors, and some musical instruments. It can also be used for entrance doors, windows, exterior cladding, garden furniture, children's playgrounds, fencing, and so on (Table 6.2). This wood has a potential for replacement of tropical wood species and also to gradually replace wood chemically protected with biocides. However, it must be highlighted that the types of thermally modified wood produced to date have not always been the most suitable material for permanently wet exposures in contact with the terrain or water.

6.3 Chemically modified wood

6.3.1 Principles, methods and technology of chemical wood modification

6.3.1.1 Principles of the chemical modification of wood

The aim of the chemical modification of wood is to improve its selected properties and increase its resistance to biological pests and abiotic effects (Rowell, 1983, 2005). In a wider meaning, this includes wood modifications connected with (1) the introduction of various chemicals into its inner parts or on its surfaces and (2) removing certain extractives from its cells. According to Norimoto and Gril (1993), or also by Reinprecht (2008), the term 'chemical modification of wood' represents the application of such chemical substances which alone usually do not have direct biocidal, fire-retardant or weather-stabilization effects, but wood in connection with them obtains higher durability and also specifically better physical and mechanical properties.

Modifying chemical substances can be located in wood (Figure 6.5) in:

- lumens of cells (I.) → *passive modification*;
- cell walls, though without covalent bonds with the wood components (II.a) → *passive modification*;
- cell walls where they are chemically bound to the structural components of wood, usually in reactions with its —OH groups (II.b, II.c) → *active modification.*

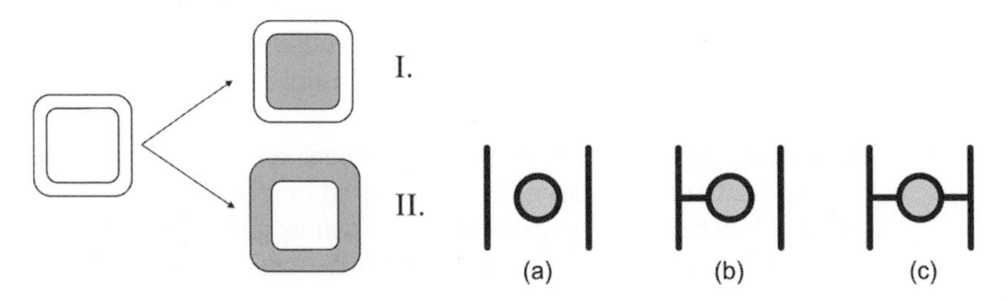

Figure 6.5 Location of substances in cells of wood at the passive (I., II.a) and active (II.b, II.c) chemical modification processes

Source: Norimoto, M. and Gril, J. (1993) Structure and properties of chemically treated woods. In: *Recent Research on Wood and Wood-based Materials*, pp. 135–154. Elsevier Barking, UK. Copyright © 1993, Elsevier Science Publishers Ltd and The Society of Materials Science. Reproduced by permission of Elsevier

According to Hill (2006), the term 'chemical modification of wood' is even narrower, and only includes its active modification.

6.3.1.2 Methods of the chemical modification of wood

Modification of the lumens of wood cells is a method during which the big pores (lumens) in wood are filled with various types of synthetic or natural substances:

- amino resins
- phenolic resins
- furfuryl-alcohol resins
- silicon compositions
- epoxides
- acrylates
- styrene
- vegetable oils
- saccharides
- waxes.

In wood, molecules of the mentioned substances can penetrate from the lumen of one cell to the lumen of neighbouring cells through opened pits in cell walls of tracheids, through perforations in vessels, and so on. Some of them may also penetrate to the cells walls of wood, depending upon their dimensions and polarity (Reinprecht, 1993; Hill, 2006).

Modifying substances located in the lumens of wood cells affect the wood's properties via the following mechanisms:

- they create a film on the S_3 layers of cell walls in contact with the lumens, which mechanically prevents the entry of degrading agents (hyphae, enzymes, etc.) into the cell walls whilst also slowing sorption and desorption processes;
- they form a compact filling in the lumens that prevents the entry of degrading agents deeper into the wood, whilst thermoset fillings (amino resins, phenolic resins, epoxides) also significantly improve the strength of the wood.

Generally, substances present only in lumens of wood improve mainly its strength, often also increasing its resistance to biological and abiotic damage, while the kinetics of hygroscopicity and swelling slow, however, without any significant improvement in its long-term dimensional stability (Schneider et al., 1991).

Modification of the cell walls of wood is a method in which small molecules of polar modifying chemicals with a diameter below 80 nm can penetrate the cell walls of wood. However, only polar molecules smaller than 1 nm are able to penetrate inside the elementary fibrils among macromolecules of crystalline cellulose. The small polar molecules of modifying chemicals (or carrier solvents) should have a splitting effect during which the cell walls of wood markedly swell and subsequently they penetrate to the interior of cell walls and create hydrogen bonds with polysaccharides and lignin. The transport of molecules of the modifying substance into the cell walls of wood (capillary transport to the dry cell walls based on a splitting effect, and/or diffusion transport to the wet cell walls based on a concentration gradient) is slow and sometimes takes several days or even years.

In the active modification of wood, the ability of the modifying substance to create covalent bonds with polysaccharides and lignin depends upon its reactivity (i.e. presence of suitable functional groups), the addition of a catalyst and also upon the technological conditions, such as temperature and pressure.

6.3.1.3 Technologies of the chemical modification of wood

In the chemical modification of wood, the first step of technology is very similar to the technology used in the chemical protection of wood. The first step is deep modification carried out using vacuum–pressure or dipping techniques, and surface modification by painting or spraying (Section 5.3). In the second step of the modification technology, the chemical structure of the modifying agents has to be in more cases additionally changed directly in the modified wood or on its surfaces. So, in the second step some molecules of modifying agents chemically react at increased temperatures or in the presence of catalysts (1) only mutually (i.e. in passive modification processes in lumens or also in cell walls of wood) and (2) with —OH or other functional groups of wood (i.e. in active modification processes in cell walls of wood) (Section 6.3.2).

6.3.2 Substances intentionally or randomly reacting with wood components

Modifying substances introduced to the cells walls of wood either intentionally chemically react and create covalent bonds with its structural components (active wood modification) or do not chemically react, possibly only randomly react with them (passive wood modification).

6.3.2.1 Substances intentionally reacting with wood components

The functional —OH groups of wood significantly participate in its hygroscopicity and dimensional instability during changes in the climate. Thanks to the —OH groups, the equilibrium moisture content of wood in moist air increases, whilst its increased moisture content is a necessary assumption for the activity of wood-boring insects ($w_{min} \cong 10\%$; $w_{opt} \cong 20–30\%$) and wood-decaying fungi and moulds ($w_{min} \cong 20\%$; $w_{opt} \cong 30–80\%$). Several negative properties of wood can be eliminated by blocking —OH groups of wood, when with a changed molecular structure it becomes more difficult to be identified as a source of nutrition for wood-decaying fungi (Kumar, 1994; Hill, 2006).

Modifying substances able to chemically react with wood should meet the following requirements:

- good penetration ability into the microcapillaries of the cell walls of wood;
- sufficient reactivity with —OH groups of wood in a neutral or slightly alkaline environment;
- sufficient reactivity with —OH groups of wood at temperatures below 120 °C, if simultaneous thermal modification of the wood is not required;
- creation of stable covalent bonds with wood components, mainly by replacing its —OH groups for less polar functional groups.

The replacement of —OH groups of modified wood affects its sorption properties in two ways:

- the spatial (steric) effect of new functional groups in the micropores of the cells walls of wood (new functional groups are larger than the original —OH groups) decreases the size and volume of the pores, and so the molecules of bound water penetrate them with more difficulty and in a lesser amount;
- the hydrophobicity effect of new functional groups, which are unable to create hydrogen bonds with water molecules, also decreasing the wood's hygroscopicity.

The most important modifying substances able to intentionally react with wood, and usually replacing its polar —OH groups, are those which enter into the following reactions:

- esterification
- urethanization
- acetalization
- etherification.

6.3.2.1.1 Esterification of wood (e.g. acetylated wood)

The esterification of wood is carried out using linear or cyclic anhydrides of carboxylic acids or using carboxylic acids themselves, halides of carboxylic acids, β-propionolactone, and so on (Equations 6.7 and 6.8).

Mechanisms of the reactions of wood with anhydrides of carboxylic and dicarboxylic acids:

- during reaction with —OH group of wood, the linear chain of anhydride splits and releases carboxylic acid (e.g. acetic acid is created from the anhydride of acetic acid, if $R = CH_3$):

$$\text{Wood}-\text{OH} + (\text{R}-\text{CO})_2\text{O} \longrightarrow \text{Wood}-\text{O}-\text{CO}-\text{R} + \text{R}-\text{COOH}$$

(6.7)

- during reaction with —OH group of wood, the cyclic chain of anhydride opens and binds fully to wood (e.g. the reaction of the anhydride of succinic acid and wood):

$$\text{Wood}-OH + (CH_2CO)_2O \longrightarrow \text{Wood}-O-CO-CH_2-CH_2-CO-OH \tag{6.8}$$

Acetylated wood is produced by reaction of wood with a liquid anhydride of acetic acid; that is, acetic anhydride, $(CH_3{-}CO)_2O$. However, it can also be produced by reactions in the gas phase, either using vapours of acetic anhydride or ketene $CH_2{=}C{=}O$. Molecules of acetic anhydride with a diameter of 0.7 nm first penetrate the cells walls, including gaps in elementary fibrils with a diameter of 1 nm, where they then react with —OH groups of the wood with or without the presence of a catalyst (pyridine, dimethylformamide, etc.). The most reactive are —OH groups of lignin, followed by —OH groups of hemicelluloses; the least reactive are —OH groups of cellulose. The acetylation of wood is well performed at a temperature of 110–140 °C, during several tens of minutes. Acetic acid is a side product that, in high concentrations above 30%, acts as an inhibitor of the further acetylation of wood, causing odour, and also having corrosive effects (Rowell et al., 1990).

In the past, the industrial production of acetylated wood was carried out discontinuously in the USA and the former USSR, as well as in Japan under the commercial name of α-Wood. At the beginning of the 21st century the production of acetylated wood fibre took place in Sweden under the trade name of A-Cell. In 2003 in Arnhem, the Netherlands, the production of acetylated Accoya solid wood commenced in a 4-m-long reactor with a diameter of 0.8 m, volume of 2300 l and a capacity of 0.9 m^3 of wood/batch; for example, in 2007 with a production of 24,000 m^3. Of course, the costs for producing acetylated wood must be adequately balanced by its specific very good properties, whilst an important role is also played here by the replacement of the classic biocide protection of wood with a more ecologically friendly modification process.

Other types of esterified wood have only been prepared under more-or-less laboratory conditions to date. Searched were anhydrides of propionic acid $(CH_3{-}CH_2{-}CO)_2O$, butyric acid $(CH_3{-}(CH_2)_2{-}CO)_2O$, pentanoic acid $(CH_3{-}(CH_2)_3{-}CO)_2O$, hexanoic acid $(CH_3{-}(CH_2)_4{-}CO)_2O$ and heptanoic acid $(CH_3{-}(CH_2)_5{-}CO)_2O$, as well as cyclic anhydrides of succinic, phthalic or maleic acid (Suttie et al., 1997, 1998; Li et al., 2001). Their reaction with wood proportionally slows with an increase in their molecular weight.

6.3.2.1.2 Urethanization of wood

The urethanization of wood takes place during an addition reaction of its —OH groups with isocyanates (Rowell & Ellis, 1984):

$$\text{Wood}\vdash OH + R{-}N{=}C{=}O \longrightarrow \text{Wood}\vdash O{-}\overset{\overset{\large O}{\|}}{C}{-}NH{-}R \tag{6.9}$$

Dry wood reacts with aliphatic and aromatic isocyanates easily in the presence of pyridine or acidic catalysts. Isocyanates react with lignin approximately 10 times faster than with polysaccharides.

6.3.2.1.3 Acetalization of wood

The acetalization of wood is a cross-linking process where two –OH groups of two polysaccharide macromolecules connect via an aldehyde molecule, firstly creating a hemiacetal bond and then an acetal bond:

hemiacetal bond

$$\text{Wood}\vdash OH + H{-}\overset{\overset{\large O}{\|}}{C}{-}H \longrightarrow \text{Wood}\vdash O{-}\overset{\overset{\large OH}{|}}{CH_2}$$

acetal bond (6.10)

$$\text{Wood}\vdash O{-}\overset{\overset{\large OH}{|}}{CH_2} + HO{-}\dashv\text{Wood} \longrightarrow \text{Wood}\vdash O{-}CH_2{-}O{-}\dashv\text{Wood} + H_2O$$

Acetal bonds are less stable than ether bonds and they hydrolyse easily.

Formaldehyde, CH_2O, is the most reactive aldehyde. It reacts with wood in the presence of acidic catalysts (HCl, HNO_3, SO_2, etc.) or under the influence of γ-radiation.

Glyoxal, OHC—CHO, is a difunctional aldehyde. Its reaction with wood only takes place in the presence of catalysts ($ZnCl_2$, $MnCl_2$). The addition of glycol increases the reactivity of glyoxal as well as its degree of cross-linking in wood, which is reflected in a significant improvement in the dimensional stability of the wood.

6.3.2.1.4 Etherification of wood

The etherification of wood is understood to be the incorporation of alkyl or alkyl-aryl groups into the structural components of wood, creating stable ether bonds —C—O—C—. One of the known etherified types of wood is alkylated wood created by the reaction of its —OH groups with dimethyl sulphate (methylation), butyl chloride (butylation), and so on. Etherified wood is also prepared using epoxides (epoxidation of wood) or other substances.

Alkylated wood is prepared by wood's reaction with dimethyl sulphate in the presence of NaOH, with methyl iodide in the presence of Ag_2O, or with halogen derivatives of carbohydrates in pyridine. These woods can have a disturbed molecular structure and worsened mechanical properties. This is mainly caused by aggressive semi-products (H_2SO_4, HCl, etc.) created during alkylation reactions.

Epoxidized wood is created by the reaction of the wood's —OH groups with monomeric epoxides CH_2=(O)=CH—R; for example, ethylene oxide (R = H), propylene oxide (R = CH_3), butylene oxide (R = CH_2CH_3) or epichlorohydrine (R = CH_2Cl). During the opening of an epoxide cyclic chain, apart from creation an ether bond, a new —OH functional group is also created and it can react with another epoxide molecule.

Epoxides are also applied as low molecular weight, linear precondensates ending with an epoxide ring, CH_2=(O)=CH—R_1 (R_1 = precondensate). Their cross-linking curing process takes place at room temperature in the presence of polyamines (diethylamine, trimethylamine, etc.). They also can penetrate the cell walls of wood, but it depends upon the size of their molecule as well as upon the polarity of the solvent used, which should be able to swell the wood (e.g. acetone). When penetrating cell walls, epoxides can enter into reactions with —OH groups of lignin and polysaccharides (Equation 6.11), but only in the presence of acidic or alkaline catalysts (Kumar, 1994). Epoxide precondensates are also applied in reinforcing of damaged wood (Reinprecht & Varínska, 1999).

$$\text{Wood}\,|\!-\!OH + R-CH\overset{O}{\diagup\!\diagdown}CH_2 \longrightarrow \text{Wood}\,|\!-\!O-CH_2-\underset{}{\overset{OH}{\overset{|}{CH}}}-R \tag{6.11}$$

1,3-Dimethylol-4,5-dihydroxy-ethyl-urea (DMDHEU – Belmadur) wood is produced from the solid wood and DMDHEU molecules able to chemically react with two —OH groups of wood; that is, DMDHEU molecules cross-linking the wood's lignin–polysaccharide macromolecules:

$$\text{Wood}\,|\!-\!OH + \underset{\text{DMDHEU}}{HOH_2C\text{-}N(\text{ring: CH(OH)-CH(OH)}, C{=}O)N\text{-}CH_2OH} \longrightarrow \text{Wood}\,|\!-\!O\text{-}CH_2\text{-}N(\text{ring: CH(OH)-CH(OH)}, C{=}O)N\text{-}CH_2\text{-}O\!-\!|\,\text{Wood} \tag{6.12}$$

The DMDHEU is usually applied to wood in a 10–20% water solution, and it reacts with the wood's components in the presence of a suitable catalyst (e.g. $MgCl_2$, $AlCl_3$ or citric acid) at a temperature of 100–150 °C (Van de Zee et al., 1998). Although DMDHEU monomers have a certain fungicidal efficiency themselves, their application is manifested more as an active chemical modifying protection of wood, because after their reaction with —OH groups of wood there is a more apparent increase in resistance to fungal decay that is also due to its decreased hygroscopicity (Verma, P. et al., 2009).

Similar to the etherification processes is the aminification process; that is, when the $—NH_2$ or —NH groups of the modification agent enter into reactions with the HO—wood groups. Hauptmann et al. (2015) analysed chemical interactions between the amino acid tricine (*N*-[tri(hydroxymethyl)methyl]glycine) and the wood components of several broadleaved species. Using FTIR and nuclear magnetic resonance spectroscopic analyses they detected reactions of the amino group of this agent only with hemicelluloses of the wood, creating xylosylamine structures.

6.3.2.2 Substances not reacting or randomly reacting with wood components

Furfuryl-alcohol resins, amino resins, phenolic resins, silicones, polyethylene glycols, metal alkoxide gels, and acrylate or vinyl monomers and polymers are able to penetrate the lumens of wood cells, and some of them also penetrate to the cell walls. These modifying substances can often cross-link in wood via intramolecular reactions such as polycondensation, polyaddition or polymerization. In this way, spatial polymer networks of a modifying substance are created in the wood, located in the lumens of cells or also in the microcapillaries of cell walls. Some of these substances sporadically can create covalent bonds with the lignin–polysaccharide matrix of wood; for example, low molecular weight amino resins, phenolic resins or silicones in the form of intermolecular reactions, creating a 'modifying-substance–wood' network (Hill, 2006). After intramolecular cross-linking reactions, the modifying substance permanently spatially blocks the lumens of cells or also the micropores in cells walls, where its presence influences several properties of the wood. For example, the presence of amino and phenolic resins mainly increases the hardness and compression strength of wood (Reinprecht, 1993).

6.3.2.2.1 Furfuryl-alcohol resins, amino resins and phenolic resins

Furfuryl-alcohol resins were already being used in the 1970s for the industrial modification of wood. For example, in the USA the production of WISTI-wood was based on the initial impregnation of wood using an initiator (e.g. $ZnCl_2$), then furfuryl alcohol and, in final cross-linking of furfuryl alcohol in the creation the furfuryl-alcohol resins directly in the wood. Furfuryl alcohol is obtained from hydrolysed wastes in the production of sugar cane, corn cobs or molasses. In the 1990s, research cooperation between the Canadian company Woodtech Inc., the Swedish company Trätek and the Norwegian company NIFS led to the development of new Wood Polymer Technologies (WPT). Products from such modified wood are known under the name Kebony Products DA (Hill, 2006). Several types of wood are used for the production of Kebony 30 (WPG = 10–50%) and Kebony 100 (WPG = 70–100%), mainly well-impregnable beech.

Furfuryl-alcohol resins, in the form of furan polymers, are also partially located in cell walls in the modified wood, where they can also be randomly cross-linked with wood components. The curing of furfuryl-alcohol resins in wood is a complicated chemical process that is constantly being improved; for example, by also using microwave heating (Treu et al., 2007).

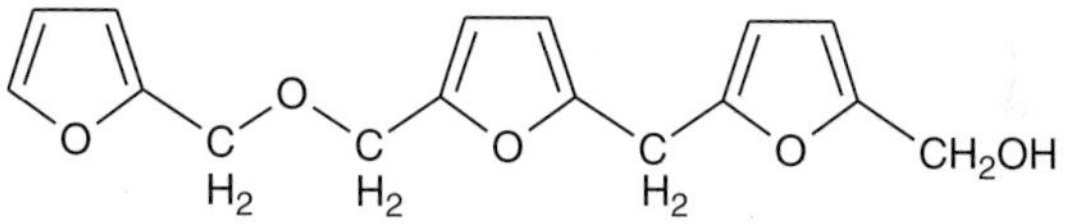

Furfuryl-alcohol resin

Amino resins of a urea-formaldehyde (UF) type were used for the production of URALLOY modified wood. Amino resins of melamine-formaldehyde (MF) and urea-melamine-formaldehyde (UMF) types are applied in water solutions, similar to UF resins. They cure in wood in a neutral or slightly acidic environment, usually at an increased temperature of 100–140 °C for a period of 2–6 h. Faults with MF resins include low stability of their solutions and higher prices (around three times more expensive than UF resins) and, therefore, they are also used in a mixture with UF resins.

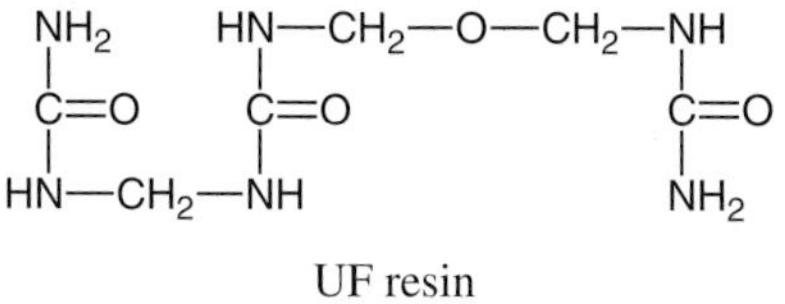

UF resin

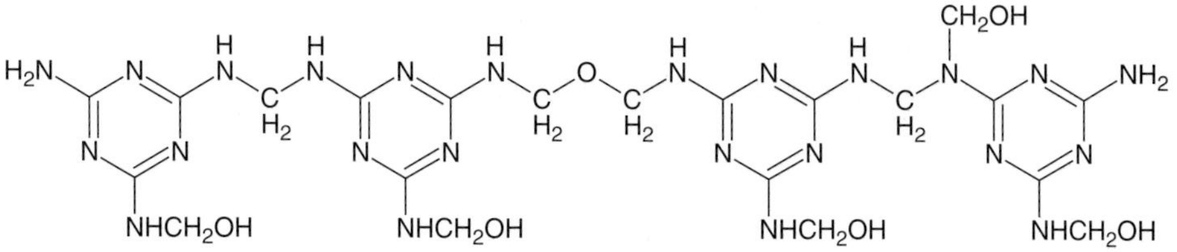

MF resin

PF resins are suitable for the production of densified wood such as IMPREG (non-compressed wood produced by gluing of impregnated veneers in a hot press at low pressure) and COMPREG (compressed wood produced by gluing of impregnated veneers in a hot press at high pressure). They are applied as water or ethanol solutions, and in wood they cure via polycondensation reactions at a high temperature of 135–165 °C, or at a normal temperature of around 20 °C using acidic catalysts. Lower molecular weight PF precondensates can also penetrate to the cell walls of wood (Furuno et al., 2004). Apart from PF resins, aqueous resorcinol-formaldehyde precondensates are also suitable for wood modification. Compared with phenol, resorcinol is 7.75 times more reactive, which facilitates its being cured at a normal temperature of around 20 °C by adding formaldehyde or paraformaldehyde.

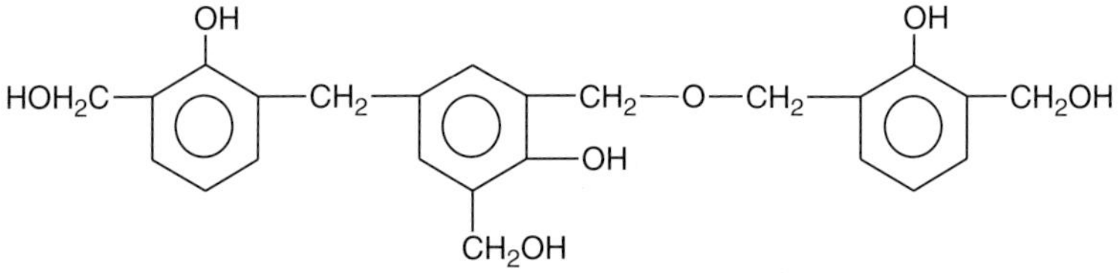

OH OH

HOH_2C—[ring]—CH_2—[ring]—CH_2—O—CH_2—[ring]—CH_2OH

—OH

CH_2OH

PF resin

6.3.2.2.2 Silicon compositions

Silicon compositions are substances derived from silicon dioxide SiO_2 or silicic acid H_4SiO_4. They are mainly applied in the hydrophobization of stone and concrete. In last years are used for wood modification – mineralization. In lumens of wood cells or also in cells walls create: (1) inorganic-silicon networks and SiO_2 gels (e.g. silicates, chlorosilanes, alkoxysilanes), or (2) organic-silicon networks (e.g. silicones, organosilanes).

Inorganic silicates, known as water glass (a mixture of Na_2SiO_3 and K_2SiO_3), are applied to wood in water solutions.

Chlorosilanes, such as tetrachlorosilane ($SiCl_4$), hydrolyse in wood, creating silicic acid and the aggressive semi-product hydrochloric acid (HCl). Silicic acid, at increased temperatures of around 100 °C and under specific conditions, is able to react with —OH groups of wood to create —Si—O—C—wood bonds.

Alkoxysilanes, such as tetramethoxysilane $(CH_3O)_4Si$, first hydrolyse in wood and subsequently, at a higher temperature and in the presence of acetic acid or another catalyst, create an SiO_2 sol–gel network with release of alcohols (ROH), but no aggressive semi-products (Mai & Militz, 2004):

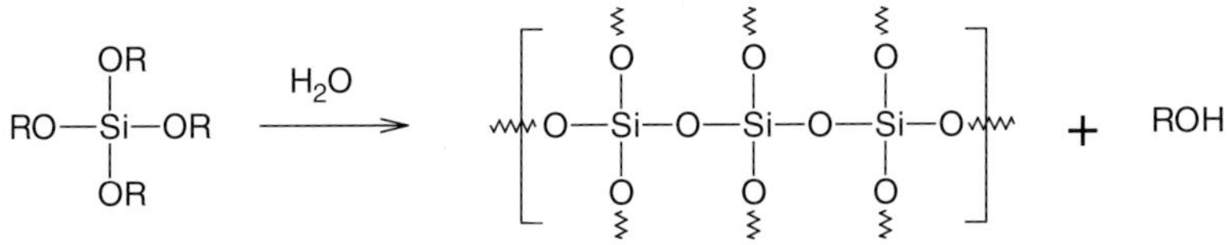

$$Si(OR)_4 \xrightarrow{H_2O} \left[\sim O{-}Si(O\sim)_2{-}O{-}Si(O\sim)_2{-}O{-}Si(O\sim)_2{-}O\sim \right] + ROH \quad (6.13)$$

Silicones are organosilicon polymers $RX_2Si{-}O{-}(SiR_2{-}O)_n{-}SiRX_2$ in which atoms of silicon bind to nonpolar carbohydrates (R) or, at the ends of the polymer chain, to reactive groups (X). Large macromolecules of silicones are applied in the form of aqueous micro-emulsions, and in wood are usually located only in lumens of its cells.

Organosilanes are organosilicon monomers with at least one Si—R′ bond, where the carbon atom from the carbohydrate (R′) binds directly to the silicon atom. The carbohydrate can be inert (R′ = methyl, *n*-propyl, etc.) or reactive (R′ = vinyl, 3-isocyanatopropyl, etc.), able to chemically react with other molecules of organosilanes or components of wood. The best known are the alkyl-trimethoxysilanes. During hydrolysis with water, they create silanols that are able to react with the —OH groups of wood:

$$RO{-}Si(R')(OR){-}OR \xrightarrow{3\,H_2O} HO{-}Si(R')(OH){-}OH + 3\,ROH \xrightarrow{-H_2O} \text{Wood}{-}O{-}Si(OH)_2{-}R' \tag{6.14}$$

However, covalent bonds between the silanols and the wood are unstable and hydrolyse easily in wet wood (Reinprecht & Grznárik, 2015). The greater stability of organosilanes in cell walls and the marked decrease in their leachability from wood can be achieved only if covalent bonds with wood create by help of reactive carbohydrate groups (R′) (e.g. vinyl or epoxide group).

6.3.2.2.3 Acrylates

Methyl acrylate (MA) and methyl methacrylate (MMA) are used for the production of Acrylic-Wood.

$CH_2{=}CH{-}COOR$	Ester of acrylic acid (MA, if $R = CH_3$)
$CH_2{=}C(CH_3){-}COOR$	Ester of methacrylic acid (MMA, if $R = CH_3$)

6.3.2.2.4 Vinyl and similar compositions

These compositions are applied to wood as nonreactive polymers or as monomers that polymerize in impregnated wood. Wood–plastic composites are produced by combining wood with polyethylene, polypropylene, polyvinyl chloride, polyvinyl acetate, polystyrene, etc.

6.3.3 Durability and other properties of chemically modified wood

6.3.3.1 Properties of acetylated and otherwise actively modified wood

Acetylated wood has excellent resistance to rot, mainly if its weight during acetylation increases minimally about 10% against white rot and about 20% against brown rot; that is, if the WPG of acetyl groups is from 10 to 20% (Hill et al., 2003). The protective mechanism against wood rotting lies in the replacement of —OH groups in wood with larger and less polar acetyl groups (Equation 6.7; Rowell et al., 2009; Rowell & Dickerson, 2014). In their presence the wood components (1) do not have, or have only limited, contact with polar large (~5–7 nm) enzymes of fungi and (2) have lower equilibrium moisture content – so even in a wet climate the wood does not have sufficient humidity for the activities of wood-decaying fungi (Alfredsen et al., 2013). However, acetylated wood resists moulds relatively less (Wakeling et al., 1992). The bacterial attack of wood by tunnelling bacteria is totally eliminated only at high acetyl content (WPG 20.9%). Acetylated wood resists termites well (Bongers et al., 2015a). For example, for acetylated pine sapwood with 22% retention of acetyls, Westin et al. (2004a) found only a 2% mass loss after a 1 year exposure to termites, whilst native wood had as much as 93% mass loss.

In terms of the relationship to water, acetylated wood has a decreased sorption capacity and increased dimensional stability (Moghaddam et al., 2016). Acetylated wood has a lower speed of sound spreading, a lower sound absorption and a lower dynamic Young's modulus of elasticity (Chang et al., 2000), which is made use of in the production of musical instruments (Yano et al., 1993). A decrease or increase in its mechanical properties is not always explicit since it strongly depends upon (1) the wood species (the influence of extractive substances upon acetylation), (2) the acetylation conditions (released acetic acid may depolymerize the hemicelluloses) and (3) the number of blocked —OH groups in the wood (wood with a higher proportion of acetyls has a lower equilibrium humidity, due to which there is also a strength increase) (Rowell & Banks, 1987; Larsson & Simonson, 1994; Jorissen et al., 2005; Bollmus et al., 2015).

Acetylated wood can be beneficially treated using coatings for exterior use; for example, the lifetime of acrylate or alkyd coatings on acetylated wood is longer than on solid wood (Bongers et al., 2005). Acetylated wooden laminates glued with reactive polyurethane adhesive have a good performance outdoors and also in various delamination tests, in contrast to the laminates bonded with PVAc or UMF adhesives (Bongers et al., 2015b). Acetylated solid wood and plywood from acetylated veneers have partially improved fire resistance (Mohebby et al., 2006).

Acetylated wood is used for facade cladding and terrace boards, but also for bridges and other structural applications. Accoya® wood is based on the acetylation of radiata pine (*Pinus radiata* D. Don). Currently, Accsys Technologies is working on the development of commercially viable acetylation processes

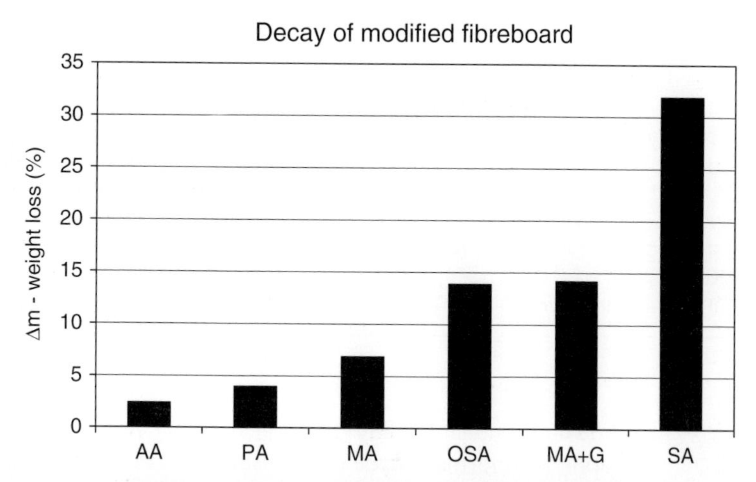

Figure 6.6 Weight losses Δm of esterified fibreboards due to 16 weeks' action of the brown-rot fungus *Gloeophyllum trabeum*. Esterification agents: AA, acetic anhydride; PA, phthalic anhydride; MA, maleic anhydride; OSA, 2-octen-1-yl-succinic acid anhydride; MA+G, maleic anhydride and glycerol; SA, succinic acid anhydride

Source: Rijckaert, V., Van Acker, J., Stevens, M. (1998) Decay resistance of high performance biocomposites based on chemically modified fibres. IRG/WP/98-40120, 11 p. Reproduced by permission of IRG/Wood Protection

for additional wood species; for example, broadleaves – alder, beech, lime and maple (Bollmus et al., 2015).

Wood esterified with other substances resists rot comparable to or not as well as acetylated wood, depending on the decay-causing fungus species and the characteristics of the wood or wood composite. An example is the more-or-less equal resistance of acetylated wood and woods esterified by other anhydride types to the brown-rot fungus *Coniophora puteana* (Hill et al., 2003). On the other hand, their efficacy against the brown-rot fungus *Gloeophyllum trabeum* was markedly worse; for example, wood fibres modified with an anhydride of cyclic succinic acid, $(CH_2CO)_2O$, succumbed to rot very easily (Figure 6.6).

Urethanized wood prepared on the basis of reactions of various isocyanates (e.g. methyl isothiocyanate, ethyl isocyanate, *n*-propyl isocyanate or *n*-butyl isocyanate) with —OH groups of wood is sufficiently resistant to attack with decay-causing fungi at WPG $\geq$15% (Williams and Hale, 1999). Urethane bonds are resistant to acidic and alkaline hydrolysis, and urethanized wood resists emissions as well.

Acetalized wood, created by cross-linking of polysaccharides with aldehydes (Equation 6.10), has decreased porosity of cell walls and, at the same time, suppresses the penetration of fungal enzymes in its cell walls. Even a relatively low retention of formaldehyde (WPG from 2 to 7%) provides its high resistance to rot (Kumar, 1994). Wood acetalized with formaldehyde molecules has a very

Table 6.3 Properties of chemically modified wood compared with untreated wood

Chemically modified wood	Property of the modified wood					
	WPG (%)	ASE (%)	$MPI_{strength-elasticity}$ (%)	$MPI_{hardness}$ (%)	DI_{decay} (%)	Durability EN 350-1
Acetylated – Accoya	15–20	75	85–115	120–130	600–900	1
Furfurylated – Kebony 100	70–100	80	40–180	200–250	500–900	1–2
DMDHEU – Belmadur	10–20	50	80–95	200	600–900	1

good dimensional stability (Fahey et al., 1987). A disadvantage of wood acetalization with formaldehyde is the disturbance of its molecular structure in the presence of aggressive acid catalysts and a decrease of its strength.

Epoxide wood resists biological pests well, and is hard, strong and dimensionally stable. Wood prepared by a reaction with butylene oxide becomes sufficiently resistant to the wood-decaying fungi *Gloeophyllum trabeum*, *Lentinus lepideus* and *Trametes versicolor* at WPG ≥17% (Hill, 2006). Wooden pegs impregnated with epoxy (at WPG 20%) and exposed in terrain well resist soil microorganisms for as long as 7 years in the northern USA, while slight rot was found in them in the southern USA (Rowell, 1984).

DMDHEU (Belmadur) wood is sufficiently dimensionally stable. For example, Belmadur beech wood has half the amount of volume swelling (i.e. from 18 to 8–9%), and it also has excellent resistance to rot (Van Acker et al., 1999; Schaffert et al. 2005). From non-durable beech wood (class 5 of durability, by criteria of EN 350-1) is created a wood that is highly resistant against fungal decay (class 1 of durability). Although Belmadur wood has slightly reduced strength, its hardness increases twofold in comparison with the native wood (Table 6.3).

6.3.3.2 Properties of furfurylated wood, silane wood and otherwise – usually only passively modified wood

Furfurylated wood contains cured furfuryl-alcohol resin in its lumens and cell walls. It resists biological attacks well; for example, Kebony 100 has high resistance to rot (class 1 of durability) and Kebony 30 with a lower amount of furfuryl-alcohol resin is also a suitable material for exteriors (class 2 of durability). It is generally true that the resistance of furfurylated wood to fungal decay, termites and marine organisms increases with a higher proportion of furfuryl-alcohol resin in the wood (Westin et al., 2004b; Slevin et al., 2015).

Furfurylated wood is very well resistant to dimensional changes. For example, the swelling of Kebony 30 is 10% and of Kebony 100 just 2%. Furfurylated wood is hard but also fragile. Thanks to its higher density, metal fasteners hold well in products made of it (Jermer and Clang, 2007). It is also very friendly in terms of ecotoxicology since it releases less volatile organic compounds (VOCs) into the atmosphere than native wood, even when it is burned (Lande et al. 2004). The aesthetics side of wood modified with furfuryl-alcohol resins is very interesting in terms of certain aims; for example, when replacing dark exotic wood species. Kebony 30 is a gold–brown colour and Kebony 100 is dark brown. Kebony wood is used for decking, garden furniture, in the boat building industry, and so on.

Wood modified with amino resins does not have apparently better biological resistance than native wood (Sailer et al., 1998). When using UF resins, the decay resistance of wood improves only with WPG $\geq$10% (Van Acker et al., 1999), or even only with WPG $\geq$30% when it was in an 8-year-long contact with soil (Westin et al., 2004a). Wood treated with methylolated melamine-formaldehyde resin, having a higher degree of cross-linking, has greater biological resistance. It is able to resist soft rot as well as decay caused by the brown-rot fungus *Coniophora puteana* fairly well (Lukowsky et al., 1998). Amino resins, to a certain extent, reduce the production of enzymes of decay-causing fungi. For example, MF resins decrease in wood the production of the *endo*-β-1,4-xylanase by the brown-rot fungus *Poria placenta* needed for damage of hemicelluloses, but they do not decrease the production of the enzyme *endo*-β-1,4-glucanase needed for depolymerization of cellulose (Ritschkoff et al., 1999). An important role in reducing the enzymatic activity of fungi is also played by formaldehyde released from amino resins, but it is disputable to what extent it is thanks to its toxic effect upon the fungi organism and to what extent it is thanks to its reaction with the fungus enzymes (Lukowsky et al., 1999).

Wood modified with amino resins retains its original colour, has decreased hygroscopicity, liquid absorbability and swelling, increased compression strength and hardness but, on the other hand, it has decreased tensile strength and it is also more fragile (Horský & Reinprecht, 1983). Amino resins in a rotten wood improve mainly its compression strength and hardness (Figure 6.7). The ability of amino resins to penetrate the cell walls of wood depends upon the method of their catalysis; that is, a highly catalysed resin is only located in the cell lumens whilst a non-catalysed resin is also able to penetrate the cell walls (Reinprecht & Makovíny, 1987; Gindl et al., 2002; see Figure 7.10).

Wood modified with phenolic resins has better resistance to rot and other biological attacks than native wood does. It has also good resistance to mechanical wear and tear and atmospheric and chemical corrosion; it also has increased strength and dimensional stability. According to Ryu et al. (1991), it is possible to make wooden products highly resistant to rot from slightly durable and non-durable wood species by modifying them with PF resins. However, when using PF resins with a molecular weight of 170, the WPG values must be at least 10% for the conifers *Tsuga heterophylla* and *Cryptomeria japonica*, and ~20% for *Fagus crenata*. An improvement in the properties of wood modified with PF resins, including its bio-resistance, is more pronounced if molecules of the

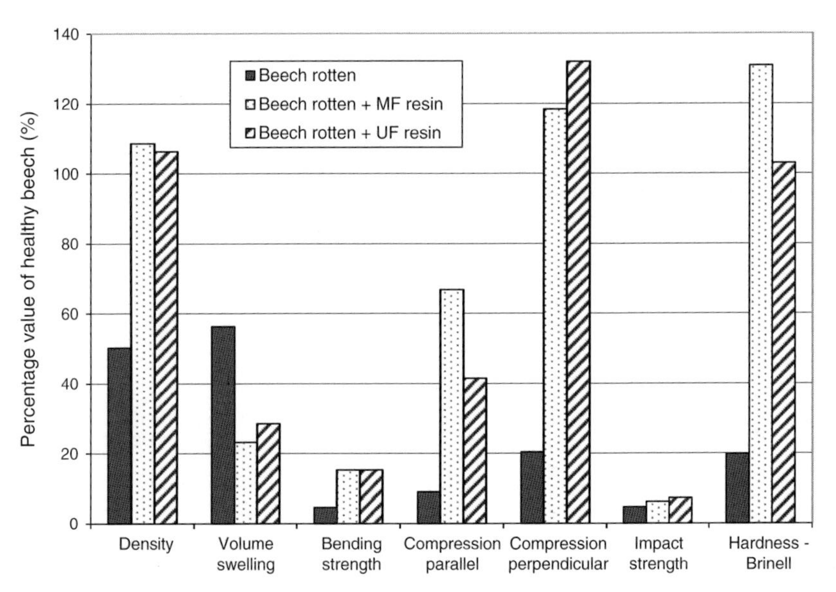

Figure 6.7 Effect of UF and MF resins at improving selected properties of rotten beech wood *Fagus sylvatica*

Source: Horský, D. and Reinprecht, L. (1986) Strengthening and dimensional stabilization of woods degraded by decaying fungi. (In Slovak, English abstract.) In *Scientific Works of DF VŠLD Zvolen 1985/1986*, pp. 63–72, Alfa Bratislava, Slovakia. Reproduced by permission of TU Zvolen

phenolic resin also penetrate the cell walls (Bicke and Militz, 2014). Ryu et al. (1993) studied the effect of seven PF resins with a molecular weight from 369 to 1143 and discovered that resistance to decay increased more significantly in wood modified using resins with a lower molecular weight, which were also able to penetrate the cell walls, where they blocked —OH groups of wood.

Silane wood sufficiently resists biological attacks if a special silicone compound – for example, with a proportion of amino (—NH_2) groups – is used in its preparation (Mai et al., 2005; Reinprecht & Grznárik, 2015). Common types of silicone-based inorganic and organic substances are usually unable to significantly improve the biological resistance of wood (Terziev & Temiz, 2005). Their combination with biocides (borates, quaternary ammonium compounds, carbamates, etc.) shows some prospects for the future with the aim that they will be fixed with silicone compound and be more resistant to leaching from wood (Mazela & Perdoch, 2012; Palanti et al., 2012; Reinprecht & Grznárik, 2014).

Wood modified with silicates (water glass) has increased resistance to water and reduced flammability. On the other hand, its biological resistance usually does not fundamentally change. Wood modified with chlorosilanes has better resistance to decay (Owens et al., 1980), but the danger is the HCl released in the wood during the modification process – its reduction can be provided by using trimethylchlorosilane $(CH_3)_3SiCl$ in pyridine, as documented Zollfrank (2001). Wood modified by alkoxysilanes is less flammable and better resistant to termites than native wood is, but its resistance to decay increases more apparently only if specific types of organosilanes are applied; for example, with $-NH_2$ groups (Tanno et al., 1998; Mai et al., 2005; Ghosh et al., 2009; Reinprecht & Grznárik, 2015). Wood modified with silicones has higher hydrophobicity thanks to methyl, ethyl and other nonpolar organic substitutes bound to silicone atoms (Hager, 1995).

6.3.4 Applications of chemically modified wood

Acetylated, furfurylated and DMDHEU woods are currently produced industrially. The properties of chemically modified woods depend upon the proportion of modifying substance in the wood, its distribution in the wood and also upon the type of its reaction with wood components. Commercially produced chemically modified woods have better dimensional stability and markedly increased resistance against decay-causing fungi and boring insects in comparison with untreated wood; however, their mechanical properties change specifically, when only hardness significantly increases (Table 6.3).

Acetylated wood (Accoya, Tricoya, Titan) is suitable for various interior functional and decorative products, but owing to its high durability is primarily recommended for using in exteriors; for example, for entrance doors, windows, facades, shingles, sound barriers, garden furniture, children's playgrounds, fencing and boat decks. Furfurylated woods (Kebony 30, Kebony 100) and woods modified with DMDHEU (Belmadur) have similar uses to acetylated wood. A disadvantage of furfurylated wood is its darker colour and significantly increased density. These chemically modified woods can also be used as an alternative to tropical wood species or wood impregnated with classical biocides. Unlike thermally modified wood, they are suitable not only for exteriors without contact the terrain (class 3 of use in accordance with EN 335), but also for very severe exposures; that is, in contact with the terrain or in the ocean (classes 4 and 5 of use).

6.4 Biologically modified wood

6.4.1 Microorganisms suppressing the activity of wood-damaging fungi and insects

Bacteria, fungi, insects and other organisms can populate and damage sterile and non-sterile wood, respectively. Fight of organisms for food and territory is normal in all ecosystems, and there are created various antagonistic and synergic relationships between them. Antagonistic relationships between organisms can be purposely used for suppressing the activity of biological pests which damage wood.

6.4.1.1 Strategy of the battle for food between organisms in wood

Every wood pest has its own strategy for success; that is, how to best populate the wood substrate and obtain a sufficient proportion of nutrients and energy to live. A good example is fungal organisms, which have three main strategies in the fight for food (Jennings & Lysek, 1999; Reinprecht & Tiralová, 2001; Schmidt, 2006):

- The production of fungicidal or other toxic substances that prevent the growth of other fungi and organisms in the wood. An example is the fungus *Oudemansiella mucida*, which produces mucidin in beech wood, by which it prevent other species of fungi from populating the wood; similarly, several microscopic fungi (moulds) (e.g. *Trichoderma viride* and *Trichoderma harzianum*) prevent the growth of wood-decaying fungi by producing viridin and other toxic metabolites (Howell, 2003; Benítez et al., 2004; Schubert et al., 2008; Gaylarde et al., 2015).
- The ability to also draw nutrients from less interesting components of wood, which they quickly populate and therefore prevent the wood from being populated by other more dangerous fungi and organisms in the future. An example is wood carbonized by fire, which is a specific source of nutrition for the fungus *Pholiota carbonica*; similarly, tannins in the heartwood of oak act as an inhibitor against several fungi and other organisms, but they in oak are particularly necessary for the activity of the decay-causing fungus *Fistulina hepatica*.
- The ability to live in extreme conditions. An example is the relatively low moisture content of wood (~18%) in which almost only the true dry-rot fungus (*Serpula lacrymans*) is able to attack wood; similarly, but the opposite case, is an extremely high moisture of wood convenient only for the activity of soft-rot fungi and anaerobic bacteria.

6.4.1.2 Biological modification of wood based on the antagonistic relationships between organisms

The biological modification of wood on the principle of antagonism is based on the ability of some bacteria, moulds and other organisms to produce biocidal substances that kill, suppress or stop the growth or enzymatic activity of organisms more dangerous to wood (e.g. wood-decaying fungi). It is basically an analogy of the already mentioned first form of strategy of the fight for food between fungal organisms (Section 6.4.1.1), mostly by giving preference to the activity of certain types of bacteria or mould over the activity of more dangerous wood-decaying fungi. During biological modification, the wood is infected with a suitable biological organism that is able to produce substances with a fungicidal, insecticidal or other biocidal effect. It is important to use such organisms that do not worsen or only slightly worsen the original quality of the wood, such as its colour and strength.

Antagonistic relationships between organisms attacking wood have been known for ~80 years. They have been used in practice for several decades (Rayner & d Boddy, 1988; Phillips-Laing et al., 2003). Several bacteria are able to release VOCs or stable toxins, or they block the activity of wood-decaying fungi in another way (Benko, 1989; Benko & Highley, 1990; Bruce et al., 2004; Singh & Chittenden, 2008).

Several antagonistic microscopic fungi known as moulds – for example, the *Trichoderma* (*T. viride*), *Penicillium* and *Chaetomium* (*Ch. globosum*) species – are able to suppress the growth and activity of wood-decaying fungi (Bruce, 1988; Verma et al., 2007; Schubert et al., 2008). More successful experiments have been carried out in the past decade or so when storing fresh timber and logs by applying a colourless mutant of the sap-staining fungus *Ophiostoma piliferum* against other sap-staining fungi, thereby preventing the undesirable colouring of stored wood (White-McDougall et al., 1998). The antagonistic fungus *Gliocladium roseum* has proven effective against moulds and fungal decay (Yang et al., 2004). This fungus and also another antagonistic fungus, *Phaeotheca dimorphospora*, were applied by Yang et al. (2007) in the production of oriented strand board from less-durable wood species in the form of infecting stored trunks or chips. The oriented strand board produced had greater resistance to moulds, but not against white-rot and brown-rot fungi. Antagonistic relationships also exist between various wood-decaying fungi (Reinprecht and Tiralová, 2001).

Lenz (2004) and other researchers illustrated an option for killing or suppressing termite activity using pathogenic viruses, bacteria, fungi and other organisms. For example, the spores of *Metarhizium anisopliae* have a repellent effect upon termites (Wang & Powell, 2004; Verma, M. et al., 2009).

Technologies for the biological modification (protection) of wood must be compatible with the application, economic, ecology and healthy questions; that is:

- What is their specificity (e.g. just against a particular species of wood-decaying fungus or against all biological pests of wood), speed and period of effectiveness?
- What is the negative impact of potentially pathogenic organisms upon the environment and human health?

6.4.2 Gene engineering for increasing durability of wood and decreasing the activity of fungal enzymes

Tree genetic engineering is becoming a routine method in forestry. It has the potential to boost global wood production in many ways – including speeding tree growth (e.g. by changes in photosynthesis), tolerance of the growing tree to abiotic stresses such as drought, but also in improving wood quantity, quality and its resistance to burning, selected pests and so on (Sutton, 1999; Sedjo, 2001; Kumar et al., 2015). The potential of production of trees with novel traits is one of the most distinct benefits of genetic transformation. Quality of harvest timber can be improved by increased resistance of living trees to drought, insects or activity of pathogenic fungi (Jia et al., 2010). For example, elm and chestnut were successfully transformed with antifungal genes to impart resistance to the chestnut blight fungus (*Cryphonectria parasitica*) and the fungus *Ophiostoma ulmi*.

Permanently increased durability of wooden products from the genetically modified trees can probably be achieved by modern biotechnology methods through modification of transportable genes or elements by virus vectors and through mutation in trees under extensive or specific cultivations of selected tree species; for example, poplars, pines, eucalyptus and spruces (Harfouche et al., 2011).

Growing of trees with a lower lignin content is important for the pulping and paper industry. However, lignin reduction can be interesting also for service life prolongation of wooden products where surfaces are exposed to sun for a long time. The amount of lignin in growing trees can be reduced by inhibition or regulation of its biosynthesis pathways with convenient genes (Hu et al., 1999). For example, the lignin ratio in transgenic Chinese white poplars was reduced by ~42% in the presence of the antisense *4CL* gene in their genome (Jai et al., 2004). Field trials with genetically modified transgenic trees are mainly in the USA and Europe. For example, in Europe in 2010 there were 25 trials for poplar, three for eucalyptus, and five in total for pine, spruce and birch (Harfouche et al., 2011).

References

Alfredsen, G., Flæte, P. O. & Militz, H. (2013) Decay resistance of acetic anhydride modified wood: a review. *International Wood Products Journal* 4(3), 137–143.

Anonymous (2002) *The Plato Technology – A Novel Wood Upgrading Technology*. Plato International BV, Arnhem, Netherlands.

Armines, J. (1986) Procédé de fabrication dun matériau ligno-cellulosique par traitement thermique et matéria obtenu par ce procédé, French Patent, No 86 14 138.

Awoyemi, L. (2006) Heat treatment of less-valuable Nigerian-grown *Ceiba pentadra* wood for improved properties. In *IRG/WP 06*, Tromsø, Norway. IRG/WP 06-40332.

Ayadi, N., Lejeune, F., Charrier, F., et al. (2003) Color stability of heat-treated wood during artificial weathering. *Holz als Roh- und Werkstoff* 61, 221–226.

Bengtsson, C., Jermer, J. & Brem, F. (2002) Bending strength of heat-treated spruce and pine timber. In *IRG/WP 02*, Cardiff, Wales, UK. IRG/WP 02-40242.

Bengtsson, C., Jermer, J., Clang, A. & Ek-Olausson, B. (2003) Investigation of some technical properties of heat-treated wood. In *IRG/WP 03*, Brisbane, Australia. IRG/WP 03-40266.

Benítez, T, Rincón, A. M., Linón, M. C. & Codón, A.C. (2004) Biocontrol mechanisms of *Trichoderma* strains. *International Microbiology* 7, 249–260.

Benko, R. (1989) Biological control of blue stain on wood with *Pseudomonas cepacia* 6253 – laboratory and field test. In *IRG/WP 89*, Lappeenranta, Finland. IRG/WP 89-1380.

Benko, R. & Highley, T. L. (1990) Selection of media for screening interaction of wood-attacking fungi and antagonistic bacteria. Part II. – interaction on wood. *Material Organisms* 25, 174–180.

Bicke, S. & Militz, H. (2014) Modification of beech veneers with low molecular weight phenol-formaldehyde for the production of plywood: comparison of the submersion and vacuum impregnation. In *The Seventh European Conference on Wood Modification*, Lisbon, Portugal, pp. 117.

Bollmus, S., Bongers, F., Gellerich, A., et al. (2015) Acetylation of German hardwoods. In *The Eighth European Conference on Wood Modification*, Helsinki, Finland, pp. 164–173.

Bongers, F., Creemers, J., Kattenbroek, B. & Homan, W. (2005) Performance coatings on acetylated Scots pine after more than nine years outdoor exposure. In *Wood Modification: Processes, Properties and Commercialisation, 2nd European Conference on Wood Modification*, Göttingen, Germany, pp. 125–129.

Bongers, F., Kutnik, M., Paulmier, I., et al. (2015a) Termite and insect resistance of acetylated wood. In *IRG/WP 15*, Vina del Mar, Chile. IRG/WP 15-40703.

Bongers, F., Meijerink, T., Lütkemeier, B., et al. (2015b) Bonding of acetylated wood. In *The Eighth European Conference on Wood Modification*, Helsinki, Finland, pp. 207–215.

Boonstra, M. J., Van Acker, J., Kegel, E. & Stevens, M. (2007) Optimisation of two-stage heat-treatment process – durability aspects. *Wood Science and Technology* 41, 31–57.

Brischke, C. & Rapp, A. O. (2004) Investigation of the suitability of Silver fir (*Abies alba* Mill.) for thermal modification. In *IRG/WP 04*, Ljubljana, Slovenia. IRG/WP 04-40275.

Bruce, A. (1998) Biological control of wood decay. In *Forest Products Biotechnology*, Bruce, A. & Palfreyman, J. W. (eds). Taylor & Francis, London, pp. 250–266.

Bruce, A., Verrall, S., Hackett, C. A. & Wheatley, R. E. (2004) Identification of volatile organic compounds (VOCs) from bacteria and yeast causing inhibition of sapstain. *Holzforschung* 58, 193–198.

Busnel, F., Blanchard, V., Prégent, J., et al. (2010) Modification of sugar maple (*Acer saccharum*) and black spruce (*Picea mariana*) wood surfaces in a dielectric barrier discharge (DBD) at atmospheric pressure. *Journal of Adhesion Science and Technology* 24(8–10), 1401–1413.

Chang, S. T., Chang, H. T., Huang, Y. S. & Hsu, F. L. (2000) Effects of chemical modification reagents on acoustical properties of wood. *Holzforschung* 54(6), 669–675.

Doi, S., Hanata, K., Kamonji, E. & Miyazaki, Y. (2004) Decay and termite durabilities of heat treated wood. In *IRG/WP 04*, Ljubljana, Slovenia. IRG/WP 04-40272.

Esteves, B. & Pereira, H. (2009) Wood modification by heat treatment: review. *BioResources* 4(1), 370–404.

Esteves, B., Domingos, I. & Pereira, H. (2005) Technological improvement of Portuguese eucalypt and pine woods by heat treatment. In *Wood Modification: Processes, Properties and Commercialisation, 2nd European Conference on Wood Modification*, Göttingen, German, pp. 91–94.

Fahey, D. S., Rowell, R. M. & Wegner, T. H. (1987) Modified woods and paper-base laminates. In *Wood Handbook – Wood as an Engineering Material*. Agricultural Handbook 72. USDA, Washington, DC, pp. 23-1–23-14.

Feher, S., Koman, S., Borcsok, Z. & Taschner, R. (2014) Modification of hardwood veneers by heat treatment for enhanced colours. *BioResources* 9(2), 3456–3465.

Furuno, T., Imamura, Y. & Kajita, H. (2004) The modification of wood by treatment with low molecular weight phenol-formaldehyde resin: a properties enhancement with neutralized phenolic-resin and resin penetration into wood cell walls. *Wood Science and Technology* 37(5), 349–361.

Gaylarde, Ch., Otlewska, A., Celikkol-Aydin, S., et al. (2015) Interactions between fungi of standard paint test method BS3900. *International Biodeterioration & Biodegradation* 104, 411–418.

Gérardin, P. (2015) New alternatives for wood preservation based on thermal and chemical modification of wood – a review. *Annals of Forest Science*, in press. doi: 10.1007/s13595-015-0531-4.

Ghosh, S. C., Militz, H. & Mai, C. (2009) The efficacy of commercial silicones against blue stain and mould fungi in wood. *European Journal of Wood and Wood Products* 67(2), 159–167.

Gindl, W., Dessipri, E. & Wimmer, R. (2002) Using UV-microscopy to study diffusion of melamine-urea-formaldehyde resin in cell walls of spruce wood. *Holzforschung* 56, 103–107.

Gonzalez-Pena, M. M., Curling, S. F. & Hale, M. D. C. (2009) On the effect of heat on the chemical composition and dimensions of thermally modified wood. *Polymer Degradation and Stability* 94, 2184–2193.

Hager, R. (1995) Waterborne silicones as wood preservatives. In *IRG/WP 95*, Helsingør, Denmark. IRG/WP 95-30062.

Harfouche, A., Meilan, A. & Altman, A. (2011) Tree genetic engineering and applications to sustainable forestry and biomass production. *Trends in Biotechnology* 29(1), 9–17.

Hauptmann, M., Gindl-Altmutter, W., Hansmann, Ch., et al. (2015) Wood modification with tricine. *Holzforschung* 69(8): 985–991.

Hill, C. A. S. (2006) *Wood Modification – Chemical, Thermal and other Processes.* John Wiley & Sons Ltd, Chichester.

Hill, C. A. S., Hale, M. D., Farahani, M. R., et al. (2003) Decay of anhydride modified wood. In *First European Conference on Wood Modification*, Ghent, Belgium, pp. 143–152.

Horský, D. & Reinprecht, L. (1983) Modification of health and damaged spruce wood with poly-condensate resins. *Drevo* 38(8), 224–226 (in Slovak, English abstract).

Horský, D. & Reinprecht, L. (1986) Strengthening and dimensional stabilization of woods degraded by decaying fungi. In *Scientific Works of DF VŠLD Zvolen 1985/1986.* Alfa Bratislava, Slovakia, pp. 63–72 (in Slovak, English abstract).

Howell, C. R. (2003) Mechanisms employed by *Trichoderma* species in the biological control of plant diseases: the history and evolution of current concepts. *Plant Disease* 87, 4–10.

Hu, W. J., Harding, S. A., Lung, J., et al. (1999) Repression of lignin biosynthesis promotes cellulose accumulation and growth in transgenic trees. *National Biotechnology* 17, 808–812.

Inoue, M., Sekino, N., Morooka, T., et al. (2008) Fixation of compressive deformation in wood by pre-steaming. *Journal of Tropical Forest Science* 20(4), 273–281.

Jämsä, S. & Viitaniemi, P. (2001) Heat treatment of wood – better durability without chemical. In *Review on Heat Treatments of Wood*, Rapp, A. O. (ed.). European Commission, Brussels, pp. 19–24.

Jennings, D. H. & Lysek, G. (1999) *Fungal Biology.* Bios, Oxford, UK.

Jermer, J. & Andersson, B.-L. (2005) Corrosion of fasteners in heat-treated wood – progress report after two years' exposure. In *IRG/WP 05*, Bangalore, India. IRG/WP 05-40296.

Jermer, J. & Clang, A. (2007) Furfurylated wood – withdrawal load for fasteners. In *IRG/WP 07*, Jackson, MS, USA. IRG/WP 07-40381.

Jai, C., Zhao, H., Wang, H., et al. (2004) Obtaining the transgenic poplars with low lignin content through down-regulation of *4CL*. *Chinese Science Bulletin* 49(9), 905–909.

Jia, Z., Sun, Y., Yuan, L., et al. (2010) The chitinase gene (*Bbchit1*) from *Beauveria bassiana* enhances resistance to *Cytospora chrysosperma* in *Populus tomentosa* Carr. *Biotechnology Letters* 32(9), 1325–1332.

Jones, D., Suttie, E., Ala-Vikari, J., et al. (2006) The commercialisation of ThermoWood® products. In *IRG/WP 06*, Tromsø, Norway. IRG/WP 06-40339.

Jorissen, A., Bongres, F., Kattenbroek, B. & Homan, W. (2005) The influence of acetylation of radiata pine in structural sizes on its strength properties. In *Wood Modification: Processes, Properties and Commercialisation, 2nd European Conference on Wood Modification*, Göttingen, Germany, pp. 108–115.

Kačík, F., Veľková, V., Šmíra, P., et al. (2012) Release of terpenes from fir wood during its long-term use and in thermal treatment. *Molecules* 17, 9990–9999.

Kamdem, D. P., Pizzi, A., Guyonnet, R. & Jermannaud, A. (1999) Durability of heat-treated wood. In *IRG/WP 99*, Rosenheim, Germany. IRG/WP 99-40145.

Kartal, S. N. (2007) Mold resistance of heat-treated wood. In *IRG/WP 07*, Jackson, MS, USA. IRG/WP 07-40358.

Kubovský, I. & Kačík, F. (2013) Changes of the wood surface colour induced by CO_2 laser and its durability after the xenon lamp exposure. *Wood Research* 58(4), 581–589.

Kubovský, I., Kačík, F. & Reinprecht, L. (2016) The impact of UV radiation on the change of colour and composition of the surface of lime wood treated with a CO_2 laser. *Journal of Photochemistry and Photobiology A: Chemistry* 322, 60–66.

Kumar, S. (1994) Chemical modification of wood. *Wood and Fibre Science* 26(2), 270–280.

Kumar, V., Rout, S., Tak, M. K. & Deepak, K. R. (2015) Application of biotechnology in forestry: current status and future perspective. *Nature Environment and Pollution Technology* 14(3), 645–653.

Laine, K., Rautkari, L., Hughes, M. & Kutnar, A. (2013) Reducing the set-recovery of surface densified solid Scots pine wood by hydrothermal post-treatment. *European Journal of Wood and Wood Products* 71, 17–23.

Lande, S., Westin, M. & Schneider, M. H. (2004) Eco-efficient wood protection. Furfurylated wood as alternative to traditional wood preservation. *Management of Environmental Quality: An International Journal* 15(5), 529–540.

Larsson, P. & Simonson, R. (1994) A study of strength, hardness and deformation of acetylated Scandinavian softwoods. *Holz als Roh- und Werkstoff* 52(2), 83–86.

Lenz, M. (2004) Biological control in termite management – the potential of nematodes and fungal pathogens. In *IRG/WP 04*, Ljubljana, Slovenia. IRG/WP 04-10521.

Li, J. Z., Furuno, T. & Katoh, S. (2001) Wood propionylation in the presence of catalysts. *Wood and Fiber Science* 33(2), 255–263.

Lovaglio, T., Rita, A., Todaro, L., et al. (2015) Determination of extractives composition in thermo-treated wood. In *The Eighth European Conference on Wood Modification*, Helsinki, Finland, pp. 75–78.

Lukowsky, D., Peek, R. D. & Rapp, A. O. (1998) Curing conditions for a low formaldehyde etherificated melamine resin. In *IRG/WP 98*, Maastricht, Netherlands. IRG/WP 98-40108.

Lukowsky, D., Buschelberger, F. & Schmidt, O. (1999) In situ testing of the influence of melamine resins on the enzymatic activity of basidiomycetes. In *IRG/WP 99*, Rosenheim, Germany. IRG/WP 99-30194.

Mai, C. & Militz, H. (2004) Modification of wood with silicon compounds – inorganic silicon compounds and sol–gel systems – a review. *Wood Science and Technology* 37(5), 339–348.

Mai, C., Donath, S., Weigenand, O. & Militz, H. (2005) Aspects of wood modification with silicon compounds: material properties and process development. In *Proceedings of the Second European Conference on Wood Modification*, Gottingen, Germany, pp. 222–231.

Mayes, D. & Oksanen, O. (2002) *ThermoWood® Handbook*. Stora Enso Timber, Finnforest, Finland.

Mazela, B., & Perdoch, W. (2012) Stabilization of IPBC in wood through the use of organosilicon compounds. In *IRG/WP 12*, Kuala Lumpur, Malaysia. IRG/WP 12-30597.

Mburu, F., Dumarcay, S., Huber, F., et al. (2006) Improvement of *Grevillea robusta* durability using heat treatment. In *IRG/WP 06*, Tromsø, Norway. IRG/WP 06-40333.

Metsä-Kortelainen, S., Antikainen, T. & Viitaniemi, P. (2006) The water absorption of sapwood and heartwood of Scots pine and Norway spruce heat-treated at 170 °C, 190 °C, 210 °C and 230 °C. *Holz als Roh- und Werkstoff* 64(3), 192–197.

Militz, H. (2002) Thermal treatment of wood – European processes and their background. In *IRG/WP 02*, Cardiff, Wales, UK. IRG/WP 02-40241.

Militz, H. (2015) Wood modification in Europe in the year 2015: a success story? In *The Eighth European Conference on Wood Modification*, Helsinki, Finland, pp. 19.

Militz, H. & Tjeerdsma, B. (2001) Heat treatment of wood by the 'Plato-process'. In *Review on Heat Treatments of Wood*, Rapp, A. O. (ed.). European Commission, Brussels, pp. 25–35.

Moghaddam, M. S., Wålinder, M. E. P., Claesson, P. M. & Swerin, A. (2016) Wettability and swelling of acetylated and furfurylated wood analyzed by multicycle Wilhelmy plate method. *Holzforschung* 70(1), 69–77.

Mohebby, B., Talaii, A., Karimi, A. & Najafi, A. K. (2006) Influence of acetylation on fire resistance of beech plywood. In *IRG/WP 06*, Tromsø, Norway. IRG/WP 06-40326.

Norimoto, M. & Gril, J. (1993) Structure and properties of chemically treated woods. In *Recent Research on Wood and Wood-based Materials*, Shiraishi, N., Kajita, H. & Norimoto, M. (eds). Elsevier, Barking, pp. 135–154.

Odrášková, M., Ráhel̆, J., Zahoranová, A., et al. (2008) Plasma activation of wood surface by diffuse coplanar surface barrier discharge. *Plasma Chemistry and Plasma Processing* 28(2), 203–211.

Owens, C. W., Shortle, W. C. & Shigo, A. L. (1980) Preliminary evaluation of silicon tetrachloride as a wood preservative. *Holzforschung* 34(6), 223–225.

Palanti, S., Feci, E., Predieri, G. & Vignali, F. (2012) A wood treatment based on siloxanes and boric acid against fungal and coleopter *Hylotrupes bajulus*. *International Biodeterioration & Biodegradation* 75, 49–54.

Petrič, M. (2013) Surface modification of wood: a critical review. *Reviews of Adhesion and Adhesives* 1(2), 216–247.

Pfriem, A., Wagenführ, A., Ziegenhals, G. & Eichelberger, K. (2005) The use of wood modified by heat-treatment for musical instruments. In *Wood Modification: Processes, Properties and Commercialisation, 2nd European Conference on Wood Modification*, Göttingen, Germany, pp. 390–397.

Phillips-Laing, E. M., Staines, H. J. & Palfreyman, J. W. (2003) The isolation of specific bio-control agents for the dry rot fungus *Serpula lacrymans*. *Holzforschung* 57, 574–578.

Podgorski, L., Chevet, B., Onic, L. & Merlin, A. (2000) Modification of wood wettability by plasma and corona treatments. *International Journal of Adhesion and Adhesives* 20(2), 103–111.

Rapp, A. O. & Sailer, M. (2001) Oil treatment of wood in Germany – state of the art. In *Review on Heat Treatments of Wood*, Rapp, A. O. (ed.). European Commission, Brussels, pp. 45–61.

Rayner, A. D. M. & Boddy, L. (1988) *Fungal Decomposition of Wood – Its Biology and Ecology*. John Wiley & Sons, Ltd, Chichester.

Reinprecht, L. (1993) Techniques of conservation and reinforcement in restoration practice – part 2. Treatments of damaged wood. In *Wood Modification, 9th International Symposium*, Poznaň, Poland, pp. 215–223.

Reinprecht, L. (1999) Changed susceptibility of the chemically and thermally degraded spruce wood to its attack by the dry rot fungus *Serpula lacrymans*. In *IRG/WP 99*, Rosenheim, Germany. IRG/WP 99-10322.

Reinprecht, L. (2008) *Wood Protection*. Handbook. TU Zvolen, Slovakia, (in Slovak).

Reinprecht, L. & Grznárik, T. (2014) Fungal resistance of Scots pine modified with organo-silanes alone and in combination with fungicides. In *The Seventh European Conference on Wood Modification*, Lisbon, Portugal, oral 6A:1, pp. 101.

Reinprecht, L. & Grznárik, T. (2015) Biological durability of Scots pine (*Pinus sylvestris* L.) sapwood modified with selected organo-silanes. *Wood Research* 60(5), 687–696.

Reinprecht, L. & Makovíny, I. (1987) Hardening of aminoplasts in modified wood with catalytic and thermic dielectric heating. In: *Wood Modification, 6th Symposium*, Poznaň, Poland, pp. 288–298.

Reinprecht, L. & Tiralová, Z. (2001) Susceptibility of the sound and the primary rotten wood to decay by selected brown-rot fungi. *Wood Research* 46(1), 11–20.

Reinprecht, L. & Varínska, S. (1999) Bending properties of wood after its decay with *Coniophora puteana* and subsequent modification with selected chemicals. In *IRG/WP 99*, Rosenheim, Germany. IRG/WP 99-40146.

Reinprecht, L. & Vidholdová, Z. (2008a) *ThermoWood: Its Preparation, Properties and Application*. Monograph, TU Zvolen, Slovakia (in Slovak, English abstract).

Reinprecht, L. & Vidholdová, Z. (2008b) Mould resistance, water resistance and mechanical properties of OHT-thermowoods. In *Sustainability through New Technologies for Enhanced Wood Durability: Socio-economic Perspectives of Treated Wood for the Common European Market*, Final Conference Proceedings of COST E37 in Bordeaux, Ghent University, Belgium, pp. 159–165.

Rijckaert, V., Van Acker, J. & Stevens, M. (1998) Decay resistance of high performance biocomposites based on chemically modified fibres. In *IRG/WP 98*, Maastricht, Netherlands. IRG/WP 98-40120.

Ringman, R., Pilgård, A., Brischke, C. & Richter, K. (2014) Mode of action of brow rot decay resistance in modified wood: a review. *Holzforschung* 68(2), 239–246.

Ritschkoff, A. C., Rätto, M., Nurmi, A., et al. (1999) Effect of some resin treatments on fungal degradation reactions. In *IRG/WP 99*, Rosenheim, Germany. IRG/WP 99-10318.

Rosenqvist, M. (2000) The distribution of introduced acetyl groups and a linseed oil model substance in wood examined by microradiography and ESEM. In *IRG/WP 00*, Kona, HI, USA. IRG/WP 00-40169.

Rowell, R. M. (1983) Chemical modification of wood. *Forest Products Abstracts* 6(12), 363–382.

Rowell, R. M. (1984) Penetration and reactivity of cell wall components. In *The Chemistry of Solid Wood*, Rowell, R. (ed.). Advances in Chemistry Series, vol. 207. American Chemical Society, Washington, DC, pp. 175–209.

Rowell, R. M. (1999) Specialty treatments. In *Wood Handbook – Wood as an Engineering Material*. USDA Forest Service, Madison, WI, chapter 19.

Rowell, R. M. (2005) Chemical modification of wood. In *Handbook of Wood Chemistry and Wood Composites*, Rowell, R. M. (ed.). CRC Press, Boca Raton, FL, chapter 14, pp. 381–420.

Rowell, R. M. & Banks, W. B. (1987) Tensile strength and toughness of acetylated pine and lime flakes. *British Polymer Journal* 19(5), 479-482.

Rowell, R. M. & Dickerson, J. P. (2014) Acetylation of wood. In *Deterioration and Protection of Sustainable Biomaterials*, Schultz, T. P., Goodell, B. & Nicholas, D. D. (eds). ACS Symposium Series, vol. 1158. American Chemical Society, Washington, DC, chapter 18, pp. 301–327.

Rowell, R. M. & Ellis, W. D. (1984) Effects of moisture on the chemical modification of wood with epoxides and isocyanates. *Wood and Fiber Science* 16(2), 257–267.

Rowell, R. M., Simonson, R. & Tillman, A. M. (1990) Acetyl balance for the acetylation of wood particles by a simplified procedure. *Holzforschung* 44(4), 263–269.

Rowell, R. M., Ibach, R. E., McSweeny, J. & Nilsson, T. (2009) Understanding decay resistance, dimensional stability and strength changes in heat-treated and acetylated wood. *Wood Material Science and Engineering* 4(1–2), 14–22.

Ryu, J. Y., Takahashi, M., Imamura, Y. & Sato, T. (1991) Biological resistance of phenol-resin treated wood. *Mokuzai Gakkaishi* 37(9), 852–858.

Ryu, J. Y., Imamura, Y., Takahashi, M. & Kajita, H. (1993) Effects of molecular weight and some other properties of resins on the biological resistance of phenolic resin treated wood. *Mokuzai Gakkaishi* 39(4), 486–492.

Sailer, M., Rapp, A. O. & Peek, R. D. (1998) Biological resistance of wood treated with water-based resins and drying oils in a mini-block test. In *IRG/WP 98*, Maastricht, Netherlands. IRG/WP 98-40107.

Schaffert, S., Krause, A. & Militz, H. (2005) Upscaling and process development for wood modification with *N*-methylol compounds using superheated steam. In *Wood Modification: Processes, Properties and Commercialisation, 2nd European Conference on Wood Modification*, Göttingen, Germany, pp. 161–168.

Scheiding, W., Kruse, K., Plaschkies, K. & Weiß, B. (2005) Thermally modified wood (TMW) for playground toys – investigations on 13 industrially manufactured products. In *Wood Modification: Processes, Properties and Commercialisation, 2nd European Conference on Wood Modification*, Göttingen, Germany, pp. 12–19.

Scheiding, W., Flade, P. & Direske, M. (2015) Determination of cracking susceptibility of three thermally modified hardwoods by advanced methods. In *The Eighth European Conference on Wood Modification*, Helsinki, Finland, pp. 88–91.

Schmidt, O. (2006) *Wood and Tree Fungi – Biology, Damage, Protection, and Use*. Springer-Verlag, Berlin.

Schneider, M. H., Brebner, K. I. & Hartley, I. D. (1991) Swelling of a cell lumen filled and cell-wall bulked wood polymer composite in water. *Wood and Fiber Science* 23(2), 165–172.

Schubert, M., Fink, S. & Schwarze, F. (2008) Evaluation of *Trichoderma* spp. as a biocontrol agent against wood decay fungi in urban trees. *Biological Control* 45, 111–123.

Sedjo, R. A. (2001) From foraging to cropping: the transition to plantation forestry, and implications for wood supply and demand. *Unasylva* 52, 24–27.
Singh, T. & Chittenden, C. (2008) In-vitro antifungal activity of chilli extracts in combination with *Lactobacillus casei* against common sapstain fungi. *International Biodeterioration & Biodegradation* 62, 364–367.
Slevin, C., Westin, M., Lande, S. & Cragg, S. M. (2015) Laboratory and marine trials of resistance of furfurylated wood to marine borers. In *The Eighth European Conference on Wood Modification*, Helsinki, Finland, pp. 464–471.
Solár, R., Reinprecht, L. & Lang, R. (2005) Resistance of untraditionally densified beech wood against rot – part II. Degradation by white-rot fungus *Trametes versicolor*. In *Drevoznehodnocujúce Huby 2005 – Wood-Damaging Fungi 2005, 4th Symposium*, Kováčová, Slovakia, pp. 81–84 (in Slovak, English abstract).
Šomšák, M. Reinprecht, L. & Tiňo, R. (2015) Effect of plasma and UV-absorbers in transparent acrylic coatings on photostability of spruce wood in exterior. *Acta Facultatis Xylologiae Zvolen* 57(1), 63–73.
Spear, M. J., Hill, C. A. S., Curling, S. F., et al. (2006) Assessment of the envelope effect of three hot oil treatments – resistance to decay by *Coniophora puteana* and *Postia placenta*. In *IRG/WP 06*, Tromsø, Norway. IRG/WP 06-40344.
Suttie, E. D., Hill, C. A. S., Jones, D. & Orsler, R. J. (1997) Assessing the bioresistance conferred to solid wood by chemical modification. In *IRG/WP 97*, Whistler, BC, Canada. IRG/WP 97-40099.
Suttie, E. D., Hill C. A. S., Jones, D. & Orsler, R. J. (1998) Chemically modified solid wood. I. Resistance to fungal attack. *Material und Organismen* 32(3), 159–182.
Sutton, W. R. J. (1999) The need for planted forests and the example of radiata pine. *New Forest* 17, 95–109.
Tanno, F., Saka, S., Yamamoto, A. & Takabe, K. (1998) Antimicrobial TMSAH-added wood-inorganic composites prepared by the sol-gel process. *Holzforschung* 52(4), 365–370.
Terziev, N. & Temiz, A. (2005) Chemical modification of wood with silicon compounds. In *Wood Modification: Processes, Properties and Commercialisation, 2nd European Conference on Wood Modification*, Göttingen, Germany, pp. 242–245.
Tjeerdsma, B. F., Swager, P., Horstman, B., et al. (2005) Process development of treatment of wood with modified hot oil. In *Wood Modification: Processes, Properties and Commercialisation, 2nd European Conference on Wood Modification*, Göttingen, Germany, pp. 186–197.
Treu, A., Larnoy, E. & Militz, H. (2007) Microwave curing of furfuryl alcohol modified wood. In *IRG/WP 07*, Jackson, MS, USA. IRG/WP 07-40371.
Van Acker, J., Nurmi, A., Gray, S., et al. (1999) Decay resistance of resin treated wood. In *IRG/WP 99*, Rosenheim, Germany. IRG/WP 99-30206.
Van de Zee, M., Beckers, E. P. J. & Militz, H. (1998) Influence of concentration, catalyst, and temperature on dimensional stability of DMDHEU modified Scots pine. In *IRG/WP 98*, Maastricht, Netherlands. IRG/WP 98-40119.
Verma, M., Brar, S. K., Tyagi, R. D., et al. (2007) Antagonistic fungi *Trichoderma* spp.: panoply of biological control. *Biochemical Engineering Journal* 37: 1–20.
Verma, M., Sharma, S. & Prasad, R. (2009): Biological alternatives for termite control: a review. *International Biodeterioration & Biodegradation* 63, 959–972.
Verma, P., Junga, U., Militz, H. & Mai, C. (2009) Protection mechanisms of DMDHEU treated wood against white and brown rot fungi. *Holzforschung* 63(3), 371–378.
Vernois, M. (2001) Heat treatment of wood in France – state of the art. In *Review on Heat Treatments of Wood*, Rapp, A. O. (ed.). European Commission, Brussels, pp. 37–44.

Wakeling, A. N., Plackett, D. V. & Cronshaw, D. R. (1992) The susceptibility of acetylated *Pinus radiata* to mould and stain fungi. In *IRG/WP 92*, Harrogate, UK. IRG/WP 92-1548.

Wang, Ch. & Powell, J. E. (2004) Cellulose bait improves the effectiveness of *Metarhizium anisopliae* as a microbial control of termites (Isoptera: Rhinotermitidae). *Biological Control* 30, 523–529.

Wang, J. & Cooper, P. A. (2005) Properties of hot oil treated wood and the possible chemical reactions between wood and soybean oil during heat treatment. In *IRG/WP 05, Bangalore, India. IRG/WP 05-40304.*

Wang, J. & Cooper, P. A. (2007) Fire, flame resistance and thermal properties of oil thermally-treated wood. In *IRG/WP 07*, Jackson, MS, USA. IRG/WP 07-40361.

Welzbacher, C. R. & Rapp, A. O. (2005) Durability of heat treated materials from industrial processes in ground contact. In *IRG/WP 05*, Bangalore, India. IRG/WP 05-40312.

Westin, M., Rapp, A. O. & Nilsson, T. (2004a) Durability of pine modified by 9 different methods. In *IRG/WP 04*, Ljubljana, Slovenia. IRG/WP 04-40288.

Westin, M., Lande, S. & Schneider, M. (2004b) Wood furfurylation process and properties of furfurylated wood. In *IRG/WP 04*, Ljubljana, Slovenia. IRG/WP 04-40289.

White-McDougall, Blanchette, R. A. & Farrell, R. L. (1998) Biological control of blue stain fungi on *Populus tremuloides* using selected *Ophiostoma* isolates. *Holzforschung* 52, 234–240.

Willems, W. (2015) Glassy polymer formation in thermally modified wood: effects on long-term moisture and durability performance in service. In *The Eighth European Conference on Wood Modification*, Helsinki, Finland, pp. 234–240.

Willems, W., Gérardin, P. & Militz, H. (2013) The average carbon oxidation state as a marker for the durability of thermally modified wood. *Polymer Degradation and Stability* 98(11), 2140–2145.

Williams, F. C. & Hale, M. D. (1999) The resistance of wood chemically modified with isocyanates I. Brown rot and acid chlorite delignification. *Holzforschung* 53(3), 230–236.

Wolkenhauer, A., Avramidis, G., Militz, H. & Viöl, W. (2008) Plasma treatment of heat treated beech wood – investigation on surface free energy. *Holzforschung* 62(4), 472–474.

Yang, D.-Q., Gignac, M. & Bisson, M-C. (2004) Sawmill evaluation of a bioprotectant against moulds, stain, and decay of green lumber. *Forest Products Journal* 54, 63–66.

Yang, D.-Q., Wang, X.-M. & Wan, H. (2007) Biological protection of composite panel from moulds and decay. In *IRG/WP 07*, Jackson, MS, USA. IRG/WP 07-10612.

Yano, H., Norimoto, M. & Rowel, R. M. (1993) Stabilization of acoustical properties of wooden musical instruments by acetylation. *Wood and Fiber Science* 25(4), 395–403.

Yildiz, S., Gezer, E. D. & Yildiz, U. C. (2006) Mechanical and chemical behaviour of spruce wood modified by heat. *Building and Environment* 41, 1762–1766.

Zanuncio, A. J. V., Motta, J. P., da Silveira, T. A., et al. (2014) Physical and colorimetric changes in *Eucalyptus grandis* wood after heat treatment. *BioResources* 9(1), 293–302.

Zauer, M., Kowalewski, A., Oberer, I., et al. (2015) Development of thermally modified European wood to substitute tropical hardwood for the use in acoustic guitars. In *The Eighth European Conference on Wood Modification*, Helsinki, Finland, pp. 96–99.

Zelinka, S. L., Ringman, R., Pilgård, A., et al. (2015) The role of chemical transport in the decay resistance of modified wood. In *The Eighth European Conference on Wood Modification*. Helsinki, Finland, p. 36–43.

Živković, V., Prša, I., Turkulin, H., et al. (2008) Dimensional stability of heat treated wood floorings. *Drvna Industrija* 59(2), 69–73.
Zollfrank, C. (2001) Silylation of solid beech wood. *Wood Science and Technology* 35(1–2), 183–189.

Standards

EN 335 (2013) Durability of wood and wood-based products – Use classes: definitions, application to solid wood and wood-based products
EN 350-1 (1994) Durability of wood and wood based products – Natural durability of solid wood – Guide to the principles of testing and classification of the natural durability of wood

7 Maintenance of Wood and Restoration of Damaged Wood

7.1 Aims and enforcement of the maintenance and the restoration of damaged wood

The aims and enforcement of the maintenance of wooden objects are connected with their service life, which is also limited by their material, design, production and exposure factors (Section 1.5). The correct and regular maintenance of wooden products and constructions satisfies or even extends their estimated service life, and therefore it becomes an integral part of their protection (Section 7.2).

The aims and enforcement of the restoration of damaged wooden objects are concerned mainly with historical artefacts, buildings and other constructions, but also with poorly protected recent wooden objects damaged in a short time by abiotic factors and/or biological pests (Chapters 2 and 3). If the technical, artistic, historical and social value of a damaged wooden object is significant and its salvage is required, a project for its restoration should be prepared as early as possible. The implementation of a restoration project for a wooden object includes several operations: (1) a thorough diagnosis of individual wooden elements in order to ascertain their current state (Section 7.3); (2) sterilization, mainly depending upon whether the biological damage is actual (Section 7.4); and (3) restoration using suitable conservation and/or reinforcement techniques (Sections 7.5 and 7.6).

7.2 Wood maintenance

7.2.1 Principles of wood maintenance in exteriors and interiors

Regular and proper maintenance of wooden products and constructions significantly extends their service life. This is of a greater importance to exterior

Wood Deterioration, Protection and Maintenance, First Edition. Ladislav Reinprecht.

uses, where there are more often favourable climatic conditions for the activity of decay-causing fungi and other pests, as well as for the degradation of the wood's surface by solar radiation, water and emissions. Maintenance of wooden constructions, similar to their building and reconstruction, requires to compliance with work safety (Bethancourt & Cannon, 2015).

The maintenance of exterior wooden products includes several regular operations:

- inspection of structural, chemical or modifying protection (e.g. from the point of view of structural protection, if the products have unplanned contact with the terrain due to vegetation or other factors and therefore move from the planned EN 335 class 3 use to an unplanned class 4 use category);
- inspection of damage (e.g. presence of fungal rot and insects galleries);
- reapplication of a weather-protecting coating depending upon the product type and exposure level (e.g. in the climate of central Europe, acrylate or alkyd coatings for wooden windows should usually be reapplied every 4–8 years (Menzies, 2013); however, vegetable oil coatings for garden furniture should be re-applied annually);
- reapplication of a biocide on the surface of the wooden product (e.g. wood shingles can be prefinished directly on the roof or façade) or also in its deeper zones (e.g. wood palisades can be treated either just in the 'terrain–air' critical zone by injecting or bandaging, or after removal from ground and cleaning by dipping or vacuum–pressure technique).

The maintenance of interior wooden products mainly includes the following regular measures:

- inspection of structural, chemical or modifying protection (e.g. whether rainwater has penetrated the truss via damaged roof covering);
- inspection of damage (e.g. rot and galleries in wooden houses – Levi, 1985);
- repair of mechanically damaged and weathered coatings (e.g. new varnish on parquet floorings should be applied directly to native wood – that is, after removing the original damaged varnish; new varnish on furniture can be applied as well to a sanded layer of the original varnish);
- reapplication of biocide or fire retardant on the surface of the wood (e.g. usually, if the original protection has over time expired due to evaporation, sublimation or changes in its molecular structure by oxidation it is necessary to retreat beams and other elements in trusses using a healthy, safe chemical protection substance).

7.2.2 Principles of the fight against the active stages of wood pests

A greater fight is ongoing with those biological pests that are already active in wood. It is necessary to find and destroy them at the outbreak of their activities. It is subsequently recommended that immediate temporary measures are taken and then, as soon as possible, to apply long-term action to eliminate pests and to secure the wooden object against further damage or even a collapse.

A *temporary fight against biological pests* can be carried out by the owner of the building or other construction after consulting with a specialist:

- removing a local source of humidity (e.g. repairing the covering or insulation in roof);
- removing of significantly biologically damaged wood and other damaged components of the object without disturbing its statics;
- local treatment of wood using biocides.

The *long-term fight against biological pests* should only be carried out by a professional, preferably a specialist or an organization specializing in performing several consecutive measures:

- survey and assessment of actual state of the wood – wood diagnostics (Section 7.3);
- liquidation of biological pests in wood and its surroundings using suitable physical or chemical sterilization methods (Section 7.4);
- renewal of damaged wooden artefacts and parts of the wooden construction by applying convenient conservation agents (Section 7.5);
- restoration of damaged wooden construction in connection with strengthening or replacement of markedly damaged wooden elements (Section 7.6);
- structural protection of wooden construction, focusing upon minimizing the risk of its damage during restoration (e.g. temporary supporting of a damaged ceiling), as well as decreasing the risk of its wetting in the future (Chapter 4);
- chemical or modifying protection of wooden elements, applied if other measures related to the construction do not exclude the possibility of their repeated biological damage (Chapters 5 and 6).

Fighting with wood-damaging fungi is based on permanently keeping wood humidity to below 20%. However, if the wood is already affected by fungal organisms, decreasing its humidity to below 20% is more difficult for the following reasons: (1) wood in the rot process is decomposed to carbon dioxide (CO_2) and water, which makes it wetter; (2) several wood-decaying fungi are able to bring water to the wooden product even from great distances via a system of surface mycelia, quite often through masonry (e.g. by rhizomorphs which are typical for the true dry-rot fungus *Serpula lacrymans*); and (3) mould growth on wood surfaces prevents the free evaporation of water from its inner zones.

In situations when the activity of fungal organisms occurs, it is insufficient just to create a climate in which the equilibrium moisture content of the wood would be slightly under 20%, but the wood must be thoroughly dried throughout the whole cross-section and, at the same time, it is necessary to remove all development stages of the fungal organism located in the wood, on its surface and in its close vicinity. It is recommended that fruiting bodies and surface mycelium of the decay-causing fungi are thoroughly separated from wood (or from masonry and other materials) prior to the use of drying and sterilization process (e.g. by microwaves) and disposed of away from the building so they are no longer the source of further infestation. Several sterilization methods are used for annihilation of decay-causing fungi and moulds; however, the selection of the particular method depends upon the particular situation; that is, whether the object is movable/unmovable, new/a cultural heritage, surface finished/unfinished, cross-sections of wooden components are large/small, and so on.

The fight against fungal organisms should also include subsequent preventative chemical protection of the sterilized wood and new wood (or also masonry and other materials in the vicinity if necessary) using a suitable fungicide. Chemical protection also serves as insurance if the sterilization was not carried out thoroughly.

A very difficult fight is ongoing with the true dry-rot fungus *Serpula lacrymans*. A general principle applies, in accordance with which not only all visibly attacked wood should be removed from the object and subsequently safely destroyed (burned, transferred for humification outside the built-up area, etc.), but also healthy wood within a distance of at least 1 m from the edge of the visible damage. The surrounding plaster, grouting and mineral fillings below rotten wooden floors should also be removed and disposed of. When disposing of these materials and fungi, increased attention should be paid to thorough sterilization and chemical protection of wood, masonry and plaster remaining in the object.

Fighting with the wood-boring insects is a more difficult task than fighting against wood-damaging fungi, because it is based on decreasing wood humidity to below 10%. This condition is not usually achieved in roofs or ceilings in a normal climate. The fight against wood-boring insects active in wood usually follows the following principles:

- It is suitable to remove all the bark from the surface of the wooden product, which will prevent the activities of the next generations of some insect species; for example, of the violet tanbark beetle (*Callidium violaceum*), bark beetles (*Scolytidae* family) and other species that create galleries more or less only in the sapwood under the bark. However, it should also be remembered that removing bark has no influence upon the activity of the house longhorn beetle (*Hylotrupes bajulus*), wood-worm beetles (*Anobiidae* family) or termites.
- It is suitable to sterilize and also treat wooden artefacts and constructions using insecticides with a preventive effect. Sterilization and preventive protection of wood may also be carried out within one process; that is, using a stable insecticide with an eradicating and preventative effect. When destroying larvae of *Hylotrupes bajulus*, opinion exists that it only rarely attacks wood over 60 years old with already-evaporated terpenoids (Kačík et al., 2012). However, this wood should also be treated by sterilization and subsequently with an insecticide for safety reasons.
- It is suitable to chemically protect wooden constructions (e.g. trusses) by spraying, with the aim of not only treating the surface of the wood, but also

gaps in the joints. Apart from spraying, injecting insecticides into insect holes is also recommended. In Europe, it is best to carry out treatment in April/May, before insects swarm. Oral insecticides, nowadays mainly hormonal, are considered to be the optimum solution.

7.3 Diagnosis of damaged wood

By diagnosis of wooden products and structures we obtain information about their current status – quality, damage, static parameters, safety class, and so on – needed for their assessment and realization of regular maintenance or restoration works (UNI 11119 2004, ASTM D245-00 2006; Dietsch & Kreuzinger, 2011; Cruz et al., 2015). Damage in wooden products must be discovered early and objectively, focusing upon determining its type, degree, scope and cause.

Diagnostic methods for the defectoscopy of wooden products and constructions are classified in accordance with the following criteria (Reinprecht & Štefko, 2000).

- Criterion of wooden material (wood, adhesive, coating, etc.) analysis:
  - direct methods (i.e. structural analysis);
  - indirect methods (i.e. analysis of physical, mechanical and biological properties).
- Criterion of testing mode:
  - sensory methods;
  - instrumental methods.
- Criterion of wood material destruction during its testing:
  - non-destructive techniques (NDT),
  - semi-destructive techniques;
  - destructive techniques.
- Criterion of the place of damages discovery:
  - in-situ methods (directly in the place of wood exposure);
  - in-vitro methods (after transferring a sample or the whole wooden product to a laboratory).

For in-situ surveys, the use of cheap sensory methods is recommended first, usually visual and sonic, and then other suitable instrumental methods (usually non-destructive) are used as needed (Brashaw et al., 2009; Niemz & Mannes, 2012; Cavalli & Togni, 2013; Kilic, 2015; Martínez et al., 2015). According to Bodig (2001), the value V of the state of the wood in a construction is a combination of its density and other properties, damage done by biotic and abiotic agents, and climatic conditions; that is, $V = f(D, B, M, T, G, C, \ldots)$, where D is density, B is biodegradation, M is moisture, T is temperature, G is geometry, C is constraining conditions.

Not only can diagnostics identify faults in wood, but they can also determine the species and age of the wood. The species of wood is usually determined via microscopic analysis and its age via the ^{14}C radioisotope or dendrochronological method.

7.3.1 Sensory diagnostic methods

Sensory methods are subjective, but their advantages are speed and low price (Piazza & Riggio, 2006, 2008). In practice they are used for initial assessment of damaged wood (Table 7.1). Commonly, damage of wood by fungi and insects is recognized by these methods – determining their scope and level, including the species of decay-causing fungus or boring insects (Bravery et al., 2010). They are very important for localization and depth measurement of cracks in timbers of large-span constructions, mainly if they appear in the glueline (Dietsch & Kreuzinger, 2011). Sensory methods are applied for investigation of wooden artefacts (furniture, musical instruments, etc.), wood in-built inside buildings (trusses, ceilings, log cabins, floors, stairs, etc.), as well as in constructions and

Table 7.1 Sensory methods for detecting the damage and the quality of wood

Visual methods
- decay-causing fungi (fruit body and mycelium of the fungus, type of rot, water droplets, etc.),
- staining fungi and moulds (changes in colour of the wood, mouldy growth on the wood, etc.),
- wood-boring insects (holes and gallery dust falling from the wood, etc.),
- weather influences (changes in colour of the wood, creation of plastic texture and cracks in the wood, etc.)
- fire (changes in colour and structure of the carbonized wood, etc.),
- effects of chemicals (defibrillation and changes in colour of the wood, etc.).

Olfactory methods
- decay-causing fungi and moulds (typical smell)
- fire (leakage of gases, carbonized wood)
- increased humidity (musty odour in the object)

Tactile methods
- surface defects (plastic textures, cracks, etc.),
- rots, galleries and other defects even deeper in the wood (changes in the hardness and integrity of the wood)
- increased moisture content

Sonic methods
- wood-boring insects (sound signals of larvae when attacking wood or imago when mating – typical for house longhorn beetle, common furniture beetle, termites)
- inner rot and cavities (typical acoustic response to tapping with a hammer, etc.)
- static faults (typical acoustic anomalies during dynamic stress to a wooden construction; e.g. by wind).

products exposed to an exterior environment (bridges, utility poles, pergolas, garden furniture, etc.).

Simple tools and dogs are often used within the sensory methods (Levi, 1985). The tools usually include portable lighting, a magnifying glass, a tape measure, a protractor, a small hammer, a claw hammer, a knife, nails, a chisel, a saw, an increment borer for sampling, a camera, a mobile phone and, depending upon the situation, also an extendable ladder. Small samples taken with a wood extractor (hollow drill) can be subsequently analysed in a laboratory. For identification of the true dry-rot fungus (*Serpula lacrymans*), the use of *rothound* dogs is successful (Hutton, 1994).

The sensitivity of sonic methods can be improved by using a stethoscope when examining sound signals emitted by wood-boring insects. Special instruments for recording sound signals from an insect in a wooden structure have also been developed, and data are subsequently processed by a computer and compared with the standard signals for individual insect species (Scheffrahn et al., 1993; Schmidt et al., 1995; Hyvernaud et al., 1996; Fujii et al., 1999; Zorovič & Čokl, 2015).

7.3.2 Instrumental diagnostic methods

Instrumental methods are designed for more accurate analysis of the type, degree, scope and causes of faults in wooden objects (Kasal, 2008). Damaged wood has an altered molecular, anatomical, morphological and geometric structure that is reflected in changes to its properties – mainly optical, electrical, acoustic, thermal, mechanical, chemical and biological. Changes to the structure and properties of damaged wood are usually identified using several physical, strength, chemical and biological methods (Table 7.2) and their combination.

More objective analyses of wooden constructions can be achieved by multiple built-in sensors emitting signals about environmental conditions and properties of wood which are continually processed (Sandak, J., et al., 2015). With the aim to obtain the most objective and comparable results, the measurement conditions around and in wood – relative humidity, moisture content, temperature, and so on – should be standardized by using various devices (Íňiguez-Gonzáles et al., 2015). Generally, more precise information about the object being analysed can be achieved only by a combination of several types of instrumental methods (e.g. Feio & Machado, 2015; Liang et al., 2016).

7.3.2.1 Optical methods

Optical methods analyse changes in the structure of damaged wood and also the presence of biological pests in the attacked wood (e.g. Schmidt, 2006; Schwarze, 2007). They may also be used for ascertaining the micro-distribution of preservatives, modifying and conservation substances in the lumens and cell walls of wood.

Table 7.2 Instrumental methods for detecting the damage and the quality of wood

Method	Device	Basic principle	Detection
Optical	Light microscope	Enlarging an image, dyeing	Changes in wood structure, dyeing of fungal hyphae
	Electron microscope	Enlarging an image (metal coating)	More detailed changes in wood structure
	Endoscope	Transfer of image through optical fibre	Damage in larger cavities, in ceilings, etc.
	Colorimeter	Visible spectrum absorption, 400–700 nm	Degree of colour change due to ageing
	Holograph	Optoelectronic	Humidity, cracks, rot
	Infrared (IR), Fourier transform IR and near-IR	Absorption of IR radiation by functional groups of materials	Humidity, analysis of rot and attack by weather
Electrical	Vitamat	Conductivity	Active rot
	Dielectric	Frequency, permittivity	Humidity, analysis of rot
	EIS	Electrical impedance spectroscopy	Gradient of humidity
Ultrasound	Pundit, Arborsonic, Sylvatest	Velocity of ultrasound waves (option to state the modulus of elasticity of wood)	Analysis of internal rot, cracks and galleries
Acoustic tomography	Fakopp 2D	Velocity of sound waves across the wood fibre	Analysis of internal rot (localization, degree, scope)
Radiographic	X-ray radiography Computed tomography	Absorption of X- or γ-rays with energy 59.54 keV (^{241}Am)	Internal structure of the wood (knots, cracks, rot, insect galleries, etc.)
Electromagnetic	Radar	Short pulses into the wood	Metals (nails, screws, etc.)
	Microwave sensor	Microwaves into the wood	Knots, cracks, etc.
	Nuclear magnetic resonance tomography	Waves with the energy of a photon between two spin levels	Humidity map of the wood, indication of internal rot
Thermography	Thermograph	Radio-frequency heating of wood	Different colour of wood knots, cavities, etc.

(*continued*)

Table 7.2 (*Continued*)

Method	Device	Basic principle	Detection
Strength	Drilling and fractometer	Drilling of wood sample and its analysis in bending	Rot, chemical corrosion
	Pilodyn	Penetration of a thin needle	Surface rot, galleries, etc.
	Resistograph	Resistance against a thin drill bit	Internal rot, cavities, etc.
Chemical	Gas detector	Concentration of CO_2, volatile organic compounds	Fungi, moulds, fire
	Chromatographs and spectroscopes	Sampling, separation and analysis of wood components	Type and degree of rot
Biological	DNA analyser	Sequence of A, G, C, T	Type of decay fungus

The *light microscope* allows determination of the initial stages of wood rot, its type and the species of wood-decaying fungus (Rapp, 1998; Messner et al., 2003). From analysing the gallery dust of wood, it is possible to determine the species of wood-boring insect (e.g. Reinprecht, 1997; Bravery et al., 2010).

The *polarized-light microscope* is suitable for monitoring changes in the crystalline cellulose; that is, its decrease in the wood is linked with a decrease in birefringence. For example, within brown rot in the wood of broadleaves there is a decrease in crystalline cellulose that is more pronounced in libriform fibres than in vessels, and in the coniferous wood it is more pronounced in early wood than in late wood. Using this method, Wilcox (1968) documented that, in the initial stages of brown rot, crystalline cellulose is mainly decomposed in the S_2 layer of cell walls, whilst in the S_1 and S_3 layers it decomposes during the later stages of rot.

The *fluorescence microscope* is suitable for identification of the initial phases of wood rot when weight decrements are still below 1–3% (Krahmer et al., 1982). After dyeing with acridine orange, rotten wood is an orange–red colour whilst healthy wood is a yellow–green colour. Fluorescent labelling and quantification of fungal biomass have been extensively reported by Spear et al. (1999).

Scanning electron microscopy gives a more detailed view of the anatomical structure of wood cells damaged by decay-causing fungi or other factors (e.g. Reinprecht & Lehárová, 1997; Tiralová & Mamoňová, 2005; Capano et al., 2015; Kim et al., 2015).

Image acquisition for assessment of wood surfaces using a scanner together with a USB interface offering sufficient optical resolution of 1200 dpi or more. This method using a proper algorithm is convenient for evaluation of mould and staining fungi growth on wood surfaces (Van den Bulcke et al., 2005).

Endoscopes are used for in-situ analysis of wooden constructions; for example, when detecting rot and galleries inside beams in ceilings without the need of their dismantling (Janotta, 1984). Bores for an endoscope with a diameter of up to ~10 mm can be closed with a seal and subsequently restored (Kasal et al., 2003). Endoscopic analysis can be documented using a photographic or video record.

Colorimetry is based on the CIE-$L^* a^* b^*$ colour system in which three parameters are measured: L^* is a lightness from black (0) to white (100), a^* is a shade between red (+) and green (−), and b^* is a shade between yellow (+) and blue (−). Colour changes in wood occur in rot, mould, weather, chemical, thermal and some other types of damage (e.g. Tolvaj et al., 2000; Pandey, 2005; Reinprecht & Hulla, 2015). Significant changes in the colour of damaged wood occur if the total colour difference $\Delta E = (\Delta L^{*2} + \Delta a^{*2} + \Delta b^{*2})^{0.5} \geq 3$.

Holography is a non-destructive method for detecting cracks, local rot and disturbed joints; for example, between lamellas in a glued composite, between veneers in plywood, or between solid wood and veneer in furniture.

IR spectroscopy and its modifications, such as FTIR and NIR, are physical–chemical methods combined with optical devices equipped with a camera and computer. They are based on the absorption of light of a specific wavelength by individual molecules or functional groups in molecules of the wood substrate. For example, increases in the carbonyl and carboxyl groups in the lignin during its oxidation by white-rot fungi (Kirk, 1984), as well as other changes in the molecular structure of the wood resulting from its weathering (Körner et al., 1992; Pandey, 2005; Kačík et al., 2014; Sandak, A., et al. 2015), or its long-term exposure underground and underwater (Solár et al., 1987; Cha et al., 2014; Capano et al., 2015; Pecoraro et al., 2015) can be determined.

7.3.2.2 Electrical methods

Electrical methods work on the principle of measuring electrical resistance, specific surface conductivity, specific volume conductivity or relative permittivity. Measuring devices are usually equipped with two electrodes that are stabbed into the wood or are gradually inserted into pre-drilled openings whilst continuously recording electrophysical values. They are used for detecting humidity and internal rot in wood utility poles, log cabins, and so on (Kučera & Bucher, 1988; Makovíny & Reinprecht, 1990). In the interval of wood moisture from 6 to 25%, their instrumental error has to be less than 2% (Said, 2004).

7.3.2.3 Ultrasound and other acoustic methods

The velocity of the transfer of ultrasound waves with a frequency from 20 kHz to 1 GHz through wood depends upon its density, humidity, direction and incline of fibres, the presence of knots, cracks, cavities, rot and other anomalies and defects. In rotten wood, acoustic waves spread more slowly than in healthy wood. Results of acoustic methods can be used for an indirect determination of bending and other mechanical properties of inspected wood (Ross et al., 1997; Huang et al., 2003; Reinprecht & Pánek, 2013; Dackermann et al., 2014).

Ultrasound devices of the one- (1-D) or two-dimensional (2-D) type have two or more probes; that is, a wave transmitter and a wave receiver(s). Probes are placed on the wood surface from two opposite sides (Figure 7.4c – see Section 7.3.2.7) or from more sides. The dynamic modulus of elasticity of wood ($MOE_{dyn} = v_L^2 \rho$) is determinable from its density ρ and the velocity of ultrasound waves passing along the fibres of the wood v_L. Using the ultrasound method, various stages of rot can be identified in wood (Arita et al., 1986; Wilcox, 1988; Marčok et al., 1997; Ross et al., 1997; Reinprecht & Pánek, 2013; Tippner et al., 2016), and also galleries caused by wood-boring insects (Prieto, 1990).

An *impulse hammer* of the 1-D or 2-D type generates impulse waves via an exciter on one side of the wood utility pole, beam, and so on. The waves are sensed by a detector on the opposite side of the product or in several places around its circumference. Impulse waves spread more slowly if there are cavities, cracks, rots or galleries in the wood.

Acoustic emissions transmitted from the wood during a mechanical stress are also a good indicator of its initial damage. Noguchi et al. (1992) documented that rotten wood transmits acoustic signals even when under minor pressure.

7.3.2.4 Radiographic methods

Radiographic methods – 1-D, 2-D or three-dimensional (3-D) – work on the principle of increasing the intensity of the passage of X-rays (X-ray radiography, computed tomography (CT)), γ-rays (CT) or neutrons (neutron radiography) through wood having a lower density (Van den Bulcke et al., 2011). So, they apply also to wood damaged by fungal rot, insect galleries or cracks. Radiography can also determine the moisture content of wood, as well as localize knots or metal items inside the wood.

Gamma-ray 1-D densitometry measures the absorption coefficient of weakening γ-radiation from a ^{241}Am radioisotope after its passage through the wood. This method allows the monitoring of, for example, the kinetics of wood decay in its cross-section (Figure 7.1).

Two-dimensional X-ray radiography uses X-rays that, after passing through the wood, are trapped on black-and-white film or converted into a computer image. Places with a more intensive passage of radiation through the damaged less-dense areas of wooden materials are blacker, as long as they are of same thickness. Radiography performed at predefined time intervals also gives us an image of the kinetics of wood decay (Bucur et al., 1997) or about the movement of insect larvae in the wood with the aim of ascertaining the efficiency of insecticide or thermal sterilization (Graham & Eddie, 1985). Today, investigation of wooden constructions is made by mobile battery-charged X-ray tubes with digital detectors (Wedvik et al., 2015).

Two-dimensional and 3-D CT work on the principle of analysing an image of the investigated wooden product from various angles (Figure 7.2). The source

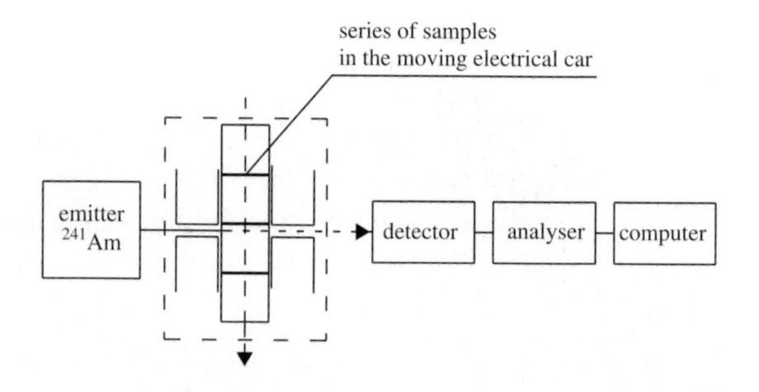

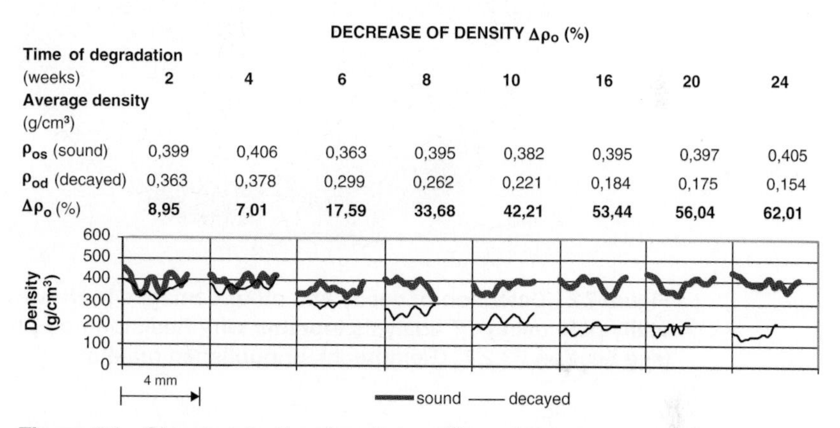

DECREASE OF DENSITY $\Delta\rho_o$ (%)

Time of degradation (weeks)	2	4	6	8	10	16	20	24
Average density (g/cm³)								
ρ_{os} (sound)	0,399	0,406	0,363	0,395	0,382	0,395	0,397	0,405
ρ_{od} (decayed)	0,363	0,378	0,299	0,262	0,221	0,184	0,175	0,154
$\Delta\rho_o$ (%)	**8,95**	**7,01**	**17,59**	**33,68**	**42,21**	**53,44**	**56,04**	**62,01**

Figure 7.1 Changes in the density profiles of Norway spruce wood as a result of rot caused by the white-rot fungus *Trametes versicolor* – detected on samples with a thickness of ~4 mm in the longitudinal direction using γ-ray densitometry

Source: Reinprecht, L., Novotná, H. and Štefka, V. (2007) Density profiles of spruce wood changed by brown-rot and white-rot fungi. *Wood Research* 52(4), 17–28. Reproduced by permission of Pulp and Paper Research Institute Bratislava

of X-rays or γ-rays circulates around the wooden product simultaneously with a detector for recording the intensity of rays that pass through the wood. A computer transforms the value of weakening rays into a scanning value (pixel) that corresponds with a certain elementary volume (voxel) in the monitored product. The size of this elementary volume is given by the parameters of the CT system (matrix) and the detected thickness of the wood. The resulting image of the scanned wooden product is displayed on a screen. Recently, special micro-CT scanners have been developed for analysis of wood and other materials (Dierick et al., 2015).

Along with a 2- or 3-D display of the wooden product, CT analysis determines its decreased density in precisely determined places, from which it is possible to identify, for example, the degree of rot (Reinprecht & Šupina, 2015), the presence of insect galleries, the wood species (Stelzner & Million, 2015), and

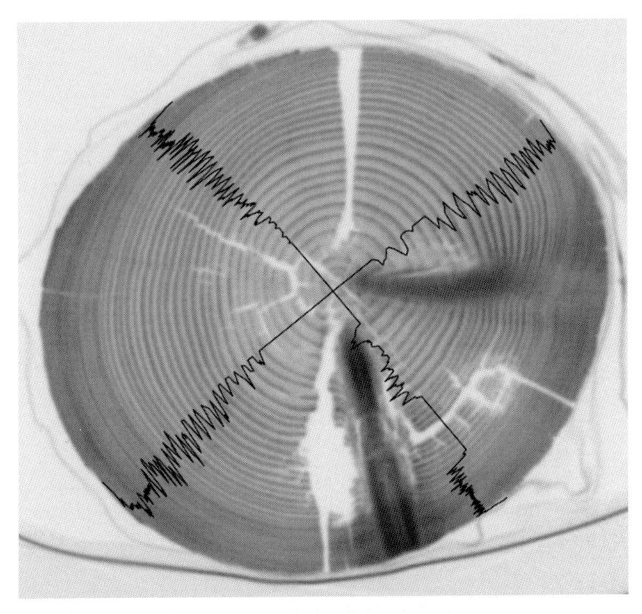

Figure 7.2 Detection of decay and other damages in Norway spruce utility pole using CT analysis together with Resistograph profile analysis (see Section 7.3.2.7) (Reinprecht, unpublished results)

so on. Apart from rot and galleries, CT analysis also allows the monitoring of cracks, metal nails and screws, layers of adhesives and coatings. CT analysis can also be applied in archaeological dendrochronology without damage of wooden objects (Bill et al., 2012). Rigon et al. (2010) undertook a feasibility study of the use of X-ray microtomography on musical instruments. CT analysis is suitable also for stating the distribution of preservatives and conservation substance in wood (Kučerová & Lisý, 1999), for monitoring the diffusion of water vapour through coatings on the wood's surface, or for monitoring other transport processes important in the protection and conservation of the wood.

7.3.2.5 Electromagnetic methods

Nuclear magnetic resonance (NMR) tomography gives information about the distribution of substances containing hydrogen atoms (water, cellulose, etc.) in the cross-section of the wood. Some types of wood degradation are related to a significant increase in its humidity, such as active rot. The NMR tomography method allows a healthy zone of wood to be distinguished from a wet freslly rotten zone. Water distribution is also important when analysing archaeological wood exposed underground for a long time (Cole-Hamilton et al., 1990). For example, Cha et al. (2014) using NMR spectra in the investigation of 800-year-old archaeological bamboos, found in the Yellow Sea, obtained information about the extensive degradation of polysaccharides and only a partial change of lignin.

7.3.2.6 Thermal imaging methods

Materials, including wood, with a temperature over −273 °C emit thermal radiation in the infrared region of the spectrum. After passing through wood, this radiation is captured on a thermal imaging camera. If the wood contains rotten cavities, insect galleries without dust or other empty areas, the flow of thermal radiation increases. On a thermal imaging record, the intensity of the flow of radiation is indicated by a typical colour – red means a high thermal flow and blue indicates a low thermal flow.

7.3.2.7 Strength methods

Rotten or otherwise damaged wood has a lower density, and then also a lower strength and hardness. These properties can be evaluated in vitro on small samples in a laboratory. However, in practice, in-situ methods are more suitable and are performed using portable devices such as:

- *Fractometer* – analyses the bending characteristics of samples taken from wood with an increment borer (Figure 7.3, Table 7.3).
- *Resistograph* – states the resistance of wood against a drill bit with a diameter of 2–3 mm (e.g. Rinn et al., 1996; Martínez et al., 2015; Reinprecht & Šupina, 2015). Using a servomotor, the drill bit penetrates through a cross-section of a wooden product at a velocity of 10–500 mm/min to a maximum depth of 400–1000 mm (Figure 7.4).

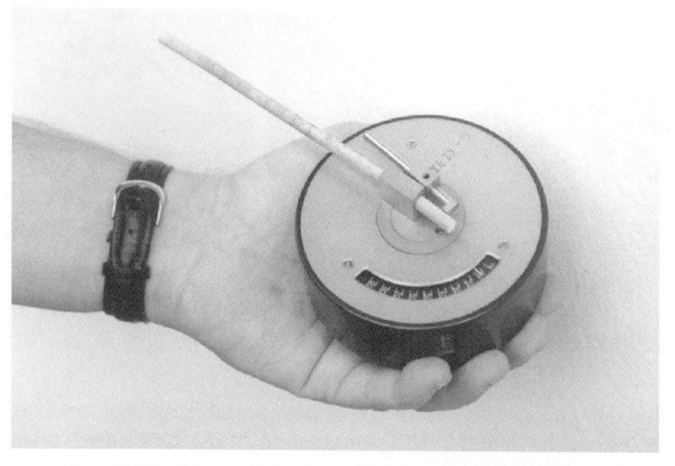

Figure 7.3 Fractometer – a manual tool for ascertaining the bending characteristics of wood

Table 7.3 Fractometer data for determining the type of wood rot

Record of fractometer		Wood property		
Bending moment	Angle of deformity	Strength	Stiffness	Type of rot
Large	Small	High	High	None (healthy wood)
Large	Large	High	Low	White – delignification
Small	Small	Low	High	Brown
Small	Large	Low	Low	White – erosive

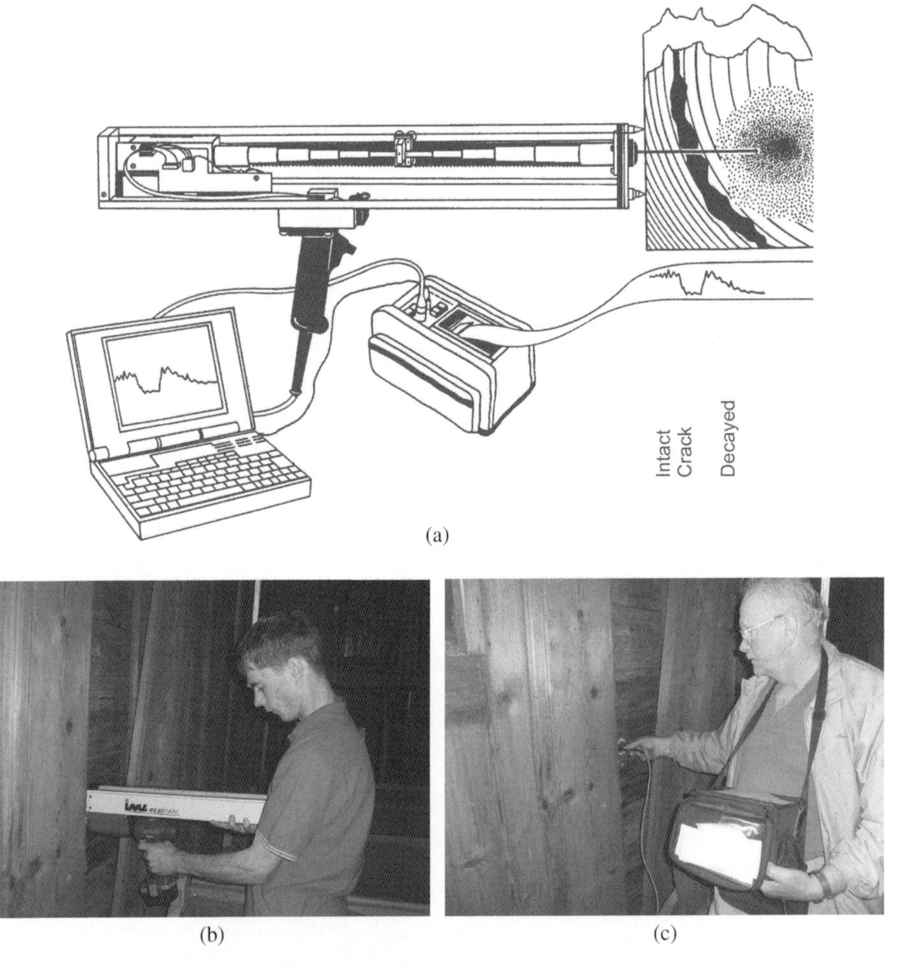

(a)

(b)

(c)

Figure 7.4 Resistograph – scheme of device, its operation and record of decreased resistance of rotten wood to the penetration of the drill bit (a), and an example of its use for analyses of biological damage in a wooden wall of the historical church in Trnovo (b) together with application of the ultrasonic device Pundit (c)

- *Pilodyn* –a penetration tester that records the penetration depth of a steel thin pin into a wooden product. There are various types of Pilodyns: 6J, 4JR or 18J. The 6J Forest Pilodyn shoots a pin with a diameter of 2.5 mm and length of 40 mm using a constant work of 6 J, to a maximum wood depth of 40 mm (i.e. if the wood does not give any resistance). A pin penetrates deeper into less dense, more rotten or with more gallery-damaged wood (e.g. Clarke & Squirrell, 1985; Görlacher, 1987; Kasal & Anthony, 2004).
- *Pin pushing and screw withdrawal resistance meter* – graphically records the development of the penetration of a steel pin into wood or the withdrawal of a screw from wood. The *y*-axis shows the force (newtons) exerted on the pin or screw, while the *x*-axis shows the movement of the pin or screw in the wood at a depth from 0 to 120 mm (Kloiber et al., 2014; Martínez et al., 2015).
- *Baumann hammer* – analyses the hardness of wood based on the penetration of a steel ball with a diameter of 20 mm into its surface layers.

7.3.2.8 Chemical and physical–chemical methods

Chemical and physical–chemical methods indicate (1) the qualitative and quantitative changes in the molecular structure of damaged wood and (2) the qualitative and quantitative profiling of gas, liquid or solid metabolites created during fire, rot and other degradation processes in wood.

Changes in the molecular structure of damaged wood can be determined using classical chemical or physical–analytical methods. For example, in the early stages of brown rot the average degree of polymerization of cellulose quickly decreases while its whole proportion in wood decreases only minimally. Changes in the molecular structure of wood can be identified quicker and usually also more accurately by various physical–chemical analytical methods; for example, by spectrometry (IR, NIR, FTIR, ultraviolet (UV), mass) and chromatography (gas, liquid), and sometimes in their mutual combination or in combination with some other specific analyses (Sandak, A., et al., 2015).

Analysis of metabolisms of fungi and other pests – that is, the microbial volatile organic compounds emitted by wood-decaying fungi (Korpi et al., 1999; Ewen et al., 2004; Evans et al., 2008) and biocontrol fungi (Stoppacher et al., 2010), such as monoterpenes, alcohols, ketones, benzoic acid or furanes – gives information (1) about their presence in buildings and (2) about relations among wood-inhabiting organisms and the possibilities of their use for biological modification in the protection of wood (Section 6.5.1).

7.3.2.9 Biological methods

Since the 1980s, it has been possible to accurately determine the species of wood-decaying fungus using biological–molecular methods. These are based upon the existence of the specific enzyme (protein) composition and nucleic acids of the particular wood pest, as well as upon the typical composition of fatty acids and other metabolites produced by this organism.

Protein methods are implemented using various techniques:

- The sodium dodecyl sulphate polyacrylamide gel electrophoresis technique uses proteins extracted from the fungus cell. They are placed on polyacrylamide gel with the aim to separate them according to size using electrophoresis. The detection of proteins separated according to the size of their molecules is carried out visually using dyes. For example, in this way, *Serpula lacrymans* can be separated from similar species of fungi (e.g. *Serpula himantioides*) based on a different protein composition (Palfreyman et al., 1991; Schmidt, 2000).
- Immunology techniques work with fragments or extracts from the fungi mycelium. They are injected into a test animal organism (mouse, rat, etc.). This organism then produces a specific antiserum that is investigated using various methods (e.g. immunofluorescence). This method facilitates the identification of various species of fungi and also the determination of the initial stages of wood rot (Clausen et al., 1991; Clausen & Kartal, 2003).

Deoxyribonucleic acid (DNA) methods are based on molecular genetics. Every organism (i.e. also every species of decay-causing fungus) has a specific composition of DNA. Depending upon the encoded sequence, DNA is formed by the alternating base components of adenine (A), thymine (T), guanine (G) and cytosine (C). These components are arranged on single-strand or double-strand arrays. Double-strand arrays have a specific lateral coupling of components, G–C and A–T, creating double strands. Some portions of fungal DNA can be decreased due to long-term storage of rotten wood (Børja et al., 2015). Identification of wood-decaying fungi based on the composition of nucleic acids can be carried out using several techniques, from which the most important are the following two:

- The polymerase chain reaction (PCR) technique – during which the double-strand DNA is first divided via denaturation into two separate fibres, usually by heating to 90–95 °C for ~1–5 min. In the second stage, the two denatured DNA strands being investigated undergo a hybridization reaction at 35–60 °C when their required sections are marked with primers (short sections of DNA consisting of 15–30 parts of A, G, C or T); that is, the first primer (positive orientation) binds to one DNA strand and the second primer (negative orientation) binds to the second complementary strand. In the third stage, the bordered DNA section undergoes a replicating PCR at 60–75 °C, whilst new copies of DNA in the amount of 2^n are synthesized in n cycles in a geometric sequence using the delineated DNA template; that is, within 30 s to 2 min, several tens of thousands of DNA units are created. In the fourth and final stage, the products of delineated DNA created undergo detection (e.g. separation using gel electrophoresis and subsequent visualization of fragments using a fluorescent source or UV emitter) and the detected products are compared with an accessible gene bank of

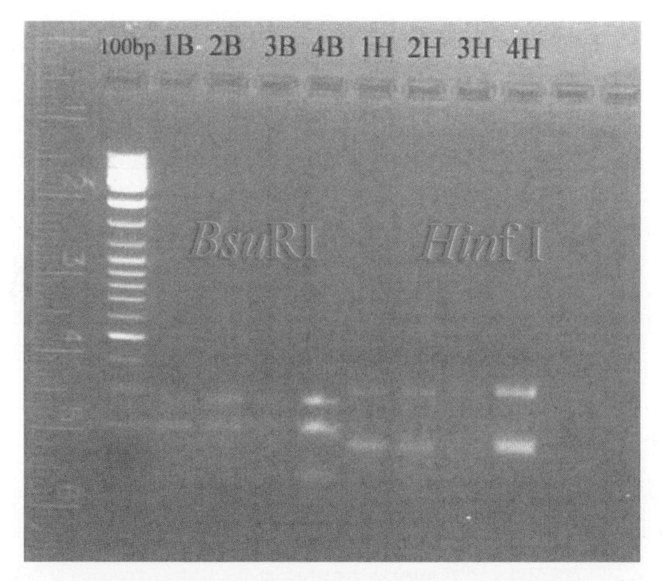

Figure 7.5 Identification of selected fungi from *Trametes* genus using PCR–restriction fragment length polymorphism technique with two restriction enzymes: *Bsu*R1 (B) and *Hin*f1 (H): 1, *Trametes versicolor*; 2, *Trametes pubescens*; 3, *Trametes sauveolens*; 4, CTB 863 A (s.v.)

Source: Bobeková, E. and Horáček, P. (2013) *Identification of Wood Damaging Fungi in Buildings with Help of Genetic Molecular Methods*. (In Slovak, English abstract). TU Zvolen, Slovakia, 66 p. Reproduced by permission of TU Zvolen

known species of fungi (www.ncbi.nlm.nih.gov – Moreth & Schmidt, 2000; Schmidt & Moreth, 2002).

- Modified PCR techniques are also suitable for analysing rotten wood without fruit bodies of fungi (i.e. only with the residues of older mycelium) or for distinguishing between genetically very closely related species of fungi. An advantage of modified PCR techniques is the high speed of analysis, accuracy and objectivity. For example, Pristaš and Gáperová (2001) used this technique for distinguishing between related species of fungi from the *Ganoderma* genus, and Bobeková and Horáček (2013) from the *Trametes* genus (Figure 7.5). The PCR–ribosomal DNA internal transcribed spacer, PCR–restriction fragment length polymorphisms of rDNA, or species-specific priming PCR techniques were similarly used to identify various isolates of wood-decaying fungi from *Coniophora* (Schmidt et al., 2002) or *Armillaria* (Potyralska et al., 2002) genera, or from so far unknown species of decaying fungi (Blanchette et al., 2004). Methods based on DNA sequence information are described in more detail by Kirker (2014).

Matrix-assisted laser desorption/ionization time-of-flight mass spectroscopy is a technique used for analysing proteins and nucleic acids. These biomolecules, or the whole cell of an organism, are embedded into crystals of matrix molecules, where they absorb energy from a laser. Thus, they are ionized and transformed into the gas phase. The ions created are accelerated in an electrical field and their flight time is recorded by a detector. The flight time of the ions analysed is expressed in the form of mass : charge (m/z) signals. Each species of wood-decaying fungus produces specific signals by which they can be distinguished from each other (Schmidt & Kallow, 2005).

Fatty acids analysis is a method based on the principle that microorganisms produce more than 200 different fatty acids. However, a certain microorganism only produces specific types of fatty acids in an accurate mass ratio. After isolation and esterification, these acids are determined qualitatively and quantitatively using gas chromatography. Diehl et al. (2003), for example, identified the wood-decaying fungi *Trametes versicolor*, *Trametes hirsuta*, *Trametes pubescens* and others in this way.

7.3.3 Diagnosing the age of wood

7.3.3.1 ^{14}C wood analysis

The basis of this method is the natural creation of the ^{14}C isotope in the atmosphere, caused by cosmic radiation, and its assimilation into plants, including wood of trees, in the form of $^{14}CO_2$ (Horský & Reinprecht, 1987; Friedrich et al., 2006). The ^{14}C half-life – that is, the time for which half of the ^{14}C atoms disintegrate into ^{12}C – is $\tau_{1/2} = 5730$ years, which is described by the kinetics of the monomolecular reaction:

$$N_A = N_{0A}\, e^{-\gamma\tau} \tag{7.1}$$

where N_A (number/kg) is the number of ^{14}C atoms in the wood at present (i.e. at real time τ), N_{0A} (number/kg) is the number of ^{14}C atoms in the wood of a growing tree for $\tau = 0$, γ is the disintegration constant and τ is the ascertained age of the wood.

7.3.3.2 Dendrochronological analysis of wood

Dendrochronological analysis is based on specific alternating early and late wood in trees growing in a defined time and climate. Trees of the same species in a defined climatic zone, forest formation and time of growth have similar increments of early and late wood. Dendrochronological analysis of a wooden product is based on measuring the width of its annual (minimum 40–60) rings using a binocular microscope or DendroCT (Bill et al., 2012) with an accuracy of 0.1–0.01 mm and a subsequent comparison of the annual ring curve with standardized reference chronologies for the same species of wood using a computer. During the growth of a tree, there are always certain climatic extremes that, in the form of very narrow or very wide annual increments, assist in finding the points of conformity with the reference chronology. For example, in

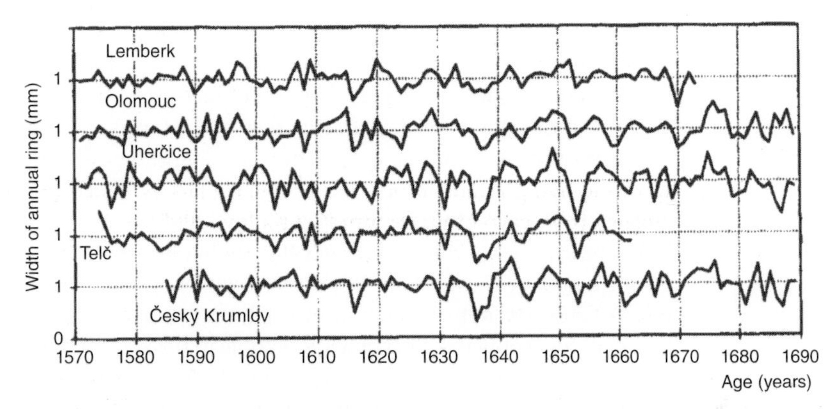

Figure 7.6 The dendrochronology of fir wood for the territory of the Czech Republic – a standardized reference chronology can be prepared based on identical, minimum increments of fir wood from various areas in the monitored climatic zone

Source: Kyncl, T. (1999) Dendrochronological dating of wood as a part of historical building's investigation in Czech Republic. (In Czech, English abstract). In: *Reconstruction and Conservation of Historical Wood 99*, pp. 15–20. TU Zvolen, Slovakia. Reproduced by permission of TU Zvolen

Germany there are well documented reference chronologies for oak wood from 360 BC to the present day, or also for pine wood from AD 799 to the present day. In the Czech Republic there are reference chronologies for fir wood from 1117 to 1996 and spruce wood from 1325 to 1998 (Kyncl, 1999).

Dendrochronological analysis is often used for dating and linking of arts; for example, triptychs (Olstad et al., 2015) and archaeological wood (Stelzner & Million, 2015). The accessibility of standardized reference chronologies for a particular time period varies and therefore they must be obtained by bridging; that is, by assembling them from annual ring curves of wood of a known age determined in accordance with archive and other data (Figure 7.6).

7.4 Sterilization of biologically damaged wood

Various physical and chemical sterilization methods are used for killing of pests in biologically damaged wood:

- heating (radiant, contact, microwave, dielectric or flame heat) – that is, for disposal of insects and fungi;
- freezing – that is, for disposal of insects and fungi;
- exposure to γ-radiation (deep sterilization), or to X-ray, UV radiation and plasma (surface sterilization) – that is, for disposal of insects, fungi and bacteria;

- exposure to ultrasound or high vacuum, for disposal of insects;
- exposure to toxic gases and vapours of liquid or solid substances with high vapour pressure, mainly for disposal of insects;
- exposure to non-toxic gases that displace air oxygen in wood, mainly for disposal of insects.

Sterilization serves for killing of pests that actively attack wood (e.g. insect larvae or fungal mycelium) or in existing higher humidity conditions represent a real danger of wood attack (e.g. insect eggs or spores of fungi). Sterilization is also a short-term protection of wood; however, when the effects of sterilization end, the wood again becomes accessible to the activities of biological pests. The long-term protection of damaged and sterilized wood is only provided by convenient structural protection measures and, if necessary, also by chemical protection using stable fungicides and insecticides.

From an ecotoxicological and health viewpoint, the physical sterilization of wood is usually preferable to the chemical one. Using chemical sterilization, it must be taken into account that toxic gases must first be removed from the wood prior to its reuse; for example, evaporate in specially ventilated rooms. However, some problems also occur in wood sterilized physically using γ-radiation or other forms of radiation, when the polysaccharides and lignin depolymerize and this wood has lower stiffness and strength and becomes more accessible to attack by moulds, decay-causing fungi and insects. Sterilization heating using high temperatures over 160 °C also causes more significant changes to the chemical structure of the wood with a negative impact upon its strength; on the other hand, polyphenols, quinones and other biologically active and/or waterproofing substances are created in the wood, which increase its durability.

7.4.1 Physical sterilization of wood

7.4.1.1 Thermal sterilization of wood: heating or freezing

Heating and freezing sterilizations are ecological and they can be used for various wooden products (furniture, flooring, artefacts) or for constructions (trusses, ceilings):

- heating connected with a decrease in the wood's humidity (unsuitable for rare artefacts)
  - radiant heating using hot air
  - contact heating
  - microwave and high frequency heating
  - flame;
- heating without decreasing the wood's humidity, in air-conditioned chambers;
- freezing of wood.

Wood-boring insects die in wood at ~55 °C after ~1 h as a result of protein coagulation and destruction of their enzymes. However, to eliminate decay-causing fungi or moulds, it is necessary to use higher temperatures, usually from 80 to 100 °C acting for 1–4 h. Only elimination of the mycelium of the true dry-rot fungus (*Serpula lacrymans*), both in wood and masonry, is slightly easier, as it dies at a temperature of 37–40 °C for 3–6 h (Mirič & Willeitner, 1984; Bech-Andersen et al., 2001; Terebesyová et al., 2010). However, it should be emphasized that the spores of wood-decaying fungi (including *Serpula lacrymans*) are much more stable than mycelia and have a short-term resistance even to temperatures exceeding 100 °C. A certain risk in the thermal sterilization of some kinds of wood at temperatures ~100 °C is the formation of attractants; that is, substances that can attract certain species of termites or other wood borers to such wood (Doi et al., 2001). In heating sterilization it must be kept in mind that at temperatures above 50 °C there is a hazard of softening of natural waxes and damaging of glues and paints on polychromed sculptures, old furniture, wood-panel paintings, and so on (Beiner & Ogilvie, 2005).

Radiant heating of wood using hot air is used in central Europe for wood trusses to eliminate the larvae of house longhorn beetle (*Hylotrupes bajulus*) and wood-worm beetles (Anobiidae). Air heated in mobile oil or gas generators to 100–130 °C is led to an enclosed truss or other object via a system of aluminium pipes (Figure 7.7). Air in the object is quickly heated to ~90 °C and wooden elements are by hot air heated to a sterilization temperature of ~55 °C for 2–20 h depending upon their cross-section. The temperature in sterilized wooden elements is indicated using sensors inserted to their centre - usually into elements with the largest cross-section (Makovíny et al., 2011). Control wood samples containing larvae of wood-boring insects cultivated in a laboratory can also be used to optimize the time of sterilization; that is, these samples are regularly opened in order to ascertain the vitality and mortality of larvae exposed to higher temperatures (Makovíny et al., 2012). Sutter (2002) states that of the larvae of the important indoor wood borers, clearly the most resistant to hot air were the larvae of the house longhorn beetle (*Hylotrupes bajulus*), when their lethal temperatures and exposure times were as follows: 50 °C/300 min; 52 °C/150 min; 54 °C/90 min; 56 °C/65 min; 60 °C/50 min. The larvae of *Anobium punctatum* and *Lyctus brunneus* were killed at more moderate conditions; for example, 56 °C/20–25 min. Hot air is also used to sterilize wood affected by the true dry-rot fungus (*Serpula lacrymans*), but not to destroy its spores (Bech-Andersen et al., 2001; Terebesyová et al., 2010).

Contact heating of wood using a flexible heater matrix is suitable for difficult-to-access wooden elements, such as walled heads of beams in ceilings.

Microwave and high-frequency heating is suitable for the sterilization of dielectrics with a proportion of water; that is, also the wet wood and biological pests in the wood. In this method, microwaves with a frequency of 300 MHz to 300 GHz produced by a magnetron (with a wavelength from 1 m to 1 mm), or high-frequency waves from 3 kHz to 300 MHz, produced by a high-frequency generator (with a wavelength from 100 km to 1 m), are transmitted into the wood. In wet wood, radiant energy interacting with water molecules is transformed into heat. Live organisms of biological pests (eggs and larvae of insects;

(a)

(b)

Figure 7.7 Sterilization of a wooden truss using hot air at 100–120 °C (a) from a heat generator (b)

mycelia of fungi) contain a high proportion of water, and so it is possible to heat and kill them in a less wet wood within several seconds to minutes.

Using a microwave emitter with a frequency of 2.45 GHz, Kisternaya and Kozlov (2007) managed to heat timber of bigger cross-sections to temperatures of 53–55 °C, which are lethal temperatures for insect larvae, in ~120–240 min, but the temperatures then had to be held for another 30 min in order for the larvae to die. Fleming et al. (2005) studied the lethal effects of microwave heating in common commercial microwave ovens on the larvae of *Bursaphelenchus xylophilus*: 100% mortality of larvae was reached at a temperature of 62 °C

(a) (b)

Figure 7.8 Sterilization of oak elements in the articular UNESCO church in Hronsek, Slovakia (a) by microwave heating eliminating larval activity of wood-worm beetle (b)

or, in case of uniform microwave heating, already at lower temperatures of 46 or 53 °C.

Microwave and high-frequency sterilization heating are mainly suitable for small wooden elements without metals, since they heat more intensively and the wood in their vicinity may carbonize and even initiate a fire. The possibility of disturbing the wood via drying cracks and damaging coatings and glued joints is also a danger during this type of heating. Makovíny et al. (2012) killed the larvae of *Hylotrupes bajulus* using microwaves at a temperature of 50 °C for 34 min. Several types of portable microwave emitters for gradual, local cross-section heating of rafters, columns and other elements of trusses or all-wooden houses, churches, etc. are used in practice (Figure 7.8).

Tanning flame of wood is not used today due to safety reasons. However, in the past, the heads of ceiling beams inserted into potentially wet walls were sterilized with a flame, in which the carbonized layer of wood was more hydrophobic and contained biocide-active tar substances (higher aromatic carbohydrates, phenols, quinones) having a long-term protection effect against pests.

Heating of wood in stationary or mobile air-conditioned chambers is used to sterilize historic furniture, musical instruments and sculptures. The aim is to preserve a constant moisture in these precious wooden items during sterilization and to not allow them to be deformed by the moisture stresses, and to avoid damaging their joints, layers of adhesives, coatings, polychrome, and so on. This sterilization method could be problematic for artefacts consisting of several materials (wood, metal, plastic, etc.) since, due to differing thermal conductivity, greater stress can occur followed by cracks and scalding. In the air-conditioned chambers, the temperature of the circulating air (usually 55 °C) and its relative humidity (modified depending upon the initial moisture of the wood and air temperature) are continuously regulated by a microprocessor. Only wooden products and elements with the same moisture content and, if possible, of the same type of wood, should be placed in the chamber. Museums currently use the Thermo Lignum® pest control treatment, which consists of accurate regulation and control of air temperature and relative humidity in special chambers, whereas inert gases may also be used (Nicholson & von Rotberg, 1996).

Freezing of wood is used to decrease its internal temperature to around −25 °C within a relatively short time (Hansen, 1992). This temperature is then retained until the biological pests die, usually for 1–7 days. During strong freezing, water crystals are formed in the pest organisms and cause their irreversible damage. Freezing must be sufficiently rapid. Shock freezing is best; that is, the freezing phase should not last longer than a few hours so the pest organisms cannot adapt to the change in temperature. Insect eggs resist lower temperatures better and, therefore, double freezing technology is sometimes used for their elimination. An example is the sterilization of the eggs of the common furniture beetle (*Anobium punctatum*) which survive a single freeze of −30 °C for even 2 days. Here, it is beneficial to use two-stage freezing with an inter-stage of heating during which time the larvae (with a lower tolerance to low temperatures) hatch from the eggs. During freezing, the original wood humidity only changes slightly and the creation of cracks is less likely than during sterilization heating. One exception is wet wooden elements that also contain free water, or whose moisture is reaching the fibre's saturation point since this water freezes easily and may create frost cracks in the wood.

7.4.1.2 Sterilization of wood using X-ray and γ-radiation

Gamma rays with a wavelength from 4×10^{-10} to 10^{-13} m (emitted from the nuclei of radioactive atoms, e.g. ^{60}Co, ^{90}Sr, ^{137}Cs) and X-rays with a wavelength from 10^{-7} to 10^{-10} m (produced using a high-voltage source in X-ray equipment) can kill living organisms. Unlike with X-rays, the penetration of γ-rays into wood is very good, which also allows them to be used for sterilizing large items throughout their whole cross-section. Doses of radiation are stated in grays (Gy). This is the amount of radiation energy absorbed per unit of mass of exposed material (1 Gy = 1 J/kg). The dose rate of radiation is the dose per time unit expressed in grays per second or per hour.

Bacteria and fungi are more resistant to γ-radiation than wood-boring insects are. In order to eliminate fungi and bacteria in wood, a dose of radiation from 1 to 18 kGy is required, depending upon the type of organism (Unger et al., 2001). For example, an 8 kGy dose of radiation can destroy all types of moulds, whilst insects can be eliminated using lower doses of 0.25–3 kGy (Urban & Justa, 1986). The required doses of radiation are lower if the wood is heated simultaneously. For example, the mycelium of the dry-rot fungus (*S. lacrymans*) at ~20 °C is killed by a dose of 2–3 kGy, but when the wood is heated to 50 °C

Figure 7.9 Radiation chamber with a ^{60}Co radioisotope for sterilization of wooden items in the Museum of Central Bohemia in Roztoky, Prague, Czech Republic

it requires a markedly lower dose of 0.5 kGy. On the other hand, when fractioning doses of radiation over a longer period, it is necessary to use higher doses. For example, in eliminating the mycelia of *Serpula lacrymans*, workers at the Nuclear Research Institute in Řež in the Czech Republic documented that the lethal dose of radiation at a dose rate of 340 Gy/h was 1.32 kGy, but at a lower rate of 22 Gy/h it would increase to as much as 3.65 kGy. When exposing wood to γ-rays, a certain decrease in its strength and in its resistance to microorganisms must be taken into account since cellulose partially depolymerizes and oligosaccharides and monosaccharides separate from hemicelluloses. However, exposed wooden products do not become radioactive and, immediately after sterilization, they can be used in interiors, or undergo reinforcement with conservation agents, protection with preservatives, and so on. Gamma radiation is very dangerous to humans and, death can be induced by a dose of just 2–5 Gy. In practice, sterilization of wooden artefacts is carried in special radiation chambers (Figure 7.9). Mobile robots with movable sources of radiation are used for sterilization of wooden trusses, ceiling, log cabins, and so on. (Teplý et al., 1986).

7.4.1.3 Sterilization of surfaces of wood using ultraviolet radiation, plasma or laser

To eliminate moulds and bacteria growing on the surface of wood and other materials, UV, plasma or laser energy field types of radiation and discharge can be used to thoroughly destroy the microorganisms to a penetration depth of ~50–200 μm.

UV emitters with a wavelength of 10–380 nm are commonly used for sterilizing the surface of wood and other materials. Sterilization effects have also plasma discharges in air, O_2, N_2, CO_2 or other environments (Shintani, 2010; Tiňo et al., 2014). Surfaces of wood pretreated with a CO_2 laser appeared even more resistant to moulds *Aspergillus niger* and *Penicillium brevicompactum* (Vidholdová et al., 2016, unpublished results). All these energy fields create radicals from the molecules of the organic substances present in the surface of wood (polysaccharides, lignin, extracts) and also in the microorganisms or enzymes acting on the surface of wood (bacteria, moulds, etc.). Radicals subsequently change the living organism of the biological wood pest via homolytic chain reactions and the pest dies.

7.4.2 Chemical sterilization of wood

Chemical sterilization of wood with toxic and nontoxic gases, and also vapours of toxic liquids and solid substances, is mainly used for eliminating wood-boring insects. Fungi organisms are apparently more resistant to toxic gases and vapours than insects are (Morrel & Corden, 1986). Gases penetrate into wood more quickly, deeper and more evenly than liquids do. The effectiveness of sterilizing biologically damaged wood with gases depends upon the type of gas, its concentration and time of action. Depending upon the size of the wooden product, the sterilization time using toxic gases is from 2 to 48 h, or using nontoxic gases from 1 to 4 weeks. The time of sterilization with gases is reduced by increasing the temperature and pressure in the environment.

The sterilization of wood was already taking place in 500–900 BC using smoke from fireplaces or using sulphur dioxide. In the middle of the 19th century, creosote vapour or a mixture of ozone and carbon dioxide (CO_2) was used. At the beginning of the 20th century, various toxic gases were used for sterilizing wood; for example, hydrogen cyanide, carbon disulphide, tetrachloromethane or, later, ethylene oxide and phosphine (PH_3) (Unger et al., 2001). Since the middle of the 20th century, other toxic gases have been gradually applied; for example, methyl bromide (CH_3Br) and sulphuryl fluoride (SO_2F_2), nontoxic nitrogen and argon, or slightly toxic CO_2. From toxic gases or a combination of toxic and less toxic gases, various commercial products are still used – usually based on ethylene oxide, CH_3Br, PH_3 or SO_2F_2 together with CO_2.

Only specialist organizations with a valid licence can work with toxic gases since they are highly toxic to humans and vitiate the environment. Some gases are chemically reactive, which may lead to damage of coatings, adhesives and, with longer exposure, of the wood itself, depending upon its humidity.

New methods of the chemical sterilization of wood with a long-term effect are also being simultaneously implemented. They are based on gradual evaporation of the molecules of carbamate, isocyanate and other toxic substances (Unger et al., 2001).

Chemical sterilization of movable wooden items is usually carried out in hermetically sealed metal and plastic chambers or in steel pressure autoclaves equipped with an anticorrosive surface or coating.

Chemical sterilization is also suitable for wooden constructions such as trusses or the interiors of churches. However, these structures must be thoroughly insulated from the ambient environment; for example, by using polyamide or polyvinylchloride foils impermeable to gases (Ashurst and Ashurst, 1996).

7.4.2.1 Toxic sterilization gases

Some toxic gases and vapours – for example, hydrogen cyanide (HCN), carbon disulphide (CS_2) or tetrachloromethane (CCl_4) – are no longer used for sterilization of wood. Other types of toxic substances are still deployed, however, they have several ecological and health limitations.

CH_3Br is a very effective gaseous insecticide against all the developmental stages of wood-boring insects. It methylates and disturbs the enzymes of insects with —SH groups. It is also toxic to the mycelium of some wood-damaging fungi (e.g. the true dry-rot fungus and moulds), disturbing their cytoplasm. It is applied in chambers when sterilizing wooden sites, timber or pallets, or directly in buildings from steel tanks. The sterilization time for insects, at a CH_3Br concentration of 20–60 g/m^3 and a temperature of 12 °C, takes from 24 to 72 h. However, the time for killing mycelia of basidiomycota is 96 h, and it is 1–2 months for destruction of their spores. CH_3Br vapour is heavier than air and should be circulated during sterilization. After completion of sterilization, the wooden product or construction must be ventilated for at least 12–72 h. Sterilization efficiency of CH_3Br increases in the presence of CO_2 (Scheffrahn et al., 1995). Working with CH_3Br must be strictly controlled since it is also highly toxic to humans with possible carcinogenicity. It also disturbs the ozone layer. In the presence of water, it partially hydrolyses to hydrobromic acid, which corrodes metals and other materials. It is still used, but it is expected that it will be banned in the near future.

Ethylene oxide (CH_2=(O)=CH_2) has a wide spectrum of sterilization effects; that is, insecticidal, fungicidal (mainly for moulds) and bactericidal. It reacts with proteins, acts as a protoplasmic poison, and it eliminates all the development stages of insects, including eggs. This gas is applied similarly to CH_3Br. It is often used in combination with CO_2, which stimulates the breathing of insects and therefore also a higher receipt of ethylene oxide. The sterilization time with the ETOX system (90% ethylene oxide and 10% CO_2) at concentrations of 150–200 g/m^3 takes 4–6 h, and with the ETOXEN system (10% ethylene oxide and 90% CO_2) at concentrations of 400 g/m^3 it is as long as 24 h. The mixture of ethylene oxide with air is explosive, and it evaporates from wood with greater difficulty. Its use has been markedly limited since the start of the 21st century due to its carcinogenicity and its negative environmental impact.

SO_2F_2 is a gas with a sufficient insecticidal effect upon larvae, pupas and imagoes of insects, but only with a slight effect upon their eggs. At the same time, it also liquidates some moulds and mycelia of decay-causing fungi (Schmidt

et al., 1997; Pfeffer et al., 2006). The sterilization time for a concentration of 20–50 g/m^3 at a temperature of 20 °C is 24–72 h, but if there is a simultaneous need to eliminate eggs this must be extended (Williams and Sprenkel, 1990). CO_2 in synergy supports its sterilizing effect. SO_2F_2 penetrates wood easily, even through coatings, and it also escapes easily from wood. This allows the time for ventilating a sterilized object to be shortened to 4–6 h. The negative environmental impact of SO_2F_2 is relatively lower than of other toxic gases. Now it is used in log fumigation as well (Barak et al., 2010).

PH_3 is a gas with a complex insecticidal effect; that is, it eliminates all the development stages of insects, including eggs. It affects the redox system of mitochondria and disturbs the breathing of insects. In high doses it also has a certain fungicidal effect. The sterilization time using PH_3 at a concentration of 2–4 g/m^3 and a temperature of 15 °C is fairly long and takes 5–10 days. A disadvantage is the heavy corrosion of metals with the creation of dark phosphides. It is also used for sterilizing damaged wood in a mixture with CO_2 or nitrogen.

7.4.2.2 Nontoxic sterilization gases

Atmospheric air contains 78.10 vol.% nitrogen, 20.94 vol.% oxygen, 0.93 vol.% rare gases and 0.03vol.% CO_2. With a shortage or total lack of oxygen, insects stop breathing and die. On the other hand, fungi can even survive in such conditions where their growth is slowed down or stopped, and their spores do not germinate (Kazemi et al., 1998).

Oxygen concentration in the air can be decreased in several ways:

- replacement of atmospheric air with an inert gas with a concentration of 96–99 vol.% (e.g. with nitrogen and argon);
- increasing the concentration of CO_2 to 60 vol.%;
- decreasing the absolute amount of oxygen in atmospheric air, either by its sorption into suitable absorbents or by a significant decrease of air pressure in deep vacuum.

Nitrogen and argon are inert gases. Their sterilizing principle lies in the displacement of oxygen from wood and decreasing its concentration to below the critical level at which wood-boring insects die (Koestler, 1992; Despot et al., 1999). The minimum critical concentration of oxygen for insects is a very low, from 0.5 to 1% vol. and it must be allowed in wood for a period of 1 to 4 weeks. Inert gases do not corrode wooden or other materials but beware, materials become dry and it is necessary to wet the sterilized environment to avoid the creation of cracks in wood. Inert gases are applied in hermetically sealable containers or in plastic, air non-permeable packaging. However, they cannot be used for the in-situ sterilization of in-built wood, e.g., damaged trusses, since there it is not possible to reach a needed 96 – 99% concentration. Argon is more effective compared to nitrogen but it is also more expensive, i.e., it is more or less only used to sterilize rare historic items.

CO_2 is a gas with an anaesthetic effect upon insects. It also suppresses the growth of fungi mycelia and prevents their spores from germinating. It penetrates wet wood with more difficulty and also desorbs more slowly. Its hydrolysis to carbonic acid is dangerous ($CO_2 + H_2O = H_2CO_3$), since in this form it attacks coatings, adhesives and other materials that are in contact with the wood. CO_2 is used in concentrations of 60–90%. It is often combined with inert nitrogen or with toxic gases such as ethylene oxide or PH_3. It is applied in chambers as well as in larger isolated spaces; for example, it is also suitable for sterilizing wooden buildings wrapped in airtight foils.

7.5 Conservation of damaged wood

Damaged wooden products and elements of constructions can be restored to a certain degree using conservation agents that stabilize them in terms of dimension and improve strength and overall appearance.

7.5.1 Natural and synthetic agents for wood conservation

Conservation agents penetrate into the wood in the form of liquid systems – solutions, dispersions or melts. The depth of their penetration depends upon the conservation technology, the permeability of the wood and the physical and chemical properties of the conservation agent, such as viscosity and surface tension. Depending upon their polarity, molecular mass and other characteristics, they are deposited either only in the lumens of cells or they penetrate also to cell walls of the wood. It is important for the person doing the conservation to know their chemical composition, chemical–physical properties and ecotoxicological parameters, as well as their relation to the wood and surface finishing, so the conservation is performed to a good quality in line with the latest knowledge.

The conservation agents should meet several of the following requirements (Reinprecht, 1998):

- good penetration into wood;
- do not change the shape of the wood during conservation;
- compatibility with preservatives, conservation agents and coatings that were applied to the wood or its surface in the past;
- minimum effect upon the wood's sorption properties;
- ensure long-term stability of the wood's shape, mainly required by wet archaeological wood whose cells should be stabilized in the swollen state;
- integrate loose or broken parts of the wood;
- reinforce the wood;
- be resistant to weather, fire and biological pests, and also protect the wood itself against a complex of these degrading factors;

- facilitate gluing, surface finishing or other technological operations with the wood that need be carried out after conservation;
- be safe to health and a minimum burden upon the environment;
- be a reversible technology process –that is, it should be possible to remove the conservation agent from the wood if its conservation effect worsens after some time or if a better agent is developed;
- have minimum impact upon the aesthetics of the treated artefact or other item.

In practice, almost no conservation agent complies with all these requirements. The type of conservation agent should be selected depending upon the particular situation; for example, whether the aim is to stabilize the shape of the damaged wood, to reintegrate and stabilize against disintegration, or mainly to reinforce it. Permanent dimensional stabilization of wood can only be achieved via the penetration of a stable conservation agent into the wood's cell walls.

The solvents for the preparation of solutions of conservation agents are selected depending upon several factors:

- effect upon the swelling and shrinkage of wood;
- boiling point and vapour tension;
- viscosity, surface tension and other physical and chemical properties which influence the transport of the conservation system (conservation agent and solvent) into the wood and its migration in the wood;
- effect upon the surface finishing of the conserved wood;
- harmfulness to health and environmental impact.

Polar solvents (e.g. water and ethanol) well swell the wood. They have a high affinity to a polar wood substance; that is, they have interactions with hydroxyl and other polar groups in the wood, which is reflected in their slowed movement in the capillary system of wood cells (Nicholas, 1972). For that reason conservation agents diluted in nonpolar solvents penetrate air-dried wood more quickly. The swelling of wood in various liquids is defined by the index of swelling (IS) rate in relation to its swelling in water, $IS_{water} = 100$; for example, $IS_{tetrachloromethane} = 13$, $IS_{toluene} = 17$, $IS_{chloroform} = 30$, $IS_{acetone} = 69$, $IS_{ethanol} = 79$, $IS_{methanol} = 90$, $IS_{ethyleneglycol} = 109$ (Mantanis et al., 1994). Such knowledge allows the selection of the optimum type of solvent for the given conservation agent and technology.

The boiling point and vapour tension of the solvent expressed as the evaporation index (EI) affect not only its ability to transport the conservation agent deeper into the wood, but also the ability of the conservation agent to migrate back to the wood's surface; for example, $EI_{diethylether} = 1$, $EI_{acetone} = 2.1$, $EI_{toluene} = 6.1$, $EI_{ethanol} = 8.3$, $EI_{xylene} = 13.5$, $EI_{water} = 80$, $EI_{turpentine} = 170$, $EI_{ethyleneglycol} = 600$ (Unger et al., 2001). Highly volatile solvents with a lower EI do not transport the conservation agent sufficiently deeply into the wood since they evaporate during the conservation, causing increased concentration and viscosity of the solution, and therefore the conservation agent mainly remains on the surface of the wood. On the other hand, lower volatility solvents with a high EI escape from the treated wood too slowly, and in the longer term they decrease the temperature of the glassy state T_g of the thermoplastic conservation agent in wood, thereby worsening the strength of the conserved wood (Carlson & Schniewind, 1990). During the slow release from the wood, these solvents also may increase the return migration of the conservation agent to wood surfaces (Payton, 1984). For the aforementioned reasons, the use of solvents with a medium evaporation index is recommended, and preferably nonpolar such as toluene or xylene.

In relation to the selection of solvent, the issue of drying and air conditioning of the conserved wood is also important. During rapid drying, the conservation agent distributes unequally in the wood and it often returns to the wood's surface.

The concentration of the conservation system is the percentage amount of a conservation agent in solution with a solvent. It is selected taking into consideration the required penetration and retention of the conservation agent in the wood, with the aim to optimally improve its properties. At lower concentrations, solutions have lower viscosity, and they penetrate more deeply and evenly into the wood, which is mainly beneficial for dimensional stabilization of the wood. On the other hand, solutions with a higher concentration of the macromolecular conservation agents have a higher viscosity, they penetrate into capillaries of wood with more difficulty and remain on the wood's surface, which is especially beneficial for surface reinforcement of wood.

Reactive monomers with double chemical bonds are applied into wood as low-viscosity liquids, and there in its pores subsequently chemically interact. Acrylate monomers, such as methyl methacrylate (MMA) or butyl methacrylate (BMA), are the reactive monomers used most. From them only more polar types can penetrate the cell walls of wood; for example, 2-hydroxyethylmethacrylate (HEMA). The subsequent polymerization of acrylate or other reactive monomers in wood is initiated chemically or via γ-radiation (Šimůnková et al., 1983; Tran et al., 1990; Adamo et al., 2015).

Chemical polymerization of monomers in wood takes place in the presence of catalysts (e.g. organic peroxides) added to the monomer. At an increased temperature, the catalyst decomposes into radicals and they initiate radical polymerization reactions in the molecules of the monomer. The type of catalyst, its quantity and temperature must be optimized in order to:

- avoid too early a polymerization of the monomer – that is, during its application into the wood, since an increase in its viscosity would slow its further penetration into the wood,
- ensure that polymerization does not take too long, since volatile monomers could evaporate from the wood during conservation.

Radiant polymerization of monomers is initiated by γ-radiation (e.g. from ^{60}Co or ^{137}Cs). The optimum dose rate for curing monomers in wood is 10^3 Gy/h, with a total radiation dose of 10–100 kGy (e.g. for curing MMA,

~50 kGy dose is required). For doses over 100 kGy there is a danger of damaging the wood's structure, mainly as a result of cellulose decomposition. An advantage of the radiation method is the even polymerization rate of the monomer throughout the whole cross-section of the treated wood. However, the temperature should be monitored since it increases with the dose rate and can cause evaporation of the monomer from the wood. Radiant polymerization of monomers is suitable for reinforcing smaller wooden artefacts. Radiation also destroys biological pests in wood (Section 7.4.1).

Polymers and other synthesized or natural agents are applied to wood in the form of solutions, dispersions or melts. Their chemical structure is (1) not changed in the preserved wood (e.g. of polyacrylates, polyethylene glycols (PEGs) and beeswax) or (2) changed after/during the conservation process due to polycondensation or polyaddition reactions (e.g. of unsaturated polyesters, epoxides and phenolic resins).

7.5.1.1 Natural agents for the conservation of damaged wood

Several natural agents were in the past used for manufacture of wooden products:

- proteins, as adhesives (bone and leather glues, casein);
- vegetable oils, as coatings (varnish, oil lacquer);
- waxes, as hydrophobizers, dimensional stabilizers and reinforcing agents of a wood surface (beeswax, montan wax, paraffin);
- plant resins, as a hydrophobizer and reinforcing agent of a wood surface (colophony, Canadian balsam, amber, copal, dammar, mastic);
- animal resins, as furniture polish (shellac);
- polysaccharides, as adhesives (starch, dextrin).

Some natural agents are also suitable for the conservation of damaged wood (e.g. Zelinger et al., 1982; Reinprecht, 1998; Unger et al., 2001). Their advantage is their low molecular weight and, therefore, good penetration into wood using a suitable solvent. However, they do not significantly improve the strength of damaged wood and just integrate it and improve its dimensional stability. Current developments in the chemical industry and new synthesized polymers mean that only a few of them are now used for the treatment of damaged wood; for example, beeswax and its combination with colophony or dammar, and also possibly shellac. Sugars, mainly saccharose, and to a lesser extent lactose or mannose, are used for treating wet archaeological wood.

Beeswax is prepared by melting honeycombs in hot water, with subsequent cleaning and whitening. It contains approximately 72% alkyl esters of fatty and wax acids (R_1) with higher molecular alcohols (R_2):

$$R_1{-}CO{-}O{-}R_2 \qquad \text{Alkyl ester of fatty or wax acid}$$

where R_1 is $C_{15}H_{31}$ and other carbohydrate radicals of acids, and R_2 is $C_{30}H_{61}$ and other carbohydrate radicals of alcohols. It also contains wax acids (~13% of $C_{24}H_{49}COOH$ to $C_{30}H_{61}COOH$), higher carbohydrates (~12% of $C_{25}H_{52}$ to $C_{32}H_{66}$), propolis, lactones, colourings and other substances. The density of beeswax is 950–970 kg/m^3. Its melting point is from 61 to 70 °C and it is fairly soft. Beeswax is soluble in toluene, petrol, turpentine and chloroform, and insoluble in water or in cold ethanol. It has good water resistance, quite good stability against ageing (but becomes fragile with age) and high resistance against acids. For conserving wood, it is mainly applied in turpentine or as a hot melt. Beeswax is suitable for treating small wooden artefacts with a less pronounced level of biological damage.

Colophony is distillation residue from turpentine balsam of coniferous wood, mainly from the *Pinus* species. It contains ~95% of terpene acids (from which ~60% are abietanic acids with conjugated double bonds that tend to be oxidized by oxygen in the air, inducing their gradual browning and embrittlement) and a small amount of phenolic compounds. Colophony is soluble in aliphatic and aromatic chlorinated carbohydrates, in acetone and in alcohols. It is insoluble in water, but under heat it dilutes in water solutions of alkalis, creating resin soaps. Colophony softens at 70–80 °C, and at a temperature exceeding 120 °C it forms a low-viscosity melt. In practice, it is usually applied diluted in ethanol and for conserving exterior wood it is applied in combination with turpentine oil and varnish.

Dammar is obtained from broadleaved, tropical wood species of South East Asia, the Diptorocarpaceae family. It consists mainly triterpenoids. It is well soluble in toluene, xylene, chloroform or in turpentine oil, but its solubility in acetone or ethanol is weaker. It softens at 85–90 °C. Dammar resists weather well since it repels water, and its chemical structure, with a minimum proportion of double covalent bonds, is only slightly disturbed by UV radiation.

Shellac is obtained from the surface of protective cover of *Lacifer lacca* larvae, which is a parasite on trees and bushes in India, southern China and Thailand. Shellac purified with water contains higher aliphatic polyhydroxyl acids (mainly aleuritic acid) and their lactones. It is insoluble not only in polar water but also in nonpolar organic solvents. It is well soluble in ethanol and other alcohols as well as in glycols. It softens at 82–88 °C. Shellac is hard but quite brittle, which predetermines it more or less only for surface reinforcement of small wooden items.

$$CH_2OH{-}[CH_2]_5{-}CHOH{-}CHOH{-}[CH_2]_7{-}COOH \qquad \text{Aleuritic acid}$$

Saccharose is a disaccharide consisting of glucose and fructose (α-D-glucopyranosyl-β-D-fructofuranose, $C_{12}H_{22}O_{11}$). It is obtained from sugar beet and sugar cane. Sucrose is very well soluble in water; for example, at 25 °C it is possible to prepare a 66% sucrose solution. As a water solution, it easily penetrates the cell walls of wood and stabilizes them in a permanently swollen state. However, for treating air-dried wood, such a stabilization effect is insignificant and even unwelcome – it results in permanent enlargement of the original dimensions of dry wood and a decrease in its strength as well as a certain growth in its hygroscopicity.

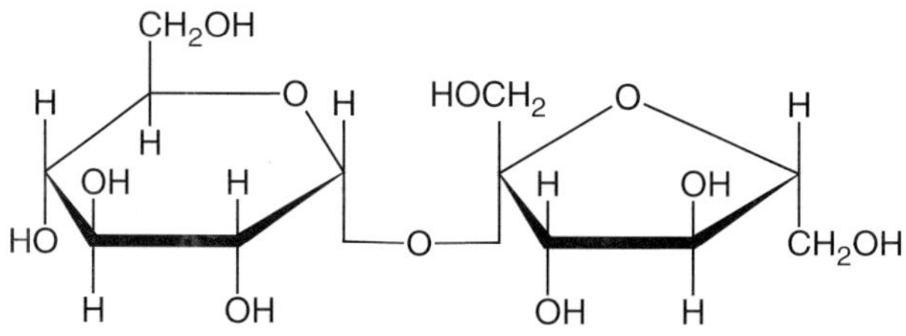

Saccharose

On the other hand, saccharose is suitable for use in the conservation of wet archaeological artefacts of a slightly lower historical value and with a lesser degree of damage (Morgós, 1999). It is applied to wet wood using diffusion technology. Wet wood treated with sucrose should be dried slowly or using vacuum sublimation to avoid the creation of cracks. The sucrose present in wet archaeological wood retains its original dimensions even after drying due to the penetration of its molecules not only into the wood's lumens and cavities, where it crystallizes, but also into the cell walls where it forms a thin, amorphous, glassy layer. The anti-shrinkage efficiency (AShE) of saccharose for treatment of wet wood is ~80–96%. It permanently increases the wood's hygroscopicity; that is, it is recommended that treated wood be kept in an environment with a relative air humidity of maximum 65–70%, so it does not receive a lot of air humidity and does not become wet on the surface.

7.5.1.2 Synthetic agents for the conservation of damaged wood

Synthetic substances for the conservation of damaged wood have been used for ~80 years. Their advantages include accessibility, acceptable price, identical chemical composition for a given type, and properties within a particular product, or the possibility to prepare polymeric substances with a special structure and properties. The majority of synthetic substances are polymers or suitable monomers able to transform into a polymer in the treated wood. The most used today include polyacrylates, unsaturated polyesters, polyurethanes and epoxides for treating dry wood, as well as PEGs for treating wet archaeological wood. More rarely used are amino resins, phenolic resins, silicones and polyvinyl derivatives.

Synthetic polymers for the conservation of wood differ in structure and modes of their structure creation (Table 7.4). Properties of the conserved wood are predetermined by their final structure in wood (Table 7.5).

Table 7.4 Structure and modes of creation of synthetic polymers used for wood conservation

Structure of polymers

- Chemical structure:
  - polyacrylates
  - epoxides
  - etc.
- Supramolecular structure:
  - thermoplasts – polymers with a linear or slightly branched structure which, in wood, soften when heated and solidify when cooled without changing the chemical structure (e.g. polyacrylates, polyvinylacetate)
  - reactoplasts/thermosets – polymers applied in wood, usually in the form of linear or slightly branched structure, which, when heated or influenced by a catalyst, irreversibly net into the form of a low-soluble to completely insoluble and infusible spatial structure (unsaturated polyesters, epoxides, amino resins, phenolic resins, etc.)

Mode of polymer creation (in factory or in wood)

- Polymerization:
  - poly-reactions, in which chemically react the same or different types of reactive monomers
  - monomers with one double bond form linear polymers or copolymers
    - vinylacetate → polyvinylacetate
    - MMA + BMA → acrylate copolymer
- Polycondensation:
  - poly-reactions of identical or differing low molecular weight substances (from which each has at least two reaction groups of types $-OH$, $-NH_2$, $-CH{=}O$, $-COOH$, etc.)
  - small molecules (H_2O, NH_3, HCl, etc.) are released from substances after their mutual reaction and the volume of the created linear or spatial polymer decreases, which can lead to shrinkage of the treated wood
    - ethylene glycol + dicarboxylic acid → polyester
    - formaldehyde + urea or melamine → amino resin
    - formaldehyde + phenol or resorcinol → phenolic resin
    - epichlorohydrin + 2,2-bis-(*p*-hydroxyphenyl)-propane → linear base of epoxide
- Polyaddition:
  - polyaddition reactions of monomers or oligomers which have at least two groups able to react, creating linear or spatial polymers
  - no small molecules are released during polyaddition; that is, there is the basic assumption that the volume and total shape balance between the reacting substances and the resulting polymer are retained, and the treated wood keeps its original shape
    - 1,6-diisocyanate-*n*-hexane + 1,4-dihydroxy-*n*-butane → polyurethane
    - ethylene oxide + ethylene glycol → PEG
    - linear base of epoxide + diethylenetriamine → cured epoxide

Table 7.5 Effect of synthetic polymer structure on the properties of conserved wood

Molecular weight of the polymer

- Greater molecular weight:
  - lower solubility → less retention by wood
  - higher viscosity of the conservation agent's solution → worse penetration into wood
  - specifically for linear polymers → a positive effect upon strength and stiffness of the conserved wood

Spatial structure of the polymer

- Linear polymer:
  - in the majority of cases, there is a suitable solvent in which it can be diluted and transported into the wood, or by which it can be extracted from the conserved wood
  - polymer chains are flexible, which creates options for their crystallization
- Branched polymer:
  - difficult or impossible crystallization
- Spatial 3-D polymer:
  - impossible crystallization
  - fragments or sections of polymer chains have low options for movement, which minimizes their solubility and swelling in solvents
  - the immovability of polymer chains is reflected in increased hardness and compression strength, as well as the brittleness of the polymer and the wood conserved with it

Degree of arrangement of the polymer chains

- Amorphous polymers:
  - special 3-D and branched polymers, or also linear polymers with spatially larger functional groups in which a sufficient amount of intermolecular connections (hydrogen bonds and van der Waal's interactions) is not created
  - polymers characterized by the glass point T_g; after this point is exceeded the movability of individual chains improves markedly; that is, the polymer is transformed from a solid glassy state into a rubber or plastic state; polymers with a lower T_g are usually more flexible; the T_g of the polymer can be decreased by adding softeners
  - chains in polymers may permanently move even in a solid (glassy) state, which is reflected favourably in decreased fragility of the polymer and also of the conserved wood
- Crystalline polymers:
  - polymers with linear or minimal branching chains, able to be arranged into crystalline structures
  - in crystalline polymers, their supramolecular structure changes when melting point T_m is reached; that is, when they transform into melt
  - a crystalline structure is expressed in an increased strength but, on the other hand, also in an increased brittleness of the polymer and the conserved wood

Polyacrylates are usually prepared by the polymerization of various esters of acrylic and methacrylic acids:

$$R_1\text{—}(CH_2\text{—}\underset{\displaystyle COOR}{\underset{|}{CH}})_n\text{—}R_2 \quad \text{Polyacrylate}$$

Polyacrylates are applied to wood as polymers or copolymers in the form of organic solutions (Reinprecht, 1998; Unger et al., 2001). The most well-known include polymethyl methacrylate, polyethyl methacrylate (e.g. Paraloid B72), polybutyl methacrylate (e.g. Solakryl BT 55), polymethyl acrylate, polyethyl acrylate, polybutyl acrylate, copolymer MMA/ethyl acrylate (e.g. Paraloid B82), copolymer methyl acrylate/ethyl methacrylate or copolymer MMA/BMA (e.g. Solakryl BMX).

Polyacrylates are usually colourless, transparent like glass, hard but also sufficiently elastic. For treating wood, they are produced in the form of granules or powder, which are well soluble in less polar organic solvents (e.g. toluene, xylene, petrol or acetone). The molecular weight of polyacrylates is relatively high (1×10^4 to 1×10^5), which is also reflected in a relatively high viscosity of their solutions. They are applied to rotten wood and wood damaged with insect galleries as 10–20% solutions with a dynamic viscosity of 5–30 mPa s. Large macromolecules of polyacrylates only remain in the lumens of the wood's cells. They can be extracted from the wood (i.e. the conservation process is reversible).

Polyacrylates have high stability against weather impacts and higher temperatures. The required properties of the treated wood can be achieved by using suitable copolymers. For example, MMA/BMA copolymer is used in conservation owing to its balance of several properties – strength, elasticity and weather resistance – which the conserved wood obtains. The MMA component of this copolymer in the polymerized state has a higher glass point T_g (105 °C) and higher hardness, strength, brittleness and resistance to UV radiation, whilst the BMA component in the polymerized state has a lower glass point T_g (20 °C) and has a greater resistance to hydrolysis in the presence of aggressive chemicals.

Monomers of acrylates are only used for conservation occasionally for smaller wooden artefacts. They polymerize directly in the wood, and this technology requires several safety measures (e.g. work with radio isotopes) and is costly. On the other hand, biological resistance of wood treated with the monomer acrylates is a good. For example, MMA applied into wood as monomer increased wood's bioresistance after its in-situ polymerization (Unger & Unger, 1995), but polyacrylate copolymers applied into wood in organic solvents only slightly increased its biological resistance (Unger et al., 1996; Tiralová & Reinprecht, 2004). The reason is a different distribution of polyacrylates in the wood, present either just in the lumens of cells (polyacrylate applied in an organic solvent) or simultaneously in the inside of the cell walls (application of an acrylate monomer and its subsequent polymerization in the wood).

Unsaturated polyesters are prepared by polycondensation of carboxylic acids and alcohols with two or more functional groups:

- dicarboxylic acids, saturated $HOOC{-}R_1{-}COOH$ (adipic acid, phthalic acid) and simultaneously unsaturated $HOOC{-}R_2C{=}CR_2{-}COOH$ (fumaric acid, maleic acid, etc.);
- dihydric alcohols $HO{-}R_3{-}OH$ (ethylene glycol, butylene glycol, etc.).

$$HO{-}[OC{-}R_1{-}CO{-}O{-}R_3{-}O{-}CO{-}R_2C{=}CR_2{-}CO{-}O{-}R_3{-}O]_n{-}H$$

Unsaturated polyester

When treating wood they are applied together with 20–45% of a reactive monomer (e.g. styrene, $CH_2{=}CH{-}(C_6H_5)$; or MMA, $CH_2{=}C(CH_3){-}COOCH_3$), which simultaneously decreases their viscosity. They are also sometimes diluted with acetone. In polymerization, cross-linking reactions of molecules of unsaturated polyester resins with molecules of styrene, as well as molecules of styrene with molecules of styrene, create spatial 3-D structures. These reactions are initiated either thermally (catalytically in the presence of dibenzoylperoxide and other catalysts at an increased temperature of 60–90 °C) or using γ-radiation (Barthez et al., 1999a).

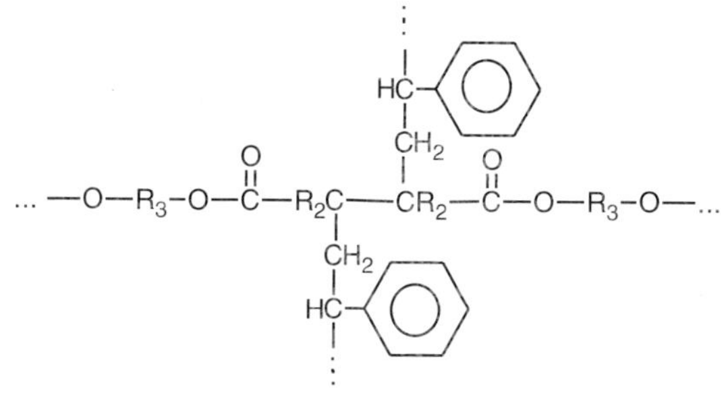

Cross-linked polyester (with styrene)

Polyesters significantly increase the strength of wood and also improve its resistance to biological pests. However, they cannot be removed from the conserved wood. They are successfully used not only for the conservation of air-dried wood, but also for the conservation of wet wood using two-level diffusion, where the water is first replaced with acetone and this solvent then with unsaturated polyester and styrene (Zelinger et al., 1982).

Epoxides are used to advantage in reinforcing of biologically damaged wood. They are applied in the form of two-component systems:

- linear epoxide base, which is a viscous, colourless, yellow or brown polymer liquid with epoxide rings $CH_2{=}(O){=}CH{-}$; it is prepared, for example, by the reaction of 2,2-bis-(*p*-hydroxyphenyl)propane $OH{-}(C_6H_4){-}C(CH_3)_2{-}(C_6H_4){-}OH$ with epichlorohydrine $CH_2{=}(O){=}CH{-}CH_2{-}Cl$;

- cross-linking agent, usually a suitable polyamine (e.g. diethylenetriamine, $NH_2{-}CH_2{-}CH_2{-}NH{-}CH_2{-}CH_2{-}NH_2$); when curing at an increased temperature, dicarboxylic acid anhydride is used as the cross-linking agent (e.g. phthalic acid anhydride);
- both these epoxide components in treated wood react via polyaddition reactions and create the spatial 3-D structure of cured epoxide.

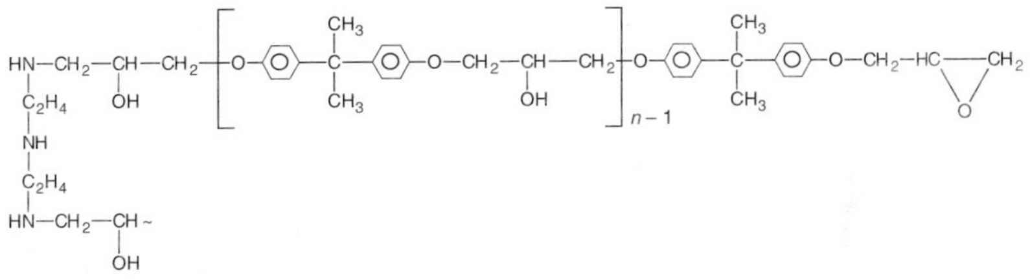

Partially cross-linked epoxide

Cured epoxides are high molecular weight polymers. They are resistant to water and to the majority of organic liquids, and do not melt under increased temperature; that is, they are not transformable into a liquid or plastic state. The use of a cross-linking agent with a smaller number of reactive functional groups decreases the total density of the cross-link, and therefore also the brittleness of the cured epoxide.

A decrease in the viscosity of epoxide systems, with the aim of their better penetration into wood, is achieved by adding organic solvents such as xylene, toluene, toluene with ethanol, and others. It is sometimes beneficial to apply a reactive solvent functioning as a cross-linking agent (e.g. butanediol diglycidyl ether). The reactive solvent becomes part of the cured epoxide and suppresses the creation of bubbles and contractions in the cured epoxide, which cannot be avoided when using nonreactive, volatile solvents. Acetone or other ketone solvents slow the cross-linking process, which is sometimes required for epoxide putties applied to the cavities of damaged wood.

The shape of the wood does not change when the epoxide resin formed by polyaddition reactions cross-links in its structure. Conserved wood has significantly increased mechanical properties (Reinprecht & Varínska, 1999). It is well resistant to bacteria, wood-decaying fungi and wood-boring insects (Unger & Unger, 1995). However, its dimensional stability does not significantly improve, because the high molecular weight epoxide resins do not penetrate to the cell walls of wood. Epoxides cured in wood are resistant to weather impacts (apart from yellowing), water, alcohols, aliphatic carbohydrates and diluted acids. However, epoxides swell in 1,2-dichloromethane or in 1,3-dioxolane. The conservation of wood with epoxides is irreversible.

Epoxide resins in the form of low-viscosity solutions are mainly used for reinforcing construction timber in exteriors and beams in log cabins, ceilings or trusses. They can also be used for the conservation of wooden artefacts if they are in a higher degree of decomposition – and they could not be sufficiently reintegrated or reinforced using polyacrylates or other reversible substances.

The excellent adhesive properties of epoxides to wood, metals and also other materials are used in various methods of repairing of wooden constructions. For example, the well-known 'beta method' (Section 7.6.2), in which epoxide is combined with steel, carbon or glass laminate rods, is used for reinforcement of wooden elements in constructions and improves their bending strength (Reinprecht, 2010).

Polyethyleneglycols (PEGs) are linear polar polymers with varying degrees of polymerization, terminated with two hydroxyl groups:

$$HO{-}[CH_2{-}CH_2{-}O]_n{-}H \qquad \text{PEG}$$

The consistency and hygroscopicity of PEGs depends upon their molecular weight. PEGs with a degree of polymerization $n < 14$ – for example, from PEG 300 (n = 6–7) to PEG 600 (n = 12–14) – are liquid and hygroscopic at room temperature. PEGs with a higher degree of polymerization $n < 20$– for example, from PEG 1000 (n = 21–24) to PEG 6000 (n = 130–140) – are solid at room temperature with a waxy consistency and a density of ~1150 kg/m^3. Their hygroscopicity is relatively low and their melting point is from 44 to 58 °C. PEGs are well soluble in water and alcohols but insoluble in less polar organic solvents. They cannot be mixed with natural nonpolar or less polar waxes, paraffin and oils.

PEGs are suitable for the conservation of wet, usually archaeological wood (Bjurhager et al., 2010). They are transported into wood with diffusion or pressure–diffusion technology, which sometimes takes several years. In such cases, it is beneficial to add a biocide with a bactericidal and fungicidal effect (preventing PEGs and treated wood from attack by bacteria and fungi), or it is also beneficial to add suitable anticorrosion substances to avoid corrosion of a metal dip tank.

In practice, two-stage technology for the conservation of wet archaeological wood using PEGs is often used. First, low molecular weight PEGs are applied, followed by higher molecular weight PEGs. In the first stage, PEGs 300–1000 penetrate the damaged/weakened cell walls of the wood where they replace the bound water. In the second stage, PEGs 1500–6000 fill the space in cell lumens and microscopic cavities in the damaged wood, where they replace free water. This effectively prevents the cells from collapsing, and wet wood permanently retains its original shape even after drying and air conditioning. Wood treated with PEGs has very good dimensional stability. However, the strength of damaged wood can only be increased by higher molecular weight types of PEGs, mainly the hard waxes PEG 2000–6000. The surface of the conserved wood remains glossy; it is usually darker than the original wood, and it is difficult to glue. Before gluing with acrylate, epoxide, isocyanate or other adhesives, the surface of conserved wood should be rid at least of a thin layer of PEG.

PEGs are also suitable for treating larger wooden objects found under water or in the ground. They were used, for example, for treating the *Wasa* wooden ship (Barkman, 1969; Håfors, 1990), as well as a number of archaeological and fossil findings (Babinski, 1999; Barthez et al., 1999b; Unger et al., 2001). Currently, PEGs application is also combined with the technology of vacuum sublimation of frozen wood (Caple and Murray, 1994; Section 7.5.3.1). Wood conservation using PEGs is a reversible process; for example, Tejedor (2010) made experiments with re-conservation of elements from *Wasa* wooden ship with alkoxysilanes.

7.5.2 Methods and technologies for the conservation of air-dried damaged wood

7.5.2.1 Conservation methods for the restoration of air-dried damaged wood

The methods of conservation of air-dried damaged wood differ upon the mode of applying a conservation agent:

- creation of surface films or glazing;
- saturation of the cell walls;
- filling of the lumens in cells;
- reintegration of the micro-damaged zones of wood;
- filling of the cavities in wood.

Surface films or glazing created on the surface of damaged wood are suitable for improving its integrity, surface hardness and aesthetics. However, this method does not evidently improve the wood's dimensional stability or strength. Classic oil film coatings or vapour-permeable acrylate water dispersions and alkyd polyester systems containing drying oils are used for surface treatment of weathered wood in outdoor structures, facades, windows, doors and decorative elements (Mosler, 2006).

Saturation of the cell walls of wood cells is a method when the conservation agent penetrates the cell walls of damaged wood. This method is not suitable for conservation of air-dried historically valuable wooden artefacts, music instruments, and so on because it permanently increases their volume. Amino or phenolic resins improve a number of wood properties, such as mechanical properties and weather resistance (Reinprecht, 1998; Gindl et al., 2003).

However, the reinforcement of damaged wood using this method cannot always be ensured since some conservation agents, such as thermoplastic PEGs and saccharose, may weaken the hydrogen bonds between the structural components in the cell walls of the wood. Such substances are therefore only mainly used for treating wet archaeological wood (Section 7.5.2.2).

Mechanical properties of conserved wood are changed specifically in relation to the chemical structure and reactivity of the conservation agents:

- Increases in wood strength are provided by conservation agents able to penetrate the wood's cell walls and subsequently create a firm polymer

network within them via chemical reactions (polymerization, polycondensation, polyaddition). These are:
- urea formaldehyde (UF), melamine formaldehyde and phenol formaldehyde polar precondensates having small molecules applied in water solutions – they harden in the cells walls of wood by cross-linking;
- monomer polar liquid acrylates – they polymerize and harden in the cell walls of wood;
- multifunctional substances reacting with —OH groups of the structural components of wood, (e.g. low molecular weight epoxides, diisocyanates, aldehydes, or anhydrides of acids) – macromolecules of lignin and polysaccharides are linked with them in the cell walls (Section 6.3).

- Decreases in wood strength in its dry state are induced by several nonreactive conservation agents applied in water solutions (mainly when treating wet archaeological wood), such as:
  - lower molecular weight PEGs ($\leq$1500);
  - saccharose;
  - various inorganic salts.

Filling of the lumens of wood cells is secured by nonpolar and less-polar conservation agents, or also with polar ones having larger macromolecules:

- synthetic polymers – polyacrylates in toluene, unsaturated polyesters with styrene, epoxides in xylene, amino- and phenol-formaldehyde resins in water, PEGs 3000–6000 in water, and so on;
- natural substances – beeswax in petrol, melted beeswax, dammar in toluene, and so on.

Dimensional stabilization of damaged wood is not ensured by this method, and only the kinetics of sorption and desorption are lengthened. On the other hand, wood with filled lumens receives less free water or other liquid, so its soaking is lower. The strengthening of damaged wood is mainly achieved if the conservation agent creates a coherent reinforcement 'skeleton' in the lumens, which mainly increases the compression strength and hardness of the wood. For this method, it is mainly those synthetic polymers that can be applied in a higher concentration that is suitable, and after their curing or evaporation of solvent they are sufficiently strong but simultaneously elastic. Epoxides and polyesters are well designed for this task. Although phenolic and amino resins are sufficiently hard and firm, they are less elastic, which causes a higher brittleness of the conserved wood. Acrylates are sometimes also suitable for this purpose despite lower firmness and hardness, since they are sufficiently elastic. Natural substances, such as beeswax, colophony or dammar, are usually unable to significantly improve the strength of damaged wood (Reinprecht, 2011).

The lumen filling method sometimes overlaps with the cell wall saturation method. The degree of improvement in the strength of the wood in this case is then the resultant of both methods. An illustrative example is the use of amino

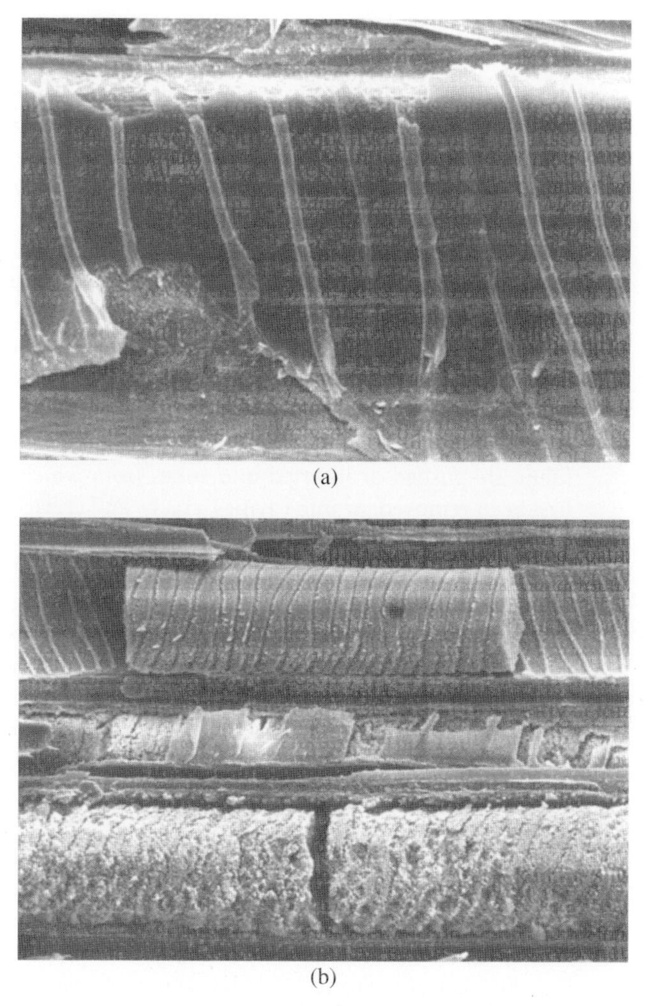

Figure 7.10 Micro-distribution of UF resin in treated lime-tree (*Tilia cordata*) wood: (a) cell wall saturation method (the use of non-catalysed UF resin); (b) the lumen filling method (the use of highly catalysed UF resin)

Source: Reinprecht, L. and Makovíny, I. (1987) Hardening of aminoplasts in modificated wood with catalytic and thermic dielectric heating. In: *Modyfikacja Drewna – Wood Modification, 6th Symposium*, pp. 288–298. Poznaň, Poland. Reproduced by permission of Faculty of Wood Technology in Poznaň

resins, which, depending upon the degree of precondensation and catalysis, either easily penetrate the cell walls (UF resins with low degree of precondensation or low degree of catalysis) or only remain in the lumens of cells (UF resins with high degree of precondensation or high degree of catalysis) (Figure 7.10).

Reintegration of the micro-damaged zones of wood is mainly used for restoring the integrity of wooden cultural sites. If insect gallery dust or pieces of rotten wood are falling from a wooden artefact, it is beneficial to treat it with a coating or deep impregnation. The conservation agent used should penetrate the wood easily and, with its good adhesive properties, it should perfectly join the disintegrating zones within the wood and also permanently retain the insect gallery dust or pieces of rotten wood inside the artefact, since these damaged pieces give it a historical value. For these purposes natural conservation agents (e.g. casein) and some synthetic polymers (mainly reversible polyacrylates) are convenient. Polyacrylates integrate wooden dust inside galleries and prevent its escape, whilst also partially reinforcing damaged cells of wood (Figure 7.11).

Filling of cavities and other spaces in damaged wood with a dimension over 1 mm can be performed with putties, solid foams and wooden fillers. This method is convenient for local repairs of wooden structures, cassette ceilings, doors, windows, floors, altars, sculptures, and so on. Putties, solid foams and fillers are pushed or inserted into zones from which the damaged wood has first been removed, or into cavities, cracks and crevices created during the use and ageing of the wooden object.

Putties are multicomponent formable systems containing a basic adhesive, cross-linking catalyst, filler, solvent and other additives, thanks to which they transform into a solid mass over time. Putties in a solid mass should have a good pliability, adhesion to wood, dimensional stability, in which their swelling and shrinkage values should be close to those of the damaged wood. Their compression strength and hardness should be equal to wood, and they should also be sufficiently flexible without cracking when exposed to external or humidity stresses. Important also is their appearance, grinding tolerance and surface finishing, whilst we must not forget their health and safety aspects and price. Good putties that are often used are epoxy resins filled with silica sand, sawdust or fractions of fine wood chips, further polyurethanes filled with sawdust, or commercial acrylates and silicones. In interiors, we can also use bone glue filled with sawdust or melted beeswax and colophony.

Solid foams are created after hardening of injected multicomponent liquid systems that contain a polymer skeleton of foam, expanding agent, catalyst, solvent and other additives. Their advantage in comparison to putties or wooden fillers is their low density, which can be important when repairing and reconstructing larger wooden objects. Today, the polyurethane foams are used primarily.

Wooden fillers are inserted into cavities or larger cracks using adhesives such as bone glue, polyvinyl acetate, epoxides, and so on. They should be made of the same wood species as the damaged wood. It is also very important to keep the same fibre orientation and the width of annual rings, the same humidity, or to use filler from older wood with a similar surface structure/patina to the wood being repaired (Mosler, 2006).

Steel, carbon or glass laminate rods can be applied to the space of a cavity as semi-flexible skeletons with the simultaneous use of the wooden fillers, putties or solid foams, mainly when reinforcing more stressed wooden elements or whole structural units (Larsen, 1994).

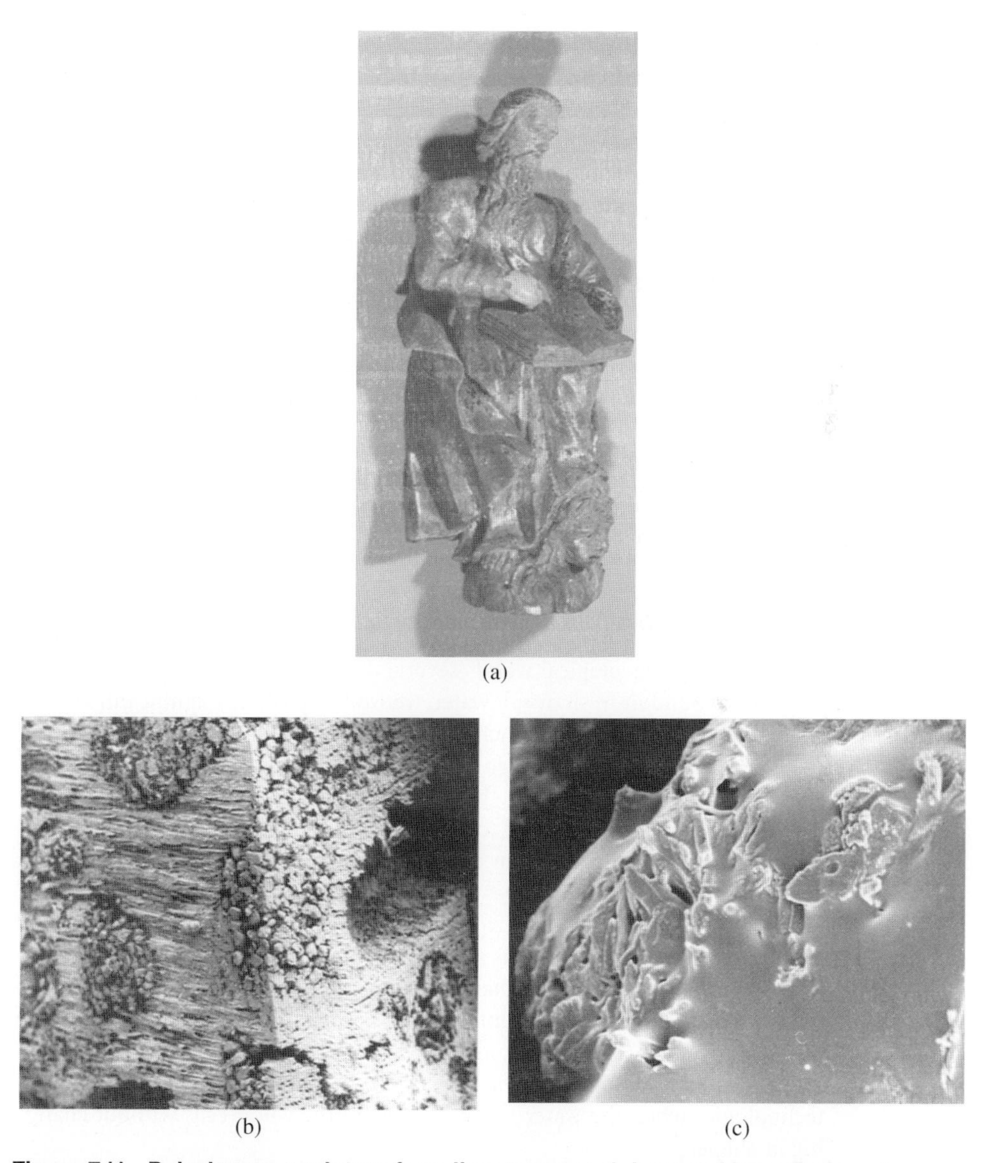

Figure 7.11 Polychrome sculpture from lime-tree wood damaged by galleries of wood-worm beetle and by brown rot (a), and consolidation of this sculpture by polyacrylate Solakryl BT 55, which strengthened and joined deteriorated wooden elements and dust in galleries (b, c)

Source: R., L. (1991) Restoration of damaged wood with polyacrylates, epoxides, fenolic resins, and amino resins. (In Slovak, English abstract). In: *Advances in Production and Application of Adhesives in Wood Industry, 10th Symposium*, pp. 312–325. VŠLD Zvolen, Slovakia. Reproduced by permission of TU Zvolen

7.5.2.2 Technologies for the conservation of air-dried damaged wood

Air-dried wood previously damaged by decay-causing fungi, insects or abiotic factors can be conserved using any pressure-free technology (e.g. coating, spraying, dipping, bandages, injecting). The vacuum–pressure technologies (vacuum impregnation in autoclaves or plastic packaging, and possibly also Bethell, Rüping or Lowry technologies – Section 5.3) can be used only if the structure of damaged wood is not markedly weakened.

The penetration depth of the conservation agent into the damaged wood depends upon the actual state of the conductive capillaries in the wood structure, predetermined by the wood species with a defined impregnability (EN 350-2), the type and degree of wood damage (penetration improves when the cells and cell walls are disturbed, but it can also decrease when there is a greater amount of fungal hyphae in the lumens), the dynamic viscosity and molecular weight of the conservation agent, as well as upon the size and time of application of the driving forces (Section 5.3).

7.5.3 Methods and technologies for the conservation of waterlogged wood

Waterlogged archaeological wood (several decades to thousands of years old) which was found in fresh or salt water, wet soil or peat, has significantly weakened and thinned cell walls. Such damaged wood, with a reduced proportion of hemicelluloses and fairly often also cellulose, has higher porosity and shrivels significantly during drying. Wet archaeological wood with high porosity dries intensively when exposed to air, decreases in volume, changes shape and cracks appear. This occurs even at moistures above the fibre saturation point (i.e. when just free water escapes from the wood). The aim of conservation is to dimensionally stabilize such wood, to reinforce and possibly reintegrate it. If the archaeological wood still has sufficient firmness and conservation is only to retain its original dimensions, then a simple stabilization technique can be used based on its long-term placement in water, or an environment with relative air humidity over 98% (Zelinger et al., 1982).

The conservation of wet archaeological wood can be carried out by special technologies when the water is removed from the wood cells without changing its shape:

- dehydration of frozen wood using vacuum sublimation;
- diffusional replacement of water in the lumens and in the cell walls of the wood with a stable, non-volatile substance;
- a combination of diffusion and sublimation technology.

7.5.3.1 Dehydration of frozen wood using vacuum sublimation

Water in the lumens and cell walls of damaged wood is first transformed into solid ice. CO_2 is used for this purpose (cooling to −79 °C). A freezer (−20 °C) may also be used if freezing of the total proportion of bound water in the cell walls is not required, or if part of the bound water has already been replaced by PEG. A vacuum of ~10 Pa is created above the frozen wood in a vacuum chamber over several hours. With a simultaneous supply of heat from heaters (30–40 °C), the sublimation of frozen water proceeds; that is, phase transformation of the ice in the wood into water vapour takes place. The shape of the wood does not change or changes much less than in classic drying. This is because the capillary action typical for liquid water does not occur, which in classic drying causes weakened cells to collapse. Wood deprived of water is very light and brittle.

Frozen wood sublimation technology can also be combined with technology for the initial replacement of a certain proportion of water in wood with non-volatile, low molecular weight PEG (300–600) or saccharose. This combination achieves the best results in the dimensional stabilization of waterlogged archaeological wood, where the AShE parameter reaches high values of around 100% (Grattan, 1982).

7.5.3.2 Diffusional replacement of water using a non-volatile substance

Water in archaeological wood is replaced with a stable, non-volatile substance using diffusion technology. Water can be replaced just in the lumens, but it is usually necessary that it also be replaced in the wood's cell walls. The AShE value depends upon the amount of conservation agent in the wood, but mainly upon its ability to also penetrate the wood's cell walls. An example is high molecular weight PEG 4000, only a small proportion of which also penetrates the cell walls, and its stabilizing effect upon maintaining the wet wood's original dimensions is weaker by half in comparison with a mixture of PEG 300 and PEG 1540 (Grattan, 1982).

Technology for the diffusional replacement of water in damaged wood mainly uses sugars (saccharose, lactose), PEGs of varying molecular weight (PEG 300 to PEG 6000), polymerizable monomers (e.g. HEMA) and also cross-linkable precondensates (e.g. melamine or phenolic resins). PEGs are also often applied in two stages. Initially, low molecular weight PEGs 300–1000 penetrate the cell walls, stabilizing their dimensions, and then higher molecular weight PEGs 3000–6000 fill the space of the lumens by which they reinforce the cells and, at the same time, prevent them from collapsing.

The conservation of wet wood is carried out using several types of diffusion technology (long-term dipping lasting several months to years, multi-spraying, bandaging, panel impregnation, and others) or sometimes also pressure–diffusion technology (Sections 5.3.5 and 5.3.6). The intensity of the diffusion of a conservation agent in wet wood $i_{diffusion}$ depends upon its diffusion coefficient in the lumens and cell walls of wood filled with water (D_L; D_{CW}), upon its concentration gradients in individual anatomical directions of wood (dc/dl), and also upon the interconnection of the wood's capillary system in the direction of diffusion (active area A_a) (Section 5.3.3). The diffusion process may be intensified by heating the conservation agent, since at an increased temperature the diffusion coefficient, solubility and the concentration gradient also increase. In practice, the diffusional exchange of water for PEG is therefore sometimes carried out at increased temperatures from 70 to 85 °C.

Today, saccharose and PEGs are mainly used for the conservation of wet archaeological wood (Hoffmann, 1991; Babinski, 1999). A typical example of the application of PEGs is the conservation of the Swedish ship *Wasa*, made of oak, which foundered in Stockholm harbour in 1628. In 1961 it was recovered and then disassembled into ~24,000 pieces. These were conserved by dipping in a mixture of PEG 1500 and PEG 4000 for 2 years. Within a special roofed structure, the 38 m long skeleton of the ship was sprayed using a 10–15% water solution of PEG 1500 for approximately 10 years, and then with a water solution of PEG 4000 with the addition of biocides H_3BO_3 and $Na_2B_4O_7 \cdot 10H_2O$ (Håfors, 1990). In Slovakia, Reinprecht (1995) treated ~800-year-old oak beams, found under wet soil in Slovenská Ľupča Castle, using pressure–diffusion technology with a water solution of PEG 600, PEG 1500 and H_3BO_3 biocide, heated to 80 °C.

7.6 Renovation of damaged wood

During renovations, a defined part of damaged wooden products or constructions is replaced or supplemented with a new part made of wood or another material. The new part is applied in the form of a filling, prosthesis, splice plate, and so on. Renovations are mainly used for restoring wooden constructions such as log cabins, bridges, trusses, ceilings or stairs, but also for restoring windows, doors or furniture. Renovation should be carried out based on a project and should include dendrologic analysis of the damaged wood, evaluation of statics and, in historically valuable objects, also the requirements of conservationists (Section 7.6.1).

Restoration of strength and stiffness of load-bearing wooden elements and whole constructions can be carried out using the following renovation methods.

- Direct reinforcement of wooden elements:
  - splicing
  - height adjustment
  - anchoring into steel brackets
  - using a prosthesis
  - sealing.
- Indirect reinforcement of wooden elements:
  - supporting
  - unloading
  - bracing.
- Reinforcing of whole or selected part of wooden construction:
  - spatial bracing of the construction
  - adding preload to the construction
  - interlinking the construction with additional elements.

7.6.1 General requirements for the renovation of wooden objects

The limited service life of wooden products and wooden constructions (Section 1.5) means that after a certain time there is a requirement for them to be renovated (i.e. using repairs, modernization or reconstruction) and possibly demolished.

- *Repair* is minor, medium or major intervention to the damaged wooden element. Repairs are intended for restoring the technical, functional and aesthetic properties of a product or construction, focusing upon ensuring safety. When repairing a wooden object, the physically worn out or biologically damaged elements are reinforced or replaced with new if necessary.
- *Modernization* serves for removing some wooden elements in the older construction in terms of their obsolete design, as well as adding new elements depending upon the current or long-term requirements of the user.
- *Reconstruction* is functional rebuilding of a wooden construction together with major interventions into its load-bearing systems necessary to remedy faults and therefore lengthen its lifetime.
- *Demolition* – represent the liquidation of wooden construction which are in a poor or critical state. In terms of wooden historic objects, it is also a set of interventions which clear a historic environment of elements with a negative artistic or hygienic impact.

Renovations of wooden objects can be carried out with more traditional and modern techniques (Brereton, 1995; Ashurst & Ashurst, 1996; Erler, 1997; Robson, 1999; Reinprecht & Štefko, 2000; Ross, 2002; Yeomans, 2003; Štefko & Reinprecht, 2004; Mosoarca & Gionen, 2013). The renovation of damaged wood is usually carried out in the form of major repairs and reconstructions based on approved projects. Regular smaller and medium repairs are carried out within maintenance of the object, often without project documentation.

7.6.1.1 Project for the renovation of wooden construction

Projects for the renovation of wooden construction are based on input researches (e.g. determination of its architectonic, history, technical state of individual wooden elements). In the technical research, the following is carried out:

- diagnostics and evaluation of the type, scope, degree and cause of faults in wood, but also possibly in other adjacent materials;
- static assessment of the wooden construction as a whole, carried out by a structural engineer in cooperation with the designer, architect and other experts (e.g. an expert for assessing the biological damage of the wood);
- documentation on the actual status of the construction, including photo-documentation and graphical annotation of wooden elements designated to be replaced and reinforced (Figure 7.12).

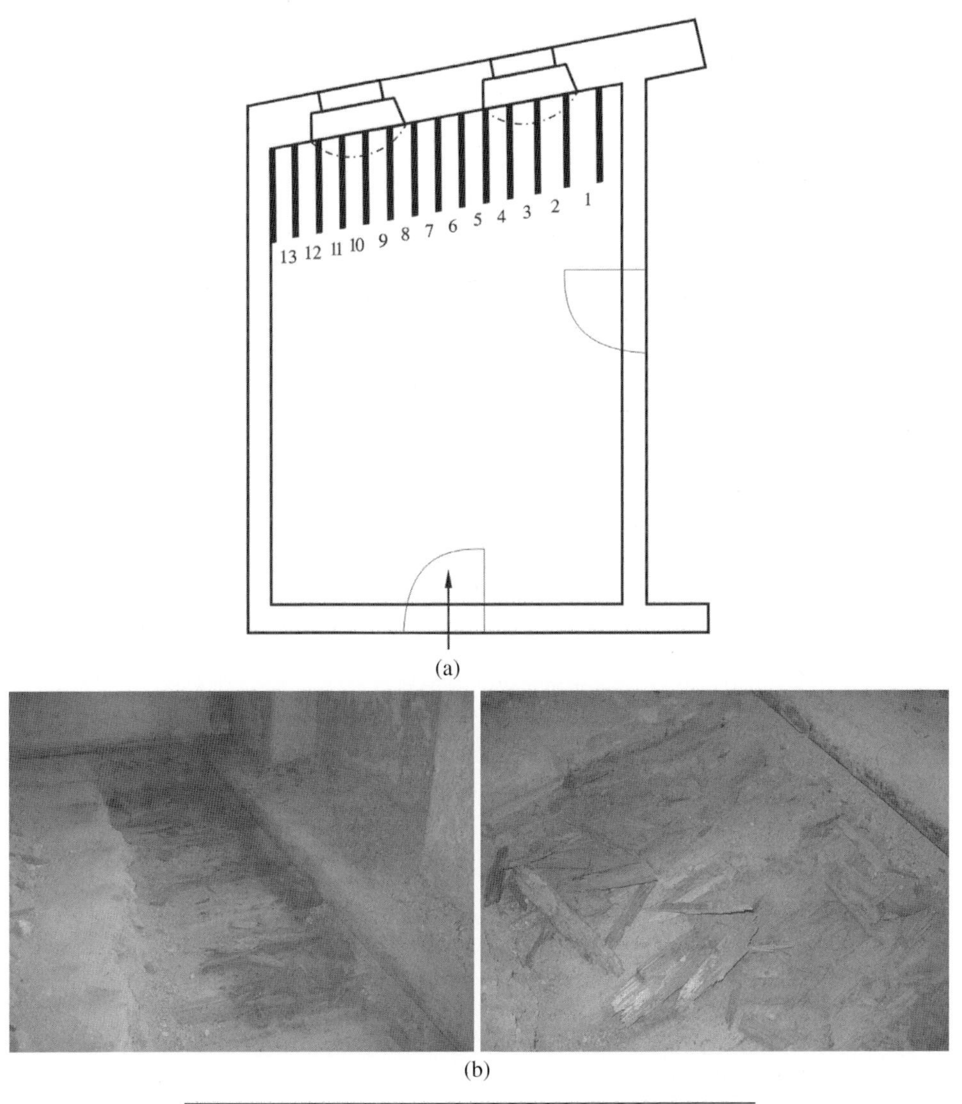

	(c) Biological damage	
Beam No.	Type	Degree
1 and 2	BR, WG	3
3 to 6	BR, WG	2 to 2-3
7	BR, WG	3
8	BR, WG	2-3
9 to 12	BR, WG	3
13	BR, WG	3-4

Figure 7.12 Drawing documentation (a) and photo-documentation (b) from diagnosis of one wooden ceiling, together with marking (c) the biological damage type (BR: brown rot; WG: wood-worm galleries) and level (1: light; 2: medium; 3: strong; 4: total)

A renovation project is usually based on faultless pre-project planning, within which relevant knowledge of the wooden construction is obtained via field surveys, written and other database resources in terms of:

- the current and future requirements for its function and service life (Section 1.5);
- the type, scope, degree and causes of wood damage (Chapters 2 and 3 and Section 7.3);
- the structural protection (Chapter 4);
- the possibilities for chemical and modifying protection (Chapters 5 and 6);
- the options for removing the causes of wood damage (Sections 7.2 and 7.4);
- the options for improving the properties of the damaged wood by conservation (Section 7.5), or improving the properties of damaged wooden elements and structural units by renovation (Section 7.6).

The static views determined by Eurocode 5 in EN 1995, taking into consideration the historic value of the wooden construction, are mainly important when renovating load-bearing wooden elements and structural units (e.g. trusses). However, when repairing non-load-bearing wooden elements (e.g. cassette ceiling soffits), the historical viewpoint is more expressly favoured.

7.6.1.2 The specifics of renovating wooden cultural heritages

The historic value of wooden objects is predetermined by their design, originality, age, decor and quite often also by the social trends of the era, which were gradually reflected in the maintenance. Increased requirements for repair and reconstruction are mainly related to those objects which have a higher degree of damage, either due to a shorter proposed lifetime (e.g. folk architecture objects) or due to neglected maintenance (e.g. religious buildings in some countries).

Major repairs and reconstructions to wooden objects of a greater historic value are carried out based on individual projects (Figure 7.13). They fully take into account the architectonic, historic and structural design, the actual state, statics and aims for the use of the wooden site in the future. For example, the maximum proportion of the original wood should be retained when renovating cultural sites.

During repair or reconstruction of the cultural heritage, the following requirements should also be sensitively considered:

- the use of the same wood species as the original one (e.g. by replacement of damaged part of wood element in truss by wooden prosthesis, or by window repair with wooden filler);
- the use of classic materials for repairing damaged wood (e.g. in joints, give preference to wooden pins with bone glue over steel pins with epoxy);

(a)

(b)

Figure 7.13 Renovation of the wooden church in Hunkovce – replacement of significantly bio-damaged beams with new ones was carried out by disassembling the log-cabin walls (a) and less-damaged beams were returned to the structure (b)

- the use of traditional techniques and methods for repairing damaged wood (e.g. when applying a replacement with a prosthesis, give preference usually to the traditional carpentry method over the 'beta method' – Section 7.6.2.1);
- give preference to temporary reversible consolidation measures over poor permanent solutions (e.g. favour temporary static securing of a structure if its repair or reconstruction cannot currently be carried out at the required level);
- retain the original structural design (e.g. it is inappropriate to change the typology of the original log cabins, ceilings or trusses).

7.6.2 Techniques for strengthening of individual wooden elements

Prior to reinforcing the load-bearing elements of a wooden structure, it is necessary to know their actual condition and, at the same time, their structural function, including forms of stresses; for example, tensile, compression, bending, shear or buckling (e.g. Brereton, 1995; Erler, 1997; Reinprecht & Štefko, 2000).

7.6.2.1 Direct reinforcement of wooden elements

The direct reinforcement of wooden elements – beams, rafters, columns and other bearing elements present in wood houses, bridges, trusses, ceilings or other constructions – is carried out whilst focusing upon regaining their mechanical properties and also taking into account the aesthetic of the renovation (Table 7.6). The following reinforcement methods are usually applied in wooden constructions:

- log cabins, half-timbered, pillared → prosthesis, sealing (filling);
- ceilings → splicing, height adjustment, anchoring to steel brackets, prosthesis, insert of carbon fibre;
- trusses → splicing, prosthesis.

Table 7.6 Options for direct reinforcing of damaged elements of a wooden object

Increasing of the cross-section of the wooden element
- splicing
- height adjustment

Insertion of lamella while keeping the cross-section of the wooden element
- carbon, Kevlar or glass fibres insertion to pre-milled grooves in the tension and/or pressure zone of the wooden element, while maintaining its original cross-section

Replacing (or supplementing) of the damaged wood
- with healthy wood or another material, while not retaining the original cross-section of the wooden element:
  - splicing, i.e. joining a healthy part of wooden element and a inserted wood or other filler with a splice from wood or steel (e.g. repairing the head of joining beam or rafter)
  - anchoring, i.e. inserting a healthy part of wooden element into a steel bracket (e.g. repairing the head of ceiling beam)
- with healthy wood or another material, while retaining the original cross-section of the wooden element:
  - prosthesis, (e.g. carpentry or 'beta' methods)
  - sealing
  - conservation with reinforcing substances (e.g. epoxides)

Additionally, the following methods are also applied in repairing wooden facades and other decorative or functional products:

- windows, doors, stairs → sealing (filling), prosthesis;
- facing, cassette soffits, and so on → sealing (filling).

Splicing of damaged wood elements is carried out using splices of wood (board, plywood, etc.) or steel (U-profile, plate, etc.), but today also of carbon, Kevlar or glass-fibre mats. They are usually attached to the wooden element from one or two sides (Figure 7.14).

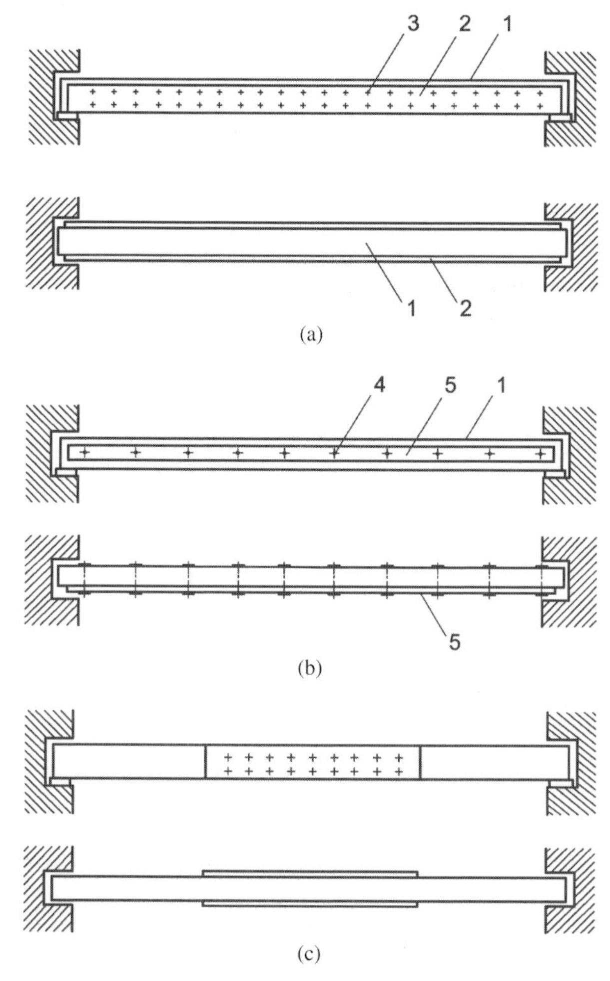

Figure 7.14 A wooden beam reinforced using the splicing method: (a) reinforcement using two wooden splices on the sides of the beam; (b) reinforcement using one steel splice on the side of the beam; (c) reinforcement using wooden splices in the central zone of the beam. (1: beam; 2: wooden splice; 3: nail; 4: bolt; 5: steel splice)

In terms of statics, it is recommended that wooden elements under bending stress are spliced of their height (i.e. from the top and bottom). Such splicing is sometimes problematic; for example, in ceilings if the floor must not be raised or the soffit must not be lowered. The reinforcement of wooden element by splicing is carried out either along its whole length or only locally; that is, in the centre, at one end or at both ends (Figure 7.14).

Wooden splices are attached to the wooden element by nails or screws with a hexagonal head. For thicker splices, bolts and other steel-connecting means are used (bulldog clips; i.e. gusset plates with double-sided spikes, etc.), or in historical structures are used wooden-connecting means (e.g. oak catches). If bolts are used for a joint, the diameter of the opening for the bolt must not be larger than the diameter of the bolt itself.

Steel splices are connected to the damaged wooden element using steel bolts, usually with a diameter of 12, 16 or 20 mm. Steel splices can bear high loads, and there is no danger of them being damaged by fungi and insects; however, their corrosion must be taken into consideration. The best are splices from a stainless steel, which have mainly been used in recent years; for example, if the head of the wooden element is placed on permanently wet masonry. In more complicated structural junctions, as well as in the requirement to join a spliced component with the surrounding wooden elements (e.g. with rafter at strengthening of a joining beam), they must also be suitably combined with other reinforcement methods. Steel splices are usually not used in historic objects, or they need to be masked.

Carbon, Kevlar and glass splicing mats are connected to the side of the wooden element that is under tensile or compressive stress, using an epoxy or other suitable glue.

Height adjustment of damaged beams is performed by polymer concretes. Epoxy, unsaturated polyester or other synthetic polymers with the addition of a mineral filler or sawdust are used in the role of the polymer concrete. It is a specific method suitable for reinforcing ceiling joists and joining beams (Figure 7.15).

Connection of the polymer concrete to the wooden element is based on its adhesive properties to wood. Polyester or epoxy resins in polymer concrete are able to partially penetrate the pores in the wood and, after curing, they are firmly anchored to the wood surface. Thus, a mechanical adhesive joint is also created along with physical and chemical adhesive joints.

Anchoring in steel brackets is applied in specific situations; for example, for reinforcing greatly damaged heads of ceiling joists or rafters. The damaged head of the beam is first removed, usually up to a distance of 0.3–1 m from its end face, and then a classical or special anchoring method is used.

Classical anchoring is the insertion of the remaining healthy part of a wooden beam into a steel bracket where it is firmly anchored with bolts, or using another suitable method (Figure 7.16). The end section of the steel bracket is suitably fixed into masonry (e.g. concreted).

Special anchoring is carried out using monolithic, load-bearing stainless steel brackets. It is mainly suitable for reinforcing wooden beams in objects of

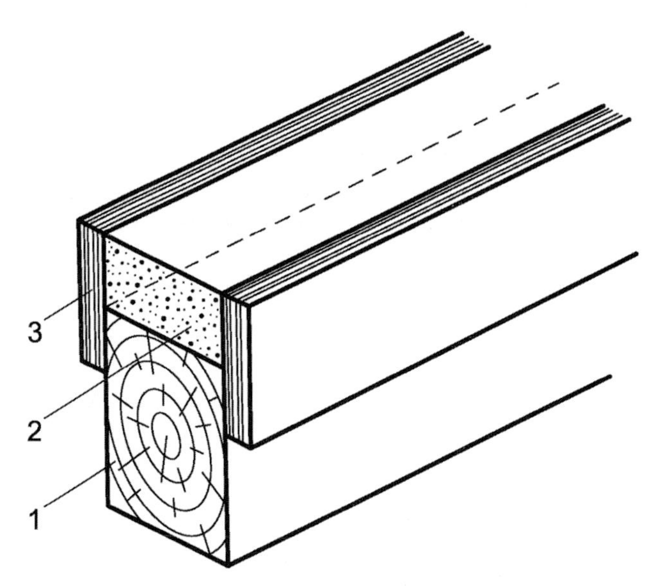

Figure 7.15 Wooden beam reinforced at the top using a layer of polymer concrete (1: beam; 2: polymer concrete; 3: temporary facing with a separation foil)

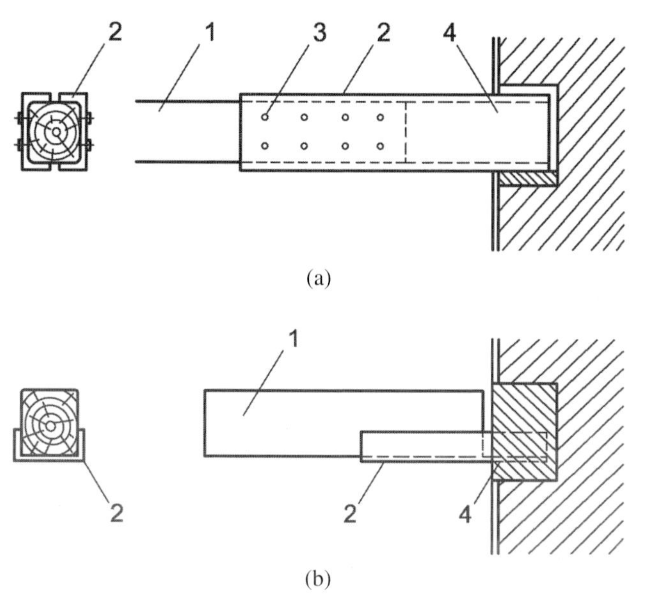

Figure 7.16 Methods (a, b) for anchoring ceiling joist with a rotten or cut head into classical steel brackets (1: ceiling joist; 2: steel bracket; 3: bolt; 4: concreting a bracket into masonry)

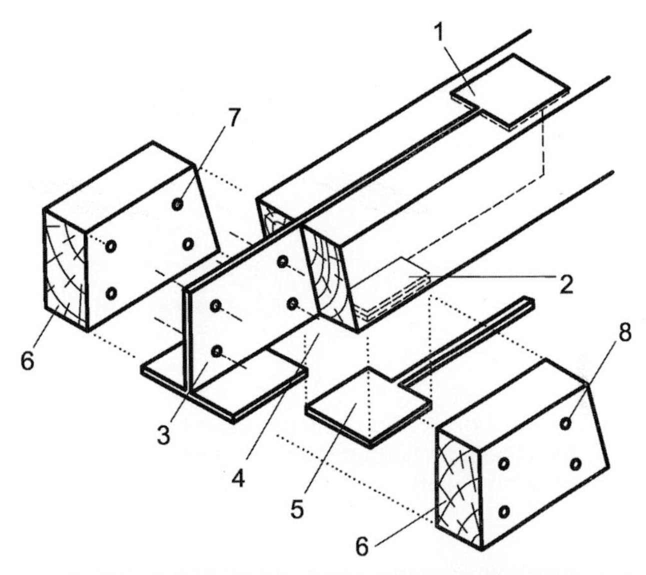

Figure 7.17 Repair of ceiling joist with a rotten head using a monolithic stainless steel bracket (1: upper holding board; 2: lower holding board; 3: horizontal load-bearing board; 4: offcut of rotten ceiling joist; 5: piece of wood for covering the bracket at the bottom (e.g. veneer); 6: new timber attached to the end of the bracket with bolts, and then the fixing points are covered with wooden inserts; 7: hole for bolt; 8: wooden insert for masking of bolt)

historic value (Brereton, 1995). Its advantage compared with classical anchoring in steel brackets is the option to perfectly mask the steel parts of the bracket using wood plates (e.g. veneers), which significantly improves the aesthetic side of the structural repair (Figure 7.17). Monolithic steel brackets with the correctly selected distance between the upper and lower plates can provide the original strength of a wooden element (Reinprecht & Štefko, 2000).

Insertion of lamella from carbon, Kevlar or glass fibres into pre-milled zones of restored wood (Figure 7.18). Lamellae are usually inserted into the tension or compression zone of the damaged wood using epoxy resin. They can also be applied from the outside of the damaged element (see splicing method). These lamellae have high tensile strength and modulus of elasticity, and they are suitable for reinforcing ceiling joists, joining beams, purlins or rafters. Inserted lamellae can be properly masked, (e.g. by applying a veneer from the same type of wood as the renovated wooden element).

Using a prosthesis is a technological operation in which the damaged part of a load-bearing wooden element is replaced or supplemented with a new part, with the aim of restoring its original strength, rigidity and aesthetics. The new part of wood (i.e. prosthesis) is identical in shape to the removed or missing part of the joining beam, rafter or other element. This means that, during replacement, the cross-section of beams and usually also the typology of the original

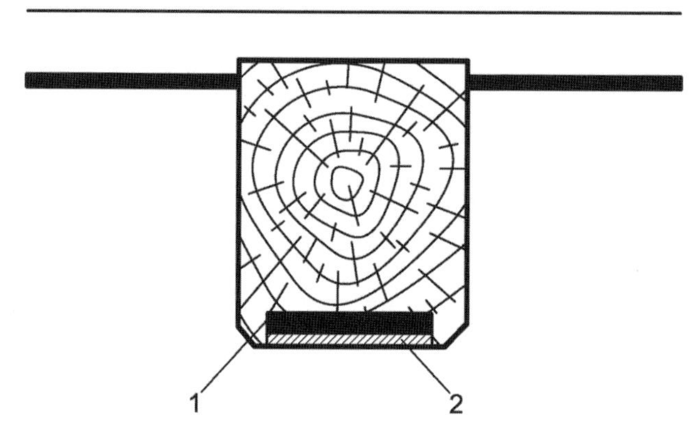

Figure 7.18 Carbon lamellae used for reinforcing a wooden beam (1: lamellae from carbon fibre; 2: veneer identical in texture and colour to the wooden beam)

joints remains unchanged. Wood for the prosthesis should be identical to the wood of the repaired element (e.g. same species and intentional patina on its surface), but other materials can also be used (e.g. other wood species or polymer concrete in combination with glass laminate rods).

Carpentry prostheses are mainly used for renovation of trusses. Replacement of damaged wood includes joining the remaining healthy part of a wooden element with a new part of wood using various types of carpentry joints (Figure 7.19). Apart from fulfilling their strengthening function, several carpentry joints may also fulfil the demanding aesthetic requirements of architects and historians. Straight and oblique plated joints secured with steel bolts are mainly used for repairing load-bearing wooden elements in trusses and other constructions. Straight plated joints are suitable for repairing elements that are under compression stress. When repairing load-bearing columns that are also under torsional and buckling stress, it is more suitable to use scissor joints. Oblique plated joints are suitable for repairing wooden elements that are under bending stress. In the case where wood is under tension and bending stress, it is recommended that the plated or oblique joints are secured using steel bulldog-type clips or oak pins. The length of plating of a wooden element with a cross-section of ~180 mm × 180 mm, by using two securing bolts, should not be shorter than 600 mm. A wooden element with carpentry prosthesis can equal the mechanical properties of the original element before damage. It is usually the case that a carpentry joint in the replacement is secured not only with bolts but also simultaneously with epoxy, polyurethane or other suitable glue (Reinprecht & Joščák, 1996).

The prosthesis of the 'beta' system was designed by Klapwijk (1978). This system is applied when repairing the heads of load-bearing elements of a greater historic value, such as ceiling joists containing carvings, paintings, polychrome, and so on. It is also suitable for repairing difficult-to-access elements, since it does not require complicated manipulation of the repaired or surrounding

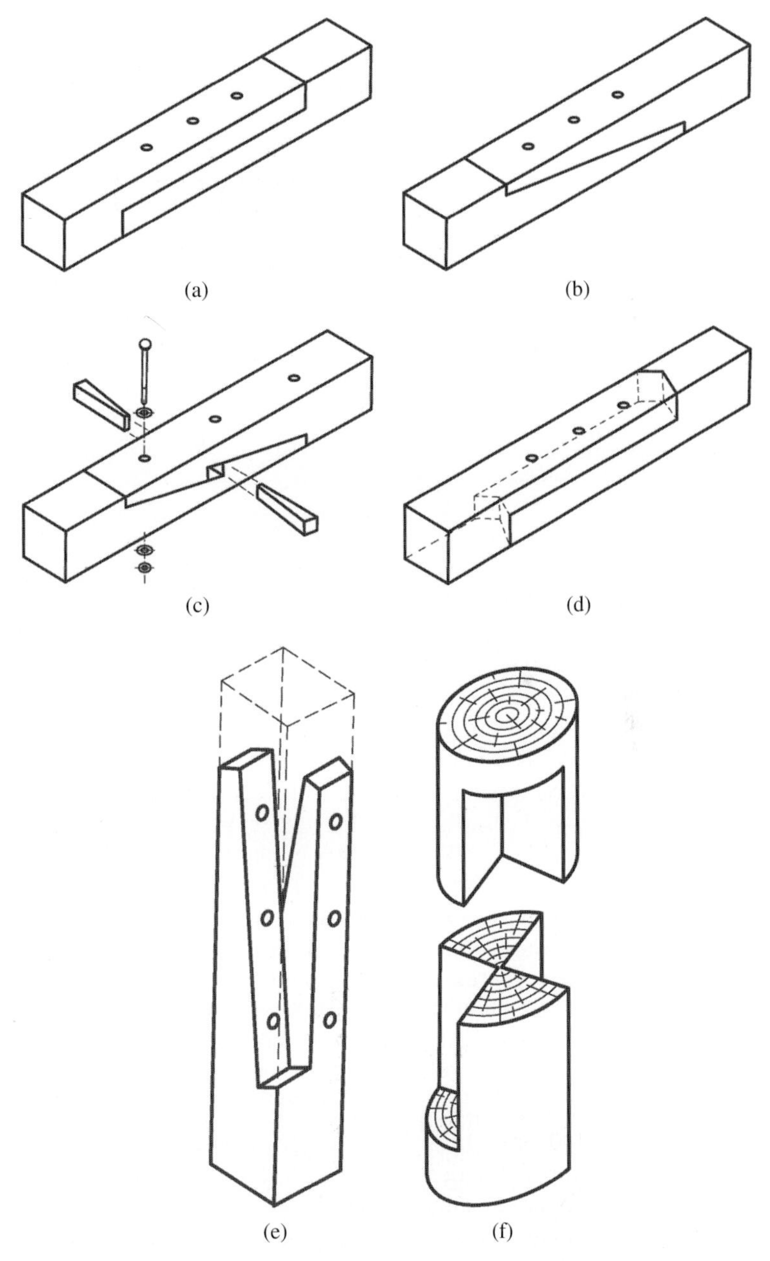

Figure 7.19 Prosthesis of wooden elements using typical carpentry joints: (a) straight plated joint (e.g. for elements under compression stress); (b) oblique plated joint (e.g. for elements under bending stress); (c) wedged inclined plated joint (e.g. for elements under tension, bending or compression stress); (d) wedged plated joint (e.g. for elements under bending stress in two directions); (e) scissor plated joint (e.g. for elements under compression stress and side movement stresses); (f) cross-plated joint (e.g. for elements under buckling stress)

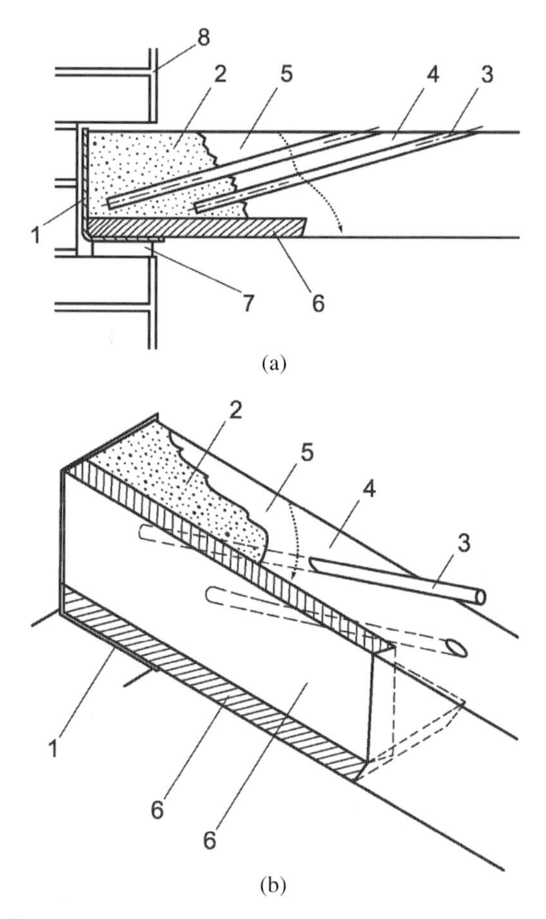

Figure 7.20 Local repair of a ceiling beam with a rotten head using the 'beta' system; that is, with a prosthesis consisting of epoxy resin, polymer concrete and reinforcing rods. (a) Side view of a substituted head of a ceiling joist. (b) Top view of the substitution method. First, the weakened zone of the wood is impregnated using low-viscosity epoxy resin; second, connecting bars are inserted into drilled openings in the remaining part of the wooden element covered with epoxy resin; third, the cavity in wood in the space of a total rot is filled with epoxy polymer concrete (1: perforated separating foil; 2: cavity (e.g. rotten beam head) replaced with epoxy resin or polymer concrete; 3: reinforcing rod of carbon fibres; 4: healthy wood; 5: damaged weakened zone in the wood; 6: new oak wood/pad and splice; 7: ventilating support; 8: masonry)

wooden elements (Ashurst & Ashurst, 1996). The damaged part of the wood is substituted using polymer concrete in combination with reinforcing rods made of profiled steel, glass laminate or carbon fibres (Figure 7.20). Polymer concrete is a mixture of a synthetic polymer (epoxy or polyester resin) and a filler (silica sand, wooden particles, etc.). The optimum mass ratio of polymer : filler

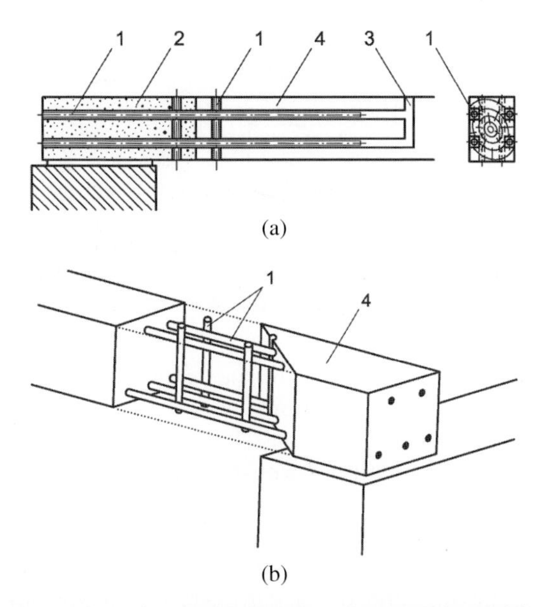

Figure 7.21 Examples of substituting the damaged head (a) or middle section (b) of the wooden element using the 'beta' system (1: reinforcing rod; 2: polymer concrete; 3: epoxy resin; 4: renovated wooden element)

depends upon a number of factors; for example, for a mixture of epoxy and silica sand it is within 1 : 3 to 1 : 7. The principle of the 'beta' system lies in the faultless connection of wood and reinforcing rods with a synthetic polymer whilst the reinforcing rods also bear the tension and bending stress (Reinprecht & Joščák, 1996). The number, length and diameter of reinforcing rods can be calculated using knowledge of the tension stress in the repaired element, the effect of transverse forces, and the adhesiveness of the polymer concrete to the reinforcing rods and the wood (Erler, 1997).

Today, the 'beta' system has undergone several modifications (Figure 7.21). Reinforcing rods can be placed longitudinally or at an incline into predrilled openings, or into milled grooves on the side of strengthened element. Solid wood can also be used instead of polymer concrete as a substitute, and is bound to the original wooden element by just a thin layer of polymer concrete. Laboratory experiments as well as long-term practical experience has shown that, over time, a polymer concrete under constant load does not change its mechanical properties and it is also well resistant to biological pests. Its fire resistance when using a silica filler is at the same level as wood.

Sealing of larger sections of damaged wood is either the replacement of a locally damaged wood with filler, or the insertion of filler into a larger cavity, crack or crevice created in the wooden element during its use and ageing. Fillers can be prepared from woods, putties or solid foams. Sealing is applied for repairs to the integrity of non-load-bearing units of objects (windows, doors, etc.) or artefacts, as well as of load-bearing elements (beams in log cabins, etc.) with the aim of restoring their compression strength.

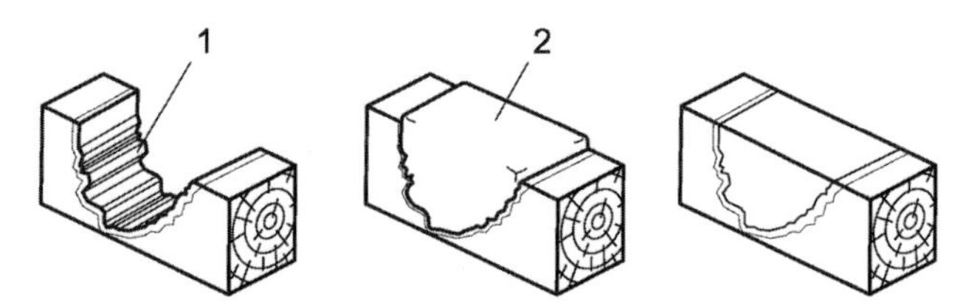

Figure 7.22 Sealing of an unevenly damaged wooden element (1: weakened zone of wood impregnated with epoxy resin, 2: inserted piece of new wood wrapped in a layer of putty from all sides)

Classical sealing is an analogy to filling of cavities in the conservation of smaller damaged sections of wood, when putties, solid foams or wooden fillers are used as well (Section 7.5.2.1).

Special sealing uses a mixture of several materials (wood, putty, glue, etc.) that are applied into a macro-cavity or other macro-space in the restored wood. It is used for repairing unevenly damaged zones of wooden element (Figure 7.22). When reinforcing load-bearing structures containing cavities, it is necessary to combine putties or wooden fillers with steel or glass laminate rods which, inside the cavities, act as semi-flexible skeletons. However, this is already analogous to the prosthesis of the 'beta' system.

7.6.2.2 Indirect reinforcement of wooden elements

The indirect reinforcement of wooden elements is the transfer of a load from them to the surrounding elements in the construction, or to newly built elements. It is carried out by supporting, relieving of the load or by bracing. Indirect reinforcements are usually temporary solutions:

- during reconstruction works;
- if the ability to use (or the load-bearing capacity) of the construction is weakened but, currently, no suitable methods for final reinforcement or no finances are available.

Supporting is used when repairing horizontal (ceiling joist, joining beam, purlin, etc.) and some inclined (rafter, brace, etc.) elements. Girders and supporting columns are used for this. A wooden or steel girder is suitable for temporarily and permanently supporting ceiling joists with rotten or otherwise damaged heads. A girder is placed under one or several ceiling joists with damaged heads, in the close vicinity of the load-bearing wall. The girder itself lies on steel, ceramic or other types of brackets, anchored to the load-bearing wall (Figure 7.23). Supporting wooden or steel columns are used during the reconstruction of ceilings and trusses (Figure 7.24).

Relieving is mainly applied in the repair of horizontal load-bearing elements. The original load acting on the damaged wooden element is transferred to the surrounding undamaged and sufficiently sized wooden elements or to added

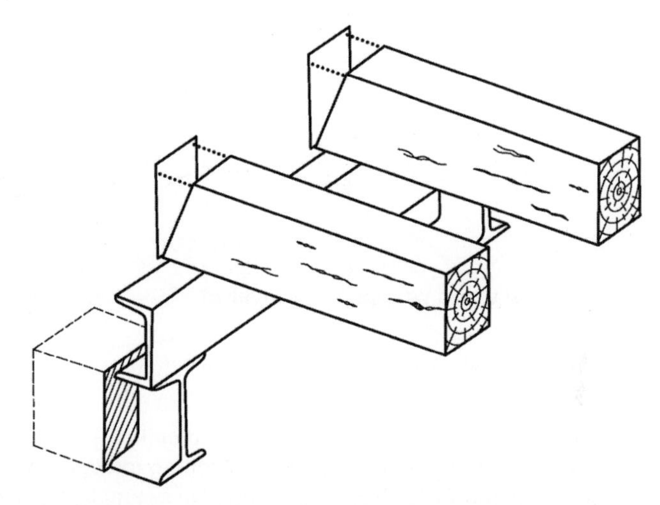

Figure 7.23 Support of ceiling joists with damaged heads using a steel girder

elements. An example is suspending a damaged ceiling joist on a new, transverse beam (Figure 7.25).

Bracing serves for strengthening the entire structural unit or individual damaged wooden elements and joints. One of the most frequently used methods for bracing is spine bracing. It is used, for example, in trusses, where bracing of the rafter with the joining beam is achieved using boards (Figure 7.26). Similarly, the bracing of two weakened opposite rafters in a roof comb can be solved.

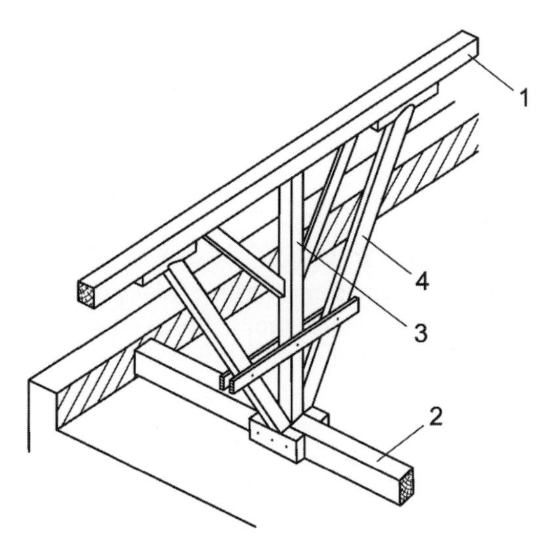

Figure 7.24 Temporary support of a damaged purlin using two supporting columns (1: purlin; 2: joining beam; 3: column; 4: supporting column)

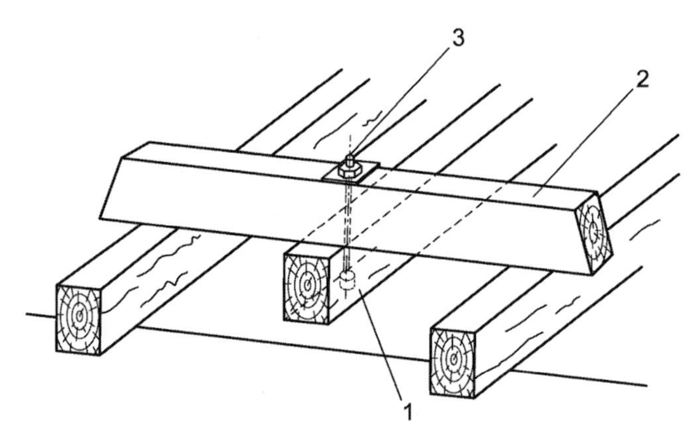

Figure 7.25 Suspending a damaged ceiling joist on a transverse beam placed on neighbouring, healthy ceiling joists (1: ceiling joist with a damaged and cut head; 2: transverse beam; 3: steel rod)

7.6.3 Techniques for strengthening of whole wooden structural units

A wooden structural unit (e.g. a log wall, a part of ceiling, a full gusset in truss) can also be comprehensively reinforced; that is, by adding new supplementary elements, or when coupled with a new supplementary load-bearing construction.

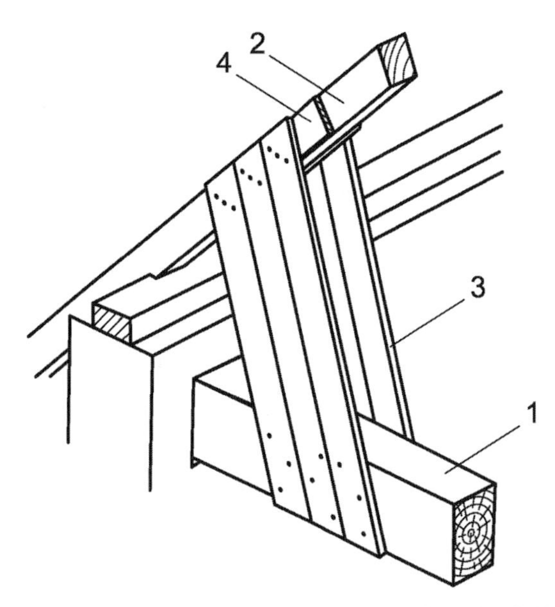

Figure 7.26 Bracing of rafter with joining beam when the full gusset in truss is weakened (1: joining beam; 2: rafter; 3: bracing boards; 4: backing plate)

7.6.3.1 Spatial bracing, tightening or stretching a wooden construction using new elements

Spatial bracing is used for renovating wooden trusses, bridges and other constructions in order to increase their functionality and stability. As required, a wooden construction is braced in a longitudinal as well as transverse direction. An illustrative example is a truss that is braced longitudinally using additional joists, saltires, roofing boards, casings as well as other components. In the transverse direction (i.e. along the plane of the full gusset in truss) the truss is braced by inserting an additional strut, crossbeam, collet, inclined brace or columns (Figure 7.27). It is not uncommon that bracing changes the original static scheme of a construction, and therefore great care should be taken that relieving of one part does not cause an increased load upon its other elements. For example, excessive sag of a joining beam is induced by additional inclined braces inserted into its centre (Figure 7.27c), or also additional columns if placed further from the joining beam supports (Figure 7.27d).

Tightening is mainly used in renovating log walls and other circumference walls; for example, using steel U-profiles (Figure 7.13b). When tightening trusses, steel rods with an adjustment of the draft article are used beneficially to trap horizontal forces. They are mounted simultaneously to the truss and to the horizontal structure; for example, to steel meshes in the ceiling (Figure 7.28).

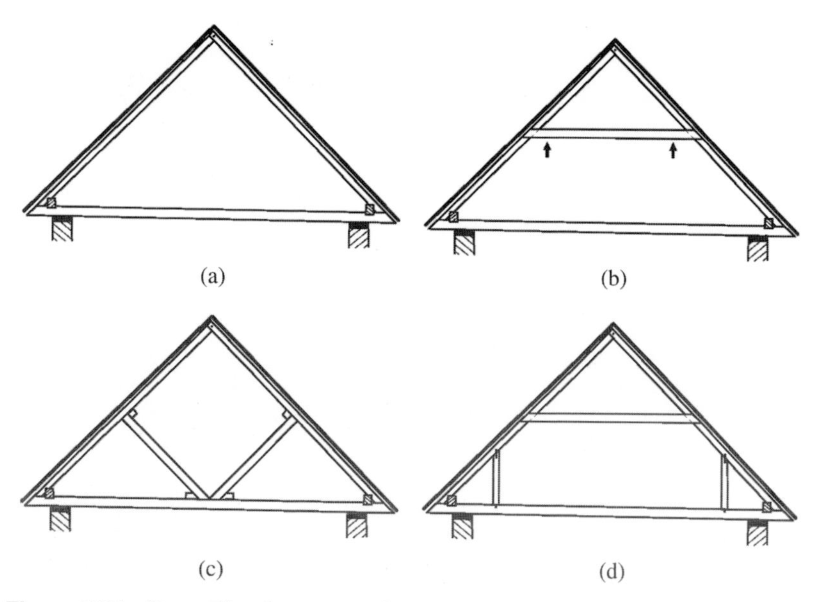

Figure 7.27 Strengthening a wooden truss using additional elements: (a) original truss; (b) strengthening of rafters using a roost; (c) supporting of rafters with inclined braces; (d) supporting of rafters with columns, close to the joining beam support

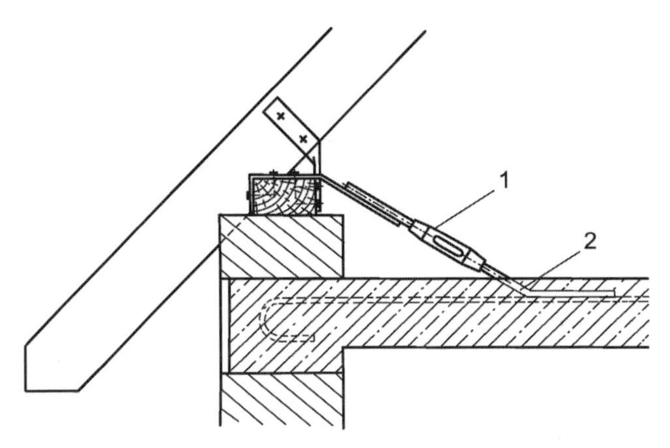

Figure 7.28 Tightening a wooden truss using a steel rod with an adjustment component (1: adjustment component; 2: steel rod welded to the steel reinforcement in a concreate ceiling and mounted to the wood rafter)

Stretching is based on the transfer of a load from the original structure or its load-bearing element (e.g. a joining beam) to new tension reinforcement, usually using a steel cable or rod (Figure 7.29).

7.6.3.2 Interlinking a wooden construction with a new-additional construction

Interlinking in practice is mainly used for strengthening wooden ceilings, using a additional layer of boards or a concrete plate (Figure 7.30).

Strengthening of wooden ceilings with a concrete plate is a wet process, in which it is necessary to take great care to avoid excess wetting of the ceiling beams and other woods. A concrete plate usually has a thickness of up to 7 cm, which can significantly increase a ceiling's load. Today, lightweight concrete is used with a smaller density of around 1500 kg/m^3 (Kanócz et al., 2014) (Figure 7.30a). From a long-term viewpoint, the ventilation of such reinforced ceilings must also be addressed, mainly due to the creation and escape of water condensate – mainly in ceilings situated under unheated areas or above spaces with high relative air humidity (Fragiacomo, 2006).

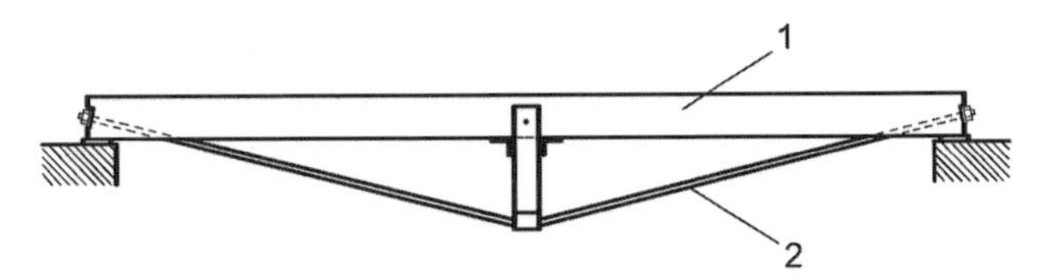

Figure 7.29 A stretching system for reinforcing a wooden beam (1: wooden beam; 2: steel cable)

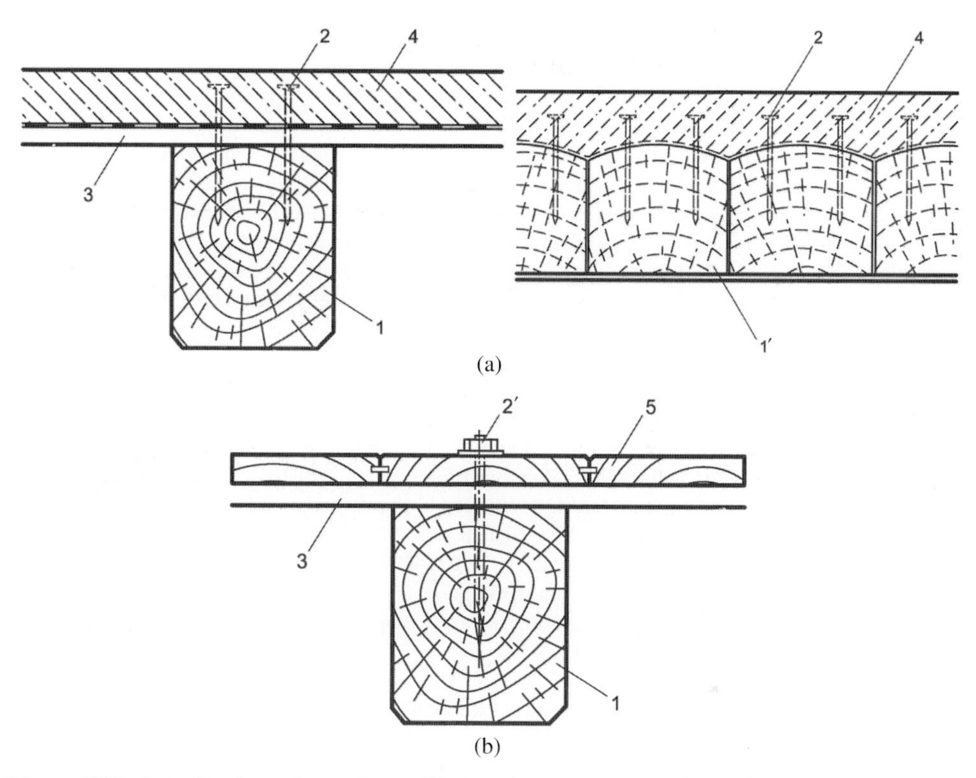

Figure 7.30 Interlinking of wooden ceilings with a concrete plate (a) or with an additional layer of new boards (b) (1: ceiling joist; 1′: attic beam; 2: steel means of anchoring to the concrete; 2′: screw to the wood; 3: wooden skeleton; 4: concrete plate; 5: layer of new boards)

References

Adamo, M., Baccaro, S. & Cemmi, A. (2015) *Radiation Processing for Bio-deteriorated Archived Materials and Consolidation of Porous Artefacts*. ENEA, Centro Ricerche Casaccia, Rome.

Arita, K., Mitsutani, S., Sakai, H. & Tomikawa, Y. (1986) Detection of decay in the interior of a wood post by ultrasonic method. *Mokuzai Kogyo* 41(8), 370–375.

Ashurst, J. & Ashurst, N. (1996) *Practical Building Conservation: Wood, Glass and Resins, and Technical Bibliography*. English Heritage Technical Handbook, vol. 5. Gower Technical, Aldershot.

Babinski, L. (1999) The two-step stabilization of the dug-out canoe from Lewin Brzeski. In *Reconstruction and Conservation of Historical Wood 99, 2nd International Symposium*. TU Zvolen, Slovakia, pp. 189–194.

Barak, A., Myers, S. & Messenger, M. (2010) Sulfuryl fluoride treatment as a quarantine treatment for emerald ash borer (Coleoptera: Buprestidae) in ash logs. *Journal of Economic Entomology* 103(3), 603–611.

Barkman, L. (1969) The conservation of the wood of the Viking war boat “Wasa”. In *Weathering of Wood, ICOMOS Symposium*, Ludwigsburg, Germany, pp. 99–108.

Barthez, J., Ramiere, R. & Tran, Q. K. (1999a) Historic dry-wood consolidation by in situ resin radio-polymerization. In *Reconstruction and Conservation of Historical Wood 99, 2nd International Symposium*. TU Zvolen, Slovakia, pp. 155–159.

Barthez, J., Hiron, X., Arnold, B., et al. (1999b) Historic and conservation treatment of the Neolithic dug-out canoes from the Bercy site in Paris. In *Reconstruction and Conservation of Historical Wood 99, 2nd International Symposium*. TU Zvolen, Slovakia, pp. 183–187.

Bech-Andersen, J., Andreasson, J. & Elborne, S. A. (2001) Quality control of microwave treatment of timber after dry rot attack. In *IRG/WP 01*, Nara, Japan. IRG/WP 01-40205.

Beiner, G. G. & Ogilvie, T. M. A. (2005) Thermal methods of pest eradication: their effect on museum objects. *The Conservator* 29(1), 5–18.

Bethancourt, J. & Cannon, M. (2015) Fall protection – structural efficacy of residential structures for fall protection systems. *Professional Safety* (May), 58–64. http://www.asse.org/assets/1/7/F3Bet_0515.pdf (accessed 10 May 2016).

Bill, J., Daly, A., Johnsen, O. & Dalen, K. S. (2012) DendroCT – dendrochronology without damage. *Dendrochronologia* 30, 223–230.

Bjurhager, I., Ljungdahl, J., Wallström, L., et al. (2010) Towards improved understanding of PEG-impregnated waterlogged archaeological wood: a model study on recent oak. *Holzforschung* 64(2), 243–250.

Blanchette, R. A., Held, B. W., Jurgens, J. A., et al. (2004) Wood destroying soft rot fungi in the historic expedition huts of Antarctica. *Applied and Environmental Microbiology* 70(3), 1328–1335.

Bobeková, E. & Horáček, P. (2013) *Identification of Wood Damaging Fungi in Buildings with Help of Genetic Molecular Methods*. TU Zvolen, Slovakia (in Slovak, English abstract).

Bodig, J. (2001) The process of NDE research for wood and wood composites. In *Proceeding of 12th International Symposium on Nondestructive Testing of Wood 2000*, NDT Test, Vol. 6/03, University of Western Hungary, Sopron, pp. 7–22.

Børja, I., Alfredsen, G., Filbakk T. & Fossdal, C. G. (2015) DNA quantification of basidiomycetous fungi during storage of logging residues. *PeerJ* 15, 3:e887. doi: 10.7717/peerj.887.

Brashaw, B. K., Bucur, V., Divos, F., et al. (2009) Nondestructive testing and evaluation of wood: a worldwide research update. *Forest Products Journal* 59(3), 7–14.

Bravery, A. F., Berry, R. W., Carey J. K. & Cooper S.E. (2010) *Recognising Wood Rot and Insect Damage in Buildings*, 3rd edn. IHS BRE Press, Bracknell, UK.

Brereton, Ch. (1995) *The Repair of Historic Buildings – Advice on Principles and Methods*. English Heritage, London.

Bucur, V., Garros, S., Navarrete, A., et al. (1997) Kinetics of wood degradation by fungi with X-ray microdensitometric technique. *Wood Science and Technology* 31(5), 383–389.

Capano, M., Pignatelli, O., Capretti, Ch., et al. (2015) Anatomical and chemical analyses on wooden artifacts from a Samnite sanctuary in Hirpinia (southern Italy). *Journal of Archaeological Science* 57, 370–379.

Caple, C. & Murray, W. (1994) Characterization of waterlogged charred wood and development of a conservation treatment. *Studies in Conservation* 39(1), 28–38.

Carlson, S.M. & Schniewind, A.P. (1990) Residual solvents in wood-consolidant composites. *Studies in Conservation* 35(1), 26–32.

Cavalli, A. & Togni, M. (2013) How to improve the on-site MOE assessment of old timber beams combining NDT and visual strength grading. *Nondestructive Testing and Evaluation* 28(3), 252–262.

Cha, M. Y., Lee, K. M. & Kim, Y. S. (2014) Micromorphological and chemical aspects of archaeological bamboos under long-term waterlogged condition. *International Biodeterioration & Biodegradation* 86(1), 115–121.

Clarke, R. W. & Squirrell, J. P. (1985) The Pilodyn – an instrument for assessing the condition of waterlogged wooden objects. *Studies in Conservation* 30(4), 177–183.

Clausen, C. A. & Kartal, S.N. (2003) Accelerated detection of brown-rot decay – comparison of soil block test, chemical analysis, mechanical properties, and immune detection. *Forest Products Journal* 53(11–12), 90–94.

Clausen, C. A., Green III, F. & Highley, T. L. (1991) Early detection of brown-rot decay in southern yellow pine using immunodiagnostic procedures. *Wood Science and Technology* 26(1), 1–8.

Cole-Hamilton, D. J., Chudek, J. A., Hunter, G. & Martin, C. J. M. (1990) NMR imaging of water in wood, including waterlogged archaeological artefacts. *Journal of the Institute of Wood Science* 12(2), 111–113.

Cruz, H., Yeomans, D., Tsakanika, E., et al. (2015) Guidelines for on-site assessment of historic timber structures. *International Journal of Architectural Heritage: Conservation, Analysis, and Restoration* 9(3), 277–289.

Dackermann, U., Crews, K., Kasal, B., et al. (2014) In situ assessment of structural timber using stress-wave measurements. *Materials and Structures* 47(5), 787–803.

Despot, R., Hrašovec, B. & Trajkovič, J. (1999) Experimental sterilization of wooden artefacts by nitrogen – N_2. In *Reconstruction and Conservation of Historical Wood 99, 2nd International Symposium*. TU Zvolen, Slovakia, pp. 145–147.

Diehl, S. V., Prewitt, M. L. & Moore Shmulsky, F. (2003) Use of fatty acid profiles to identify white-rot wood decay fungi. In *Wood Deterioration and Preservation – Advances in our Changing World*, Goodell, B., Nicholas, D. D. & Schultz, T. P. (eds). ACS Symposium Series vol. 845. American Chemical Society, Washington, DC, pp. 313–324.

Dierick, M., Van Loo, D., Masschaele. B., et al. (2014) Recent micro-CT scanner developments at UGCT. *Nuclear Instruments and Methods in Physics Research B* 324, 35–40.

Dietsch, P. & Kreuzinger, H. (2011) Guideline on the assessment of timber structures: summary. *Engineering Structures* 33, 2983–2986.

Doi, S., Kurimoto, Y., Takiuchi, H. & Aoyama, M. (2001) Effects of drying processes on termite feeding behaviour against Japanese larch wood. In *IRG/WP 01*, Nara, Japan. IRG/WP 01-10390.

Erler, K. (1997) *Alte Holzbauwerke – Beurteilen und Sanieren*. Verlag für Bauwesen GmbH, Berlin.

Evans, J. A., Eyre, C. A., Rogers, H. J., et al. (2008) Changes in volatile production during interspecific interactions between four wood rotting fungi growing in artificial media. *Fungal Ecology* 1(2–3), 57–68.

Ewen, R. J., Jones, P. R. H., Ratcliffe, N. M. & Spencer-Phillips, P. T. N. (2004) Identification by gas chromatography–mass spectrometry of the volatile compounds from the wood-rotting fungi *Serpula lacrymans and Coniophora puteana*, and from *Pinus sylvestris* timber. *Mycological Research* 108(7), 806–814.

Feio, A. & Machado, J. S. (2015) In-situ assessment of timber structural members: combining information from visual strength grading and NDT/SDT methods – a review. *Construction and Building Materials* 101, 1157–1165.

Fleming. R. M., Janowiak J., J., Kimmel, J. D., et al. (2005) Efficacy of commercial microwave equipment for eradication of pine wood nematodes and cerambycid larvae infesting red pine. *Forest Products Journal* 55(12), 226–232.

Fragiacomo, M. (2006) Long-term behaviour of timber-concrete composite beams. II: numerical analysis and simplified evaluation. *Journal of Structural Engineering ASCE* 132(1), 23–33.

Friedrich, W.L., Kromer, B., Friedrich, M., et al. (2006) Santorini eruption radiocarbon dated to 1627–1600 B.C. *Science* 312, 548.

Fujii, Y., Yanase, Y., Yoshimura, T., et al. (1999) Detection of acoustic emission (AE) generated by termite attack in a wooden house. In *IRG/WP 99*, Rosenheim, Germany. IRG/WP 99-20166.

Gindl, W., Zargar-Yaghubi, F. & Wimmer, R. (2003) Impregnation of softwood cell walls with melamine-formaldehyde resin. *Bioresource Technology* 87(3), 325–330.

Görlacher, R. (1987) Zerstörungsfreie Prüfung von Holz – ein "in situ"-vefahrten zur Bestimmung der Rohdichte (Non-destructive testing of wood: an in-situ method for determination of density). *Holz als Roh- und Werkstoff* 45(7), 273–278.

Graham, M. D. & Eddie, T. H. (1985) *X-Ray Techniques in Art Galleries and Museums*. Hilger, Bristol.

Grattan, D. W. (1982) A practical comparative study of several treatments for waterlogged wood. *Studies in Conservation* 27(3), 124–136.

Håfors, B. (1990) The role of the Wasa in the development of the polyethyleneglycol preservation method. In *Archaeological Wood – Properties, Chemistry, and Preservation*, Rowell, R. M. & Barbour, R. J. (eds). Advances in Chemistry Series, vol. 225. American Chemical Society. pp. 195–216.

Hansen, L. S. (1992) Use of freeze disinfection for the control of the common furniture beetle *Anobium punctatum*. In *IRG/WP 92*, Harrogate, UK. IRG/WP 92-1528.

Hoffmann, P. (1991) Sucrose for the stabilization of waterlogged wood – some investigation into anti-shrink-efficiency (ASE) and penetration. In *ICOM – Group on Wet Organic Archaeological Materials, 4th Conference*, Bremerhaven, Germany, pp. 317–328.

Horský, D. & Reinprecht, L. (1986) *Study of Subfossil Oak Wood*. Monograph. VŠLD Zvolen, Slovakia (in Slovak, English abstract).

Huang, C. L., Lindstrom, H., Nakada, R. & Ralston, J. (2003) Cell wall structure and wood properties determined by acoustics – a selective review. *Holz als Roh- und Werkstoff* 61(5), 321–335.

Hutton, T. (1994) Non-destructive surveying the built environment, maintenance and monitoring. In *Conservation and Preservation of Timber in Buildings*, Prague–Telč, Czech Republic, pp. 16–20.

Hyvernaud, M., Wiest, F., Serment, M. M., et al. (1996) Make ready of a detection system for insect attack by acoustical method. In *IRG/WP 96*, Guadeloupe. IRG/WP 96-10183.

Íñiguez-Gonzáles, G., Arraga, F., Esteban, M. & Llana, D. F. (2015) Reference conditions and modification factors for the standardization of nondestructive variables used in the evaluation of existing timber structures. *Construction and Building Materials* 101, 1166–1171.

Janotta, O. (1984) Die bautechnische Endoskopie, ein modernes Verfahren zur Untersuchung von Holzdecken. *Internationaler Holzmarkt* 75(1–2), 3–4.

Kačík, F., Veľková, V., Šmíra, P., et al. (2012) Release of terpenes from fir wood during its long-term use and in thermal treatment. *Molecules* 17, 9990–9999.

Kačík, F., Šmíra, P., Kačíková, D., et al. (2014) Chemical changes in fir wood from old buildings due to ageing. *Cellulose Chemistry and Technology* 48(1–2), 79–88.

Kanócz, J., Bajzecerová, V. & Mojdis, M. (2014) Experimental and numerical analysis of timber-lightweight concrete composite with adhesive connection. *Advanced Materials Research* 969, 155–160.

Kasal, B. (2008) In situ evaluation of timber structures – education, state-of-the art and future directions. In *International Rilem Conference*, SACoMaTiS, Varenna, Italy, pp. 1025–1032.

Kasal, B., & Anthony R. (2004) Advances in in situ evaluation of timber structures. *Progress in Structural Engineering and Materials* 6(2), 94–103.

Kasal, B., Drdácký, M. & Jirovský, I. (2003) Semi-destructive methods for evaluation of timber structures. In *Structural Studies – Repairs and Maintenance of Heritage Architecture VIII*, Brebia, A. A. (ed.). Advances in Architecture. WET Press, Southampton, pp. 835–842.

Kazemi, S. M., Dickinson, D. J. & Murphy, R. J. (1998) The influence of gaseous oxygen concentration on fungal growth rates, biomass production and wood decay. In *IRG/WP 98*, Maastricht, Netherlands. IRG/WP 98-10283.

Kilic, G. (2015) Using advanced NDT for historic buildings: towards an integrated multidisciplinary health assessment strategy. *Journal of Cultural Heritage* 16(4), 526–536.

Kim, J. S., Gao, J. & Daniel, G. (2015) Ultrastructure and immunocytochemistry of degradation in spruce and ash sapwood by the brown rot fungus *Postia placenta*: characterization of incipient stages of decay and variation in decay process. *International Biodeterioration & Biodegradation* 103, 161–178.

Kirk, T. K. (1984) Degradation of lignin. In *Microbial Degradation of Organic Compounds*, Gibson, D. T. (ed.). Marcel Dekker, New York, pp. 399–437.

Kirker, G. T. (2014) Genetic identification of fungi involved in wood decay. In *Deterioration and Protection of Sustainable Biomaterials*, Schultz, T. P., Goodell, B. & Nicholas, D. D. (eds). ACS Symposium Series, vol. 1158. American Chemical Society, Washington, DC, chapter 4, pp. 81–91.

Kisternaya, M. V. & Kozlov. V. A. (2007) *Wood-Science Approach to the Preservation of Historic Timber Structures*. Izd-ro KarRC RAS, Petrozavodsk, Russia (in Russian).

Klapwijk, D. (1978) Restoration and preservation of decayed timber structures and constructions with epoxides. In *Conservation of Wood in Painting and Decorative Arts*, Oxford Congress IIC, London, pp. 75–76.

Kloiber, M., Tippner, J. & Hrivnák, J. (2014) Mechanical properties of wood examined by semi-destructive devices. *Materials and Structures* 47(1), 199–212.

Koestler, R. J. (1992) Practical application of nitrogen and argon fumigation procedures for insect control in museum objects. In *Biodeterioration of Cultural Property, 2nd International Conference*, Yokohama, Japan, pp. 94–96.

Körner, I., Faix, O. & Wienhaus, O. (1992) Versuche zur Bestimmung des Braunfäule-Abbaus von Kiefernholz mit Hilfe der FTIR-Spektroskopie. *Holz als Roh- und Werkstoff* 50, 363–367.

Korpi, A., Pasanen, A-L. & Viitan, H. (1999) Volatile metabolites of *Serpula lacrymans, Coniophora puteana, Poria placenta, Stachybotrys chartarum* and *Chaetomium globosum*. *Building and Environment* 34, 205–211.

Krahmer, R. L., DeGroot, R. C. & Lowell, E. C. (1982) Detecting incipient brown rot with fluorescence microscopy. *Wood Science* 15, 78–80.

Kučera, L., & Bucher, H. P. (1988) Ein neuartiges Messgerät für Holzuntersuchungen. Schweizer *Ingenieur und Architekt Sonderdruck aus Helf* 45(3), 1243–1246.

Kučerová, I. & Lisý, J. (1999) Viewing of the impregnation substance penetration into wood with the computed tomography. In *Reconstruction and Conservation of Historical Wood 99, 2nd International Symposium*. TU Zvolen, Slovakia, pp. 167–170 (in Czech, English abstract).

Kyncl, T. (1999) Dendrochronological dating of wood as a part of historical building's investigation in Czech Republic. In *Reconstruction and Conservation of Historical Wood 99, 2nd International Symposium*. TU Zvolen, Slovakia, pp. 15–20 (in Czech, English abstract).

Larsen, K. E. (1994) *Architectural Preservation in Japan*. ICOMOS – International Wood Committee, Paris, France.

Levi, M. P. (1985) *A Guide to the Inspection of Existing Homes for Wood-Inhabiting Fungi and Insects*. North Carolina State University, Raleigh, NC.

Liang, Z., Yang, H., Xiao, Y., et al. (2016) Nanoscale surface analysis that combines probe microscopy and mass spectrometry: a critical review. *Trends in Analytical Chemistry* 75, 2–34.

Makovíny, I. & Reinprecht, L. (1990) Possibilities of wood decay detection by its electrophysical properties. In *Ochrana Dreva (Wood Protection), 8th Conference*. DT Bratislava, Czechoslovakia, pp. 70–74 (in Slovak).

Makovíny, I., Reinprecht, L., Šmíra, P., et al. (2011) Temperature fields at microwave sterilization of wood. *Acta Facultatis Xylologiae Zvolen* 53(2), 15–24 (in Slovak, English abstract).

Makovíny, I., Reinprecht, L., Terebesyová, M., et al. (2012) Control of house longhorn beetle (*Hylotrupes bajulus*) larvae by microwave heating. *Wood Research* 57(2), 179–188.

Mantanis, G. I., Young, R. A. & Rowell, R. M. (1994) Swelling of wood, part II. Swelling in organic liquids. *Holzforschung* 48, 480–490.

Marčok, M., Reinprecht, L. & Beničák, J. (1997) Detection of wood decay with ultrasonic method. *Drevársky Výskum* (*Wood Research*) 42(1), 11–22.

Martínez, R., Arriaga, F., Llana, D. F., et al. (2015) NDT to identify biological damage in wood. In *19th International Nondestructive Testing and Evaluation of Wood Symposium*, GTR FPL-GTR-239, Rio de Janeiro, Brazil, pp. 453–461.

Menzies, G. F. (2013) Whole life analysis of timber, modified timber and aluminium-clad timber windows: service life planning (SLP), whole life costing (WLC) and life cycle assessment (LCA). Report for the Wood Window Alliance, Institute for Building and Urban Design, Heriot Watt University, Edinburgh, UK.

Messner, K., Fackler, K., Lamaipis, P., et al. (2003) Overview of white-rot research – where we are today. In *Wood Deterioration and Preservation – Advances in our Changing World*, Goodell, B., Nicholas, D. D. & Schultz, T. P. (eds). ACS Symposium Series, vol. 845. American Chemical Society, Washington, DC, pp. 73–96.

Mirič, M., & Willeitner, H. (1984) Lethal temperatures for some wood-destroying fungi with respect to eradication by heat. In *IRG/WP 84*, Ronneby Brunn, Sweden. IRG/WP 84-1229.

Moreth, U. & Schmidt, O. (2000) Identification of indoor rot fungi by taxon-specific priming polymerase chain reaction. *Holzforschung* 54, 1–8.

Morgós, A. (1999) The conservation of large dried out and waterlogged archaeological wooden objects and structural elements – low molecular weight epoxy resin and sucrose, lactitol treatments (Parts I and II). In *Drewno Archeologiczne: Badania i*

Konserwacja (Archaeological Wood: Investigations and Conservations), Symposium, Biskupin–Wenecja, Poland, pp. 135–147, 149–166.

Morrel, J. J. & Corden, M. E. (1986) Controlling wood deterioration with fumigants – a review. *Forest Products Journal* 36(10), 26–34.

Mosler, J. (2006) Technology of conservatioin and restoration of historical windows. In *Okná a Dvere pri Obnove Pamiatok* (*Windows and Doors in Heritage Renovation*), Banská Štiavnica, Slovakia, pp. 94–100 (in Slovak and German).

Mosoarca, M. & Gionen, V. (2013) Historical wooden structures from Banat region, Romania. Damages: modern consolidation solutions. *Journal of Cultural Heritage* 14(3, Supplement), e45–e59.

Nicholas, D. D. (1972) Characteristics of preservative solutions which influence their penetration into wood. *Forest Products Journal* 22(5), 31–36.

Nicholson, M. & von Rotberg, W. (1996) Controlled environment heat treatment as a safe and efficient method of pest control. In *2nd International Conference on Insect Pests in the Urban Environment*, Edinburgh, UK. http://www.icup.org.uk/reports%5CICUP735.pdf (accessed 11 May 2016).

Niemz, P. & Mannes, D. (2012) Non-destructive testing of wood and wood-based materials. *Journal of Cultural Heritage* 13(3, Supplement), s26–s34.

Noguchi, M., Ishii, R., Fujii, Y. & Imamura, Y. (1992) Acoustic emission monitoring during partial copmpression to detect early stages of decay. *Wood Science and Technology* 26(4), 279–287.

Olstad, T. M., Stornes, J. M. & Bartholin, T. S. (2015) When dendrochronology corroborates art history. *European Journal of Science and Theology* 11(2), 159–169.

Palfreyman, J. W., Vigrow, A., Button, D., et al. (1991) The use of molecular methods to identify wood decay organisms. Part 1. The electrophoretic analysis of *Serpula lacrymans*. *Wood Protection* 1, 15–22.

Pandey, K. K. (2005) Study of the effect of photo-irradiation on the surface chemistry of wood. *Polymer Degradation and Stability* 90(1), 9–20.

Payton, R. (1984) The conservation of an eighth century BC table from Gordion. In Adhesives and Consolidants. International Institute for Conservation, London, pp. 133–137.

Pecoraro, E., Pizzo, B., Alves, A., et al. (2015) Measuring the chemical composition of waterlogged decayed wood by near infrared spectroscopy. *Microchemical Journal* 122, 176–188.

Pfeffer, A., Unger, W., Fröba, G. & Binker, G. (2006) Effect of fumigation with sulfuric difluoride on wood inhabiting fungi, a laboratory test. In *IRG/WP 06*, Tromsø, Norway. IRG/WP 06-30410.

Piazza, M. & Riggio, M. (2006) Limits of visual strength grading: old timber roof beams of "Ai Caduti dell' Adamello" refuge. In *Extending the Life of Bridges, Concrete, Composites, Building, Masonry, Civil Structures*. CD-Multimediale, Edinburgh, UK.

Piazza, M. & Riggio, M. (2008) Visual strength grading and NDT of timber in traditional structure. *Journal of Building Appraisal* 3(4), 267–296.

Potyralska, A., Schmidt, O., Moreth, U., et al. (2002) rDNA-ITS sequence of *Armillaria* species and a specific primer for *A. mellea*. *Forest Genetics* 9, 119–123.

Prieto, G. (1990) Detection and estimation of *Hylotrupes bajulus* L. wood damages by ultrasonics. In *IRG/WP 90*, Rotorua, New Zealand. IRG/WP 90-2350.

Pristaš, P., & Gáperová, S. (2001) About possibilities of the PCR method application in mycology. In *The PCR in Biological Research and Diagnostic. Conference*, ÚFHZ SAV Košice, Slovakia, pp. 158–163 (in Slovak, English abstract).

Rapp, A.O. (1998) Method of embedding and staining of wood after biological testing to support the identification of decay type. In *IRG/WP 98*, Maastricht, Netherlands. IRG/WP 98-20131.

Reinprecht, L. (1991) Restoration of damaged wood with polyacrylates, epoxides, phenolic resins, and amino resins. In *Advances in Production and Application of Adhesives in Wood Industry*, 10th Symposium, VŠLD Zvolen, Slovakia, pp. 312–325 (in Slovak, English abstract).

Reinprecht, L. (1995) Conservation of oak timbers from the castle Slovenská Ľupča. In *Reconstruction and Conservation of Historical Wood 95, 1st International Symposium*. TU Zvolen, Slovakia, pp. 161–168 (in Slovak, English abstract).

Reinprecht, L. (1997) *Processes of Wood Degradation*. TU Zvolen, Slovakia (in Slovak).

Reinprecht L (1998) *Reconstruction of Wooden Objects*. TU Zvolen, Slovakia (in Slovak).

Reinprecht, L. (2010) Repairing of wooden elements by beta-method using carbon, steel or beech rods. In *IRG/WP 10*, Biarritz, France. IRG/WP 10-40488.

Reinprecht, L. (2011) Possibilities for improvement of moisture and strength properties of decayed spruce wood with natural resins. *Wood Research* 56(3), 285–296.

Reinprecht, L. & Hulla, M. (2015) Colour changes in beech modified with essential oils due to fungal and ageing-fungal attacks with *Coniophora puteana*. *Drewno (Wood)* 58(194), 37–48.

Reinprecht, L. & Joščák, P. (1996) Reinforcement of model-damaged wooden elements. Part 2: restoration of wooden elements by the extension method using natural wood or epoxy–wood composite. *Drevársky Výskum (Wood Research)* 41(2), 4–55.

Reinprecht, L. & Lehárová, J. (1997) Microscopic analyses of woods – beech (*Fagus sylvatica* L.), fir (*Abies alba* Mill.) and spruce (*Picea abies* L. Karst.) in various stages of rot caused by fungi *Serpula lacrymans, Coriolus versicolor* and *Schizophyllum commune*. In *Drevoznehodnocujúce huby '97 – Wood-Damaging Fungi 97*, 1st Symposium. TU Zvolen, Slovakia, pp. 91–113 (in Slovak, English abstract).

Reinprecht, L. & Makovíny, I. (1987) Hardening of aminoplasts in modificated wood with catalytic and thermic dielectric heating. In *Modyfikacja Drewna (Wood Modification)*, 6th Symposium, Poznaň, Poland, pp. 288–298.

Reinprecht, L. & Pánek, M. (2013) Ultrasonic technique for evaluation of biodefects in wood: part 2 – in-situ and in-vitro analyses of old beams using ultrasonic and bending tests. *International Wood Products Journal* 4(1), 22–29.

Reinprecht, L. & Štefko, J. (2000) *Wooden Ceilings and Trusses – Types, Failures, Inspections and Reconstructions*. ABF–ARCH Praha, Czech Republic, (in Czech).

Reinprecht, L. & Šupina, P. (2015) Comparative evaluation of inspection techniques for impregnated wood utility poles – ultrasonic, drill-resistive, and CT-scanning assessments. *European Journal of Wood and Wood Products* 73(6), 741–751.

Reinprecht, L. & Varínska, S. (1999) Bending properties of wood after its decay with *Coniophora puteana* and subsequent modification with selected chemicals. In *IRG/WP 99*, Rosenheim, Germany. IRG/WP 99-40146.

Reinprecht, L., Novotná, H. & Štefka, V. (2007) Density profiles of spruce wood changed by brown-rot and white-rot fungi. *Wood Research* 52(4), 17–28.

Rigon, L., Vallazza, E., Arfelli, F., et al. (2010) Synchrotron-radiation microtomography for the non-destructive structural evaluation of bowed stringed Instruments. *e-PreservationScience* 7, 71–77.

Rinn, F., Schweingruber, F. H. & Schär, E. (1996) Resistograph and X-ray density charts of wood. Comparative evaluation of drill resistance profiles and X-ray density charts of different wood species. *Holzforschung* 50, 303–311.

Robson, P. (1999) *Structural Repair of Traditional Buildings*. Donhead, Shaftesbury.

Ross, P. (2002) *Appraisal and Repair of Timber Structures*. Thomas Telford, London.

Ross, R. J., DeGroot, R. C., Nelson, J. N. & Lebow, P. K. (1997) The relationship between stress wave transmission characteristics and the compressive strength of biologically degraded wood. *Forest Products Journal* 47(5), 89–93.

Said, M. M. (2004) Moisture measurement guide for building envelope applications. In *Research Report 190*. Institute for Research in Construction, NRC, Canada, pp. 1–34.

Sandak, A., Rozanska, A., Sandak. J. & Riggio, M. (2015) Near infrared spectroscopic studies on coatings of 19th century wooden parquets from manor houses in south-eastern Poland. *Journal of Cultural Heritage* 16, 508–517.

Sandak, J., Sandak, A. & Riggio, M. (2015) Multivariate analysis of multi-sensor data for assessment of timber structures: principle and application. *Construction and Building Materials* 101, 1172–1180.

Scheffrahn, R. H., Robbins, W. P., Busey, P., et al. (1993) Evaluation of a novel, hand-held, acoustic emissions detector to monitor termites (Isoptera: Kalotermitidae, Rhinotermitidae) in wood. *Journal of Economic Entomology* 86(6), 1720–1729.

Scheffrahn, R. H., Wheeler, G. S. & Su, N-Y. (1995) Synergism of methyl bromide and sulfuryl floride toxicity against termites (Isoptera: Kalotermitidae, Rhinotermitidae) by admixture with carbon dioxide. *Journal of Economic Entomology* 88(3), 649–653.

Schmidt, E., Juzwik, J. & Scheider, B. (1997) Sulfuryl floride fumigation of red oak logs eradicates the oak wilt fungus. *Holz als Roh- und Werkstoff* 55, 315–318.

Schmidt, O. (2000) Molecular methods for the characterization and identification of the dry rot fungus Serpula lacrymans. *Holzforschung* 54, 221–228.

Schmidt, O. (2006) *Wood and Tree Fungi – Biology, Damage, Protection, and Use*. Springer-Verlag, Berlin.

Schmidt, O. & Kallow, W. (2005) Differentiation of indoor wood decay fungi with MALDI-TOF mass spectrometry. *Holzforschung* 59(3), 374–377.

Schmidt, O. & Moreth, U. (2002) Data bank of rDNA-ITS sequences from building-rot fungi for their identification. *Wood Science and Technology* 36(5), 429–433.

Schmidt, O., Grimm, K. & Moreth, U. (2002) Molecular identity of species and isolates of the *Coniophora* cellar fungi. *Holzforschung* 56, 563–571.

Schmidt, R., Göller, S.T. & Hertel, H. (1995) Computerized detection of feeding sounds from wood boring beetle larvae. *Material und Organismen* 29(4), 295–304.

Schwarze, F. W. M. R. (2007) Wood decay under the microscope. *Fungal Biology Reviews* 21(4), 133–170.

Shintani, H., Sakudo, A., Burke, P. & McDonnell, G. (2010) Gas plasma sterilization of microorganisms and mechanisms of action. *Experimental and Therapeutic Medicine* 1(5), 731–738.

Šimůnková, E., Šmejkalová, Z. & Zelinger, J. (1983) Consolidation of wood by the method of monomer polymerization in the object. *Studies in Conservation* 28(3), 133–144.

Solár, R., Reinprecht, L., Kačík, F., et al. (1987) Comparison of some physico-chemical and chemical properties of carbohydrate and lignin part of contemporary and sub-fossil oak wood. *Cellulose Chemistry and Technology* 21(5), 513–524.

Spear, R. N., Cullen, D. & Andrews, J. H. (1999) Fluorescent labels, confocal microscopy, and quantitative image analysis in study of fungal biology. *Confocal Microscopy* 307, 607–623.

Štefko, J., & Reinprecht, L. (2004) *Wooden Buildings: Constructions, Protection and Maintenance*, Jaga Group, Bratislava (in Czech).

Stelzner, J. & Million, S. (2015) X-ray computed tomography for the anatomical and dendrochronological analysis of archaeological wood. *Journal of Archaeological Science* 55, 188–196.

Stoppacher, N., Kluger, B., Zelinger, S., et al. (2010) Identification and profiling of volatile metabolities of the biocontrol fungus *Trichoderma atroviride* by HS-SPME–GC–MS. *Journal of Microbiological Methods* 81, 187–193.

Sutter, H. P. (2002) *Holzschädlinge an Kultürgutern erkennen und bekämpfen*. Haupt Verlag, Bern.

Tejedor, C. C. (2010) Re-conservation of wood from the seventeenth century Swedish warship the Vasa with alkoxysilanes: a re-treatment study applying thermosetting elastomers. Master of Arts Thesis, Texas A&M University.

Teplý, J., Franěk, C., Kraus, R. & Červenka, V. (1986) Mobile irradiator and its application in the preservation of the objects of art. *Radiation Physics and Chemistry* 28(5–6), 585–588.

Terebesyová, M., Reinprecht, L. & Makovíny, I. (2010) Microwave sterilization of wood for destroying mycelia of the brown-rot fungi *Serpula lacrymans, Coniophora puteana* and *Gloeophyllum trabeum*. In *Wood Structure and Properties*, Kúdela, J. & Lagaňa, R. (eds). Arbora Publishers, Zvolen, Slovakia, pp. 145–148.

Tiňo, R., Repáňová, Z. & Jablonský, M. (2014) Effects of atmospheric plasma treatment on wood surfaces. In *57th SWST International Annual Convention – Sustainable Resources and Technology for Forest Products*. TU Zvolen, Slovakia, pp. 876–877.

Tippner, J., Hrivnák, J. & Kloiber, M. (2016) Experimental evaluation of mechanical properties of softwood using acoustic methods. *BioResources* 11(1), 503–518.

Tiralová, Z. & Mamoňová, M. (2005) Activity of brown-rot fungus *Gloeophyllum trabeum* in thermally treated wood – microscopic analysis. In *Drevoznehodnocujúce Huby '05 – Wood-Damaging Fungi 05, 4th Symposium*, Kováčová, TU Zvolen, Slovakia, pp. 65–68 (in Slovak).

Tiralová, Z., & Reinprecht, L. (2004) Fungal decay of acrylate treated wood. In *IRG/WP 04*, Ljubljana, Slovenia. IRG/WP 04-30357.

Tolvaj, J., Horváth, I., Sováti, E. & Sáfár, C. (2000) Colour modification of black locust by steaming. *Drevársky Výskum* (*Wood Research*) 45(2), 25–32.

Tran, Q. K., Ramière, R. & Ginier-Gillet, A. (1990) Impregnation with radiation – curing monomers and resins. In *Archaeological Wood – Properties, Chemistry, and Preservation*, Rowell, R. M. & Barbour, R. J. (eds). Advances in Chemistry Series, vol. 225. American Chemical Society, Washington, DC, pp. 217–233.

Unger, A., Schniewind, A.P. & Unger, W. (2001) *Conservation of Wood Artifacts*. Springer-Verlag, Berlin.

Unger, W. & Unger, A. (1995) Die biologische Korrosion von Konsolidierungsmitteln für Kunst- und Kulturgut aus Holz. *Kunsttechnology und Konservierung* 9(2), 377–384.

Unger, W., Fritsche, H. & Unger, A. (1996) Zur Resistenz von mAlmaterialien und Stabilisierungsmitteln für Kunst- und Kulturgut gegenüber holzzerstörenden insekten. *Kunsttechnology und Konservierung* 10(1), 106–116.

Urban, J. & Justa, P. (1986) Conservation by gamma radiation. *Museum* 151, 165–167.

Van den Bulcke, J., Van Acker, J. & Stevens, M. (2005) Image processing as a tool for assessment and analysis of blue stain discolouration of coated wood. *International Biodeterioration & Biodegradation* 56, 178–187.

Van den Bulcke, J., De Windt, I. & Van Acker, J. (2011) Non-destructive evaluation of wood decay. In *IRG/WP 11*, Queenstown, New Zealand. IRG/WP 11-20479.

Wedvik, B., Stein, M., Stornes, J. M. & Mattsson J. (2015) On-site radioscopic qualitative assessment of historical structures: identification and mapping of biological deterioration of wood. *International Journal of Architectural Heritage: Conservation, Analysis, and Restoration*, in press. doi: 10.1080/15583058.2015.1077905.

Wilcox, W. W. (1968) *Changes in Wood Microstructure through Progressive Stages of Decay*. US Forest Service Research Paper FPL-70. US Department of Agriculture, Forest Service, Forest Products Laboratory, Madison, WI.

Wilcox, W. W. (1988) Detection of early stages of wood decay with ultrasonic pulse velocity. *Forest Products Journal* 38(5), 68–73.

Williams, L. H. & Sprenkel, R. J. (1990) Ovicidal activity of sulfuryl floride to anobiid and lyctid beetle eggs of various ages. *Journal of Entomological Science* 25(3), 366–375.

Yeomans, D. T. (2003) *The repair of Historic Timber Structures*. Thomas Telford, London.

Zelinger, J., Šimůnková, E. & Kotlík, P. (1982) *Chemistry in Work of the Conservator*. Academia, Prague (in Czech).

Zorovič, M. & Čokl, A. (2015) Laser vibrometry as a diagnostic tool for detecting wood-boring beetle larvae. *Journal of Pest Science* 88, 107–112.

Standards

ASTM D245-00 (2006) American Society for Testing and Materials – Standard practice for establishing structural grades and related allowable properties for visually graded lumber.

EN 335 (2013) Durability of wood and wood-based products – Use classes: definitions, application to solid wood and wood-based products.

EN 350-2 (1994) Durability of wood and wood based products – Natural durability of solid wood – Guide to natural durability and treatability of selected wood species of importance in Europe.

EN 1995 (2008) Eurocode 5: Design of timber structures.

UNI 11119 (2004) Cultural Heritage – Wooden artefacts – Load-bearing structures – On site inspections for diagnosis of timber members (The Italian standard).

Index

Note: Page numbers given in italics refer to figures and numbers in bold refer to tables and boxes.

abiotic factors
  needed for activity of wood-damaging fungi 72–76, **74**
  needed for activity of wood-damaging insects 94, **95**, *96*, 97
abiotic wood degrading factors *3*, 28–30, **29**, 52
accompanying biodegradable/attracting substances **10**, 72, 96–97
acetalization 234, 236
acetylated wood 199, 234–235, 242–243, 247
  Accoya wood 221, 235, 242, **244**, 247
  A-Cell 221, 235
acetylation **10**, **20**, 234–235, 244
acid
  acetic *47*, **55–56**, 175, 223, 229, 234–235
  attacking of wood *3*, **9**, **13**, 45, 55
  carboxylic 234
  formic 115, 223, 229
  hydrochloric 45, *47*, 48, **55–56**, 240
  nitric 45–46, *47*, **55–56**
  oxalic **77**, **79**, 112, 114
  resin **10**
  salt of 45, 170
  sulphuric 45–46, *47*, 48, **55–56**
acrylates 171, 232, 241, 302
  butyl methacrylate (BMA) 291, 297
  methyl methacrylate (MMA) 241, 291, 297–298
  mono- 291–292
  poly- 292, 294, **295**, 297, *305*
activation energy 34, 37
active substances of preservatives, *see* “individual indexes” for wood treating substances
additives
  colour-stabilizing, against UV radiation 170–172
  effecting thermal decomposition of wood **39**, 42
  for wooden composites 197–205
adhesives
  bio-resistant 202
  fire-resistant 203
  for modified wood 229
  water-resistant **24**, 205
  for wooden composites 197–199, 201–203, 205
adhesion strength
  of coatings 171
  of joints in wooden composites 199, 201
aesthetic
  of damaged wood 10, 48, 52, 90
  effect of preservatives 170, 176
  of modified wood 245
  of restored wood 290, 301, 309, 313
ageing, *see* weathering
aggressive chemicals **13**, 28, 45–46, *47*, 48, 53, **56**
algae 62
alkali
  ammonium hydroxide *47*, 53, **55–56**
  attacking of wood *3*, **13**, 45, 53
  cell collapse by 53–54
  sodium hydroxide *47*, 53, **55–56**

Wood Deterioration, Protection and Maintenance, First Edition. Ladislav Reinprecht.

alkaline copper quat (ACQ) 158, **200**
alkaloids 76, 162
Alternaria alternata **90**
American Wood Protection Association (AWPA) 145, 151, 194
amino resins 48, 198, 221, 232, 238–239, 245, **295**, 301
ammonium dihydrogen phosphate **46**, 169–170, 204
ammonium sulphate 169
anhydrides of acids 234–235
Anobiidae 93, 95, 100, *101*, 166, 263
Anobium punctatum **95**, *96*, 100, *101*, 165, 281, 284
antagonism **20**, 90, 221, 247–249
anthropogenic factors
  needed for activity of wood-damaging fungi 76
  needed for activity of wood-damaging insects 97
antibiotics 65, 76
antioxidants 160, 170–171
anti-shrinkage efficiency (AShE) 294, 307
anti-splitting 131
anti-swelling efficiency (ASE) 222, **244**
anti-weathering agents 145
Antrodia vaillantii **74**, 85, **86**
ants 103, *104*, 164
archaeological wood
  conservation of 289, 292, 294, 300, 306–308
  dating of 279
  durability of 17, 49, 56, 63
artefacts 18, 56, 94, 193, 263, 265, 285
  treatment of 260, 290, 292–294, 298, 301, 321
Ascomycota
  asexual reproduction of 66
  phylum 66–67, **71**, **127**
  sexual reproduction of 66, *67*
ascospores 66, *67*
Aspergillus amstelodami **90**
Aspergillus flavus 91
Aspergillus niger *70*, 72, 75–76, **90**, 91, 286
atmospheric corrosion 28, 30, *31*, 245
attractants **19**, 94
Aureobasidium pullulans **89**
autoclave 177, *187*, 189–191, 225
autoclave impregnation of wood
  pressure-pressure-vacuum (PPV – Rüping) **178**, *188*, 189, *190*
  pressure-vacuum (PV – Lowry) **178**, *188*
  pulse pressure-diffusion 190, *191*
  station for *187*
  vacuum-atmospheric pressure-vacuum (VAV) *188*
  vacuum-pressure-vacuum (VPV – Bethell) 156, **178**, *188*, 189, *190*
azaconazole 160

Bacillus asterosporus 65, 76
Bacillus brevis 63
Bacillus subtilis 63, *64*, 65, 176
bacteria **13**, 62, **63**, 108, 152, 76
  aerobic 63
  anaerobic **19**, 49, 63, 248
  antagonistic 63, 65, 248–249
  fixing nitrogen for fungi 76
  physiology of 62–63
bacteria destroying cell walls of wood
  cavitation 63–64, **65**
  erosion 63–64, **65**
  tunnelling 63–64, **65**, 242
bacteria destroying pit membranes in cells of wood 63
bactericides **19**, 145, 152
basidia 66, *69*
Basidiomycota
  asexual reproduction of **68**
  phylum 66–67, **71**, **127**
  sexual reproduction of 66, *69*
basidiospores 66, *69*
beeswax 292–293, 302
beetles 93, 97–102, **127**
  ambrosia 97
  bark 97
benzo(*a*)pyrene 156
bifethrin 164
Biocidal Product Directive 98/8/EC 150–151
biocides **19**, *23*, **24**, 192, 197
  acute toxicity (lethal dose LD) of 149, **150**, 160, 165–167
  degradability by proteobacteria and staining fungi 150–151

historical development of *153*
marking of 146, 148
registration of 151
test methods of 151, 173, 196
bio-control, *see* modification of wood, biological
biodegradation
of cellulose **77**, **79–80**, 110–113, *116*
of hemicelluloses **77**, 113–114, *116*
of lignin **79–80**, 114–115, *116*, 117
biological factors
needed for activity of wood-damaging fungi 76
needed for activity of wood-damaging insects 96–97
biological pests
activity prevention of **10**, **129**, 142, 146
attacking wood 1, *3*, 62, **63**
changes of wood properties by 12, 117–119
conditions for activity of **19**, *73*, **74**, **95**, **127**
fights to 261–264, 280–289
resistance of wooden products to 52, 152–167, 198–200, 218, 247
biological resistance
of biologically modified wood 63, 247–250
of chemically modified wood 232, 242, *243*, **244**, 245–246
of chemically preserved wood 152–167, 173, 192–193, 196
of thermally modified wood 52, 226, *227*, 228
of wooden composites 198–200
biological wood-degrading factors *3*, **63**
bioresmethrin 164
bis-[*N*-cyclohexyldiazeniumdioxy]-copper (Cu-HDO) 157–158, **183**, 189, 191
bonds
covalent 34, 108, **109–110**, 117, 197, 233–234, 238, 241
hydrogen **8**, **11**, 54, 56, 118, 197, 234, **296**, 301
splitting of covalent **109–110**, 115
van der Waals interactions **8**, **11**, 54, 56, 118, **296**
bordered pit **5–6**, **13**, *31*, 49, *64*, 174–176, 228, 232
boric acid **19**, **150**, 155, 159, 164, 168, 170, 182, 204–205, 308
boron, organic esters of 155–156
bridges
damaging of 36, 80, 98
diagnosis of 266
lifetime of 21–22, *23*, **24**
protection of 22, 126, 137, 145, 156, 164, 242
renovation of 308, 313
thermal 140
brown rot **13**
mechanisms of **77–78**, 112–113
properties of wood with **113**, **274**
burning
speed of *38*, 40–41
of wood 34, *35*, 36, *37–39*, 40–42, **45**

^{14}C radioisotope 264, 278
calcium oxalate 70
calcites 49
Callidium violaceum 94, *99*, 263
Camponotus herculeanus 103
Camponotus ligniperdus 103
capillary
forces **11**, 137, 182
pressure 177, **178**, *179*, 180
transport **8**, *179*, 233
water 20
carbamates 159
carbolineum 157
carbon (reinforcing) rods 304, *320–321*
carbonized foam *169*
carbonyl group **8**
ceilings
damaging of 1, 85, 98, 100, 103, *310*
diagnosis of 265, *310*
lifetime of 22
protection of 126, 142, 145, 182
renovation of 308, 313, *327*
cell element of wood **4–6**
libriform fibre **5**, **15**
parenchyma **5**, **15**, 62, 88
resin canal **4–5**, 62
tracheid **2**, **5–6**
vessel **5–6**, **15**

cell wall of wood **4–8**
  cross-linked at modification 218
  decomposed at fossilization 50, *51*
  destroyed by bacteria 63–64, *65*
  destroyed by fungi 77–82
  middle lamella of **7**, *31*, **80**
  mineralization of *51*
  penetration to 172, 181, 192, 220
  permeability of, *see* penetration into cell walls of wood
  pit membrane (margo) in **8**, *64*, 65
  pit in **5–6**, **13**, *64*, 174–176, 232
  plasticized 219
  pores of 1 to 80 nm in **8**
  porosity of **8**, 243
  primary wall of **7**
  secondary wall of **7–8**
cellulose **7–9**
  amorphous **9**, **109**, 112
  biodegradation of **77**, **79**, **82**, 108–113, **109**, *111–112*, 113
  chemical degradation of **46**
  crystalline **9**, **15**, 77, *78*, **79**, **109**, 112, 268
  crystallinity index of 28
  de-crystallization of **13**
  depolymerization of *31*, 55, 65, 111–113, 245, 285
  fibrils of 31–32
  thermal decomposition of *34–35*, **46**, 223
Cerambycidae 93, 98–100
Ceratostomella pilifera **89**
Chaetomium sp. 79, 249
charcoal 36–38, *39–40*, 41, 44, *153*, 168
chemical corrosion 45, 245
chemical modification of wood, methods of
  active *231*, 233–238, 241
  filling its cell walls *231*, 232–238, 245
  filling its lumens *231*, 232, 238–241, 245
  passive *231*, 238–241
chemical protection of wood
  ecology of 149–151, 245
  legislation of 146–149
  methodology of 146–149
  quality control of 193–196
  regulation of 151–152
  technologies of, *see* technologies of wood treatment with chemicals
chemical protection of wooden composites
  preservatives for 165, 199, **200**
  technologies for 200, *201*, 202, *203*, 204–205
chemically modified wood
  biological durability of 233, 242–247
  changes in structure of 232–233
  dimensional stability of 233, 242, 244–245, 247
  fire resistance of 242, 247
  with substances, *see* "individual indexes" for wood treating chemicals
  technologies 233. *See also* technologies of wood treatment with chemicals
children's playgrounds 22, 137, 157, 187, **225**, 231, 247
chitin
  in cell walls of fungi 67, 70–71, 76, 111
  production inhibitors of 166–167
  in wing-cases of insects 97
chitosan 162–163
chlamydospores **68**
chlorinated carbohydrates *153*, 164
chromatography 35, 194–195, 275, 278
class
  of fire response 42, **43**
  penetration 146, *147*, 194
  retention 146, *147*, 194
  of wood durability 15, **16–17**, **24**, 146, *147*, **148**, *227*, 228, 244
  of wood impregnability 146, *147*, **174**
  of wood use **127**, 146, *147*, **148**, 149, 165, 172–173, 182, 187, 194, 228, 247, 261
classification
  of wood-damaging fungi 66–70, **71**
  of wood-damaging insects 93–94
climatic conditions
  humidity of wood effected by **11**
  regulation of **129**, 141–142
coatings (paints)
  acrylate 242, 261, 301
  acrylate-alkyd 138
  acrylate-polyurethane 138
  adherence to wood of *134*

alkyd 138, 170–171, 242, 261, 301
defects in film-forming *171*
film-forming **20**
pigmented 170–172
polyacrylate 138, 170–171
polyurethane 138, 170–171
transparent 170–172
vapour-impermeable 130, 138, 170
vapour-permeable 138, 301
waterproof 138, 155
water-repellent **24**
coatings, test of 196
Coleoptera, *see* beetles
colophony 292–293, 302
colorimeter **267**, 269
colour
changes in deteriorated wood **13**, 15, 28, **29**, 32, 48, 51, 66, **78**, 88, **89**, **267**, 269
of painted wood 170
stability of treated wood 171, 196, 229
of wood **4**
combustion
triangle *37*
of wood *35*, 37–38, *39*, 42, 44
computed tomography (CT) **267**, 270–271, *272*
concentration gradient 177, *179*, 182–183, 233, 307
conditioning 142, 224–225
conidia 66, **68**, *70*, 89
Coniohora puteana 71, *73*, **74**, 85, **86**, 113, 163, 173, 192, 243, 245
conservation agents
application properties of 289–292
natural 289, 292–294, 304
synthetic 289, 294–301
conservation of air-dried damaged wood
by creation of surface films 301
by filling its cavities 304
by filling its lumens 302, *303*
by its reintegration 301, 303–304, *305*
by saturation its cell walls 301–302, *303*
technologies for 306
conservation of waterlogged wood
by diffusion 306–308
by vacuum sublimation, *see* vacuum sublimation
copper
active to soft-rot fungi 154
as biocide *153*
copper azoles 158
copper bis-dimethyldithiocarbamate (CDDC) 158
copper carbonate 154
copper/chromate/arsenic (CCA) *153*, 154, 160
copper/chromate/borate *153*, 154, **183**
copper hydroxide **150**
copper hydroxide-carbonate 154, 159
copper naphtenate (CuN) *153*, 158, **200**
copper oxide 152, 154, 193
copper quat – micronized nano-form (MCQ) 154
copper-8-quinolinolate 158
copper sulphate 154, 182
corrosion of wood
under aerobic conditions 45
under anaerobic conditions 49
atmospheric, *see* erosion atmospheric
atmospheric-chemical 46
chemical 45–46, *47*, **55**
cracks
creation in wood **13**, **29**, *30*, 32, 51–52, 284, 288, 306
prevention in wooden products 129, 131, *132–133*, 142, 170, 172, 294
creosotes **19**, *23*, **82**, *153*, 156–157, **183**, 189, *190*, **200**
chemical composition of 156
efficacy against biological pests 157, 164
types by WEI 156
cultural heritage 263, 311
cyfluthrin **150**, 164
cypermethrin **150**, 164
cyproconazole 160

Daedalea quercina **74**, **84**
dammar 292–293, 302
Darcy's law 180
decay-causing fungi, *see* fungi, decay-causing
in exteriors 80, **83–84**
in interiors 85, **86**

degradation of wood 12, **13**
abiotic *3*, 28, **29**
biological *3*, 62, **63**, **77**, **79**, **82**
chemical **29**, 45
photo 28, *30–31*
radiation **29**
thermal **29**, 34, *35*, *39*
thermal-chemical **55**
degree of polymerization **8**, 53, 113, 275, 300
deltamethrin **150**, 164
demolition 309
dendrochronology 264, 278–279
density
of damaged wood 117, 119, *246*
of wood, *see* wood density
deoxyribonucleic acid (DNA) 68, 117, 163, 276–277
depolymerization (decomposition) **11**
of cellulose *31*, 55, *109*, *111–112*
of hemicelluloses 30, 55, *109*, 113–114
of lignin 30, 55, *109–110*, 114–117
design
of construction 22, *23*, 44, 312
of details 22
proposal/realization **24**, 129, 140
Deuteromycota 66–67, **71**
diagnosis of damaged wood **24–25**, 194, 264, 309, *310*
diagnostic methods
acoustic tomography **267**
biological **267**, 275–278
chemical **267**, 275
electrical **267**, 269
electromagnetic **267**, 272–273
instrumental 266–276
optical **267**, 268–269
radiographic **267**, 270–272
sensory **265**, 266
sonic **265**, 266
strength **267**, 273–275
thermography **267**, 273
ultrasonic **267**, 269–270, *274*
dialkyl-dimethyl-ammonium compound (DDAC) 151, 158–159
dichlofluanid **150**, 151
dichloro-diphenyl-trichloroethane (DDT) **150**, 151, 164
4,5-dichloro-2-*n*-octyl-4-isothiazol-3-one (DCOIT) 160–161
didecyl-dimethyl-ammonium tetraflouroborate (DBF) 155
differential thermal analysis (DTA) *34*, 35
differential thermo-gravimetric 35
diffusion
in cell walls of wood **8**, 74, 108, 110, 117, 233
coefficient 182, 192, 307–308
driving forces 174, 177, **178**, 179, 181–182
gradient 177, 179, 181–182
intensity 182
nonstationary 179, 182
resistance 140–141
stationary 182
of substances in wood 177–182, 298
of water vapour 140, 171
dimensional stability
of chemically modified wood 232–233, 236, 242, 244
improving of damaged wood **10**, 291, 290, 292, 299–300, 302, 306
of thermally modified wood 52, 223
of wooden composites *132*, 198
1,3-dimethylol-4,5,-dihydroxy-ethyl-urea (DMDHEU – for Belmadur wood) 237–238, **244**, 247
dispersion stability 192
dry-out factor (*a*) 129–130
durability increase (DI) 222, **244**

ecological risk
of moulds 91
of preservatives 149–151, 205
efficiency of chemical wood modification (ψ_{mod}) 222
elementary fibrils of cellulose **7**
emissions **24–25**, 46, 51
endoscope **267**, 269
energy
activation 34, 37
chemical 34, 36
initiation 38
light 38
radiant 281

radiation 284
thermal 38
environment
acid 72, 239
aerobic 45, *50*
anaerobic 49–50, 56
exterior 14, 22, 28, **29**, 80
inert 220, 226
interior 14, 22, 28, **29**, 52
preservative's impact on 145, 151
wet 49, 199, 231
environmental Protection Agency (EPA) 151
enzymes **10**, 108, **109–110**
catalysts for biochemical decomposition of wood 76, **109–110**
catalytic efficiency of 110
catalytic function of **13**, 108
cellobiose: quinone-oxidoreductase (CBQ) **79**, **109**, 115, *116*
cellulases 106, 111, *116*
dehydrogenase 161
dioxygenase **80**, **110**, *116*
extracellular 70, 74, 108
1,4-β-glucanases 65, **77**, **79**, **109**, *111*, 245
1,4-β-glucosidase **77**, **109**, *111*, *116*
hydrolases **77**, **79**, **82**, 108, 111–113, 176
intracellular 70, 108
laccase **80**, **110**, 115, *116*
ligninolytic **79**, **82**, 114–115, *116*
mannanolytic **77**, **109**, 114
oxidases **80**, **110**, 111, 114–115, *116*, 221
oxidoreductases **79**, **82**, 108, **109**, 111, 114–115, *116*
pectinase 176
peroxidases **10**, **79–80**, **109**, 115, *116*, 117
phenoloxidases 115, 117
xylanolytic **77**, **109**, 114, *116*, 176, 245
epoxides 203, 232, 237, 244, 292, 295, **295**, 298–300, 302, *320–322*
equilibrium moisture content 129–130, 218, 228, 233, 242, 262
Ergates faber **95**, *96*, *99*
Ernobius mollis *101*
erosion atmospheric 28, *30–31*, 32
essential oils *153*
bio-active types obtained from plants 161–162
chemical composition of 161–162
esterification 234–235, *243*
etherification 234, 236–238
ethylene oxide 237, 286–289
eurythermic organisms 75
evaporation index (EI) 290–291
extractives **7–10**
corrosion of metals with 48
effective to biological pests **9–10**, 30, 72, 161–162
inorganic 71
leaching out of wood 170

factor method 21–22
failures
during usage 21
in project 21
in technology 21
fenoxycarb **150**, 166
Fenton reagent 72, **77**, **79**, 108, *112*, 113–115, 154
fibreboards (FB) *2*, 196
chemical protection of **200**, 203, *243*
degradations of **43**, 197–198, *243*
fibre saturation point (FSP) **11–12**, **55**, 190, 284, 306
fibril
elementary **7**
macro **7**
micro **7**
fibrillation **29**
Fick's first low 182
Fillers, wooden 304
film-forming coatings 170–171
fire **13**, **29**, 34, 36–37, *40*, 44, 53
classes of response of wood and other materials to **43**
design concepts 142–143
hazard 45, 146
phases of *44*
protection 126, 143, 192
resistance 43–44, 128, 142–143, 193, 198, 242
risk of a construction to 142

fire (*Continued*)
- safety in buildings 142
- safety equipment 44, 142
- section **129**, 142
- suffocation at 35–36

fire retardants
- ammonium salts 48, 204
- degradation of wood by **13**, **46**
- Epsom's salt 48
- Glauber's salt 48
- halides 169
- intumescent paints (creating solid foams) 167–168
- meltable borates 168, 170, 192
- meltable water glass 168
- nano-compounds 192–193
- natural (charcoal) 41
- potassium carbonate 48, 169
- salts of acids – phosphoric, sulphuric, hydrochloric 169–170
- sodium carbonate 48, 169
- testing of 42–44, 196
- types of 167–170
- using of **20**, 37, 42–43, 142–143, 145, 197, 199–202

fire retardants, chemical mechanism of
- activation of endothermic dehydration reactions 168–170
- creation of covalent bonds with oxygen 168–169

fire retardants, physical mechanism of
- dilution of flammable gases with inflammable gases 168–169
- mass insulation 167–168
- thermal insulation 167–168

Fistulina hepatica 248

fixation
- of boron 155–156
- of modifying substances **10**
- of preservatives **10**, 155–156, 158, 182
- of synthetized organic preservatives 158–159

flame
- attacking of wood *3*, 37
- spreading of 40–43, 229

flammable
- gas 34–36, 38, *39*, 41–42, 44
- material **12**, *37*, 40, 44, 128

flammability 14, 42–43
- of materials **43**, 128, 143, 157, 198, 246
- test 42–43

flavonoids **9–10**, 72, 161–162

flow
- gradient of 177, 179–180
- intensity of 180
- nonstationary viscous 179–180
- of preservatives in wood 177–181
- stationary viscous 180

flufenoxuron **150**, 166–167

foil, vapour-impermeable 126, *185*

formaldehyde 236, 240, 245, **295**

Formicidae, *see* ants

Forstner drill bit 194

fossilization **29**, 49–50, **56**
- by carbonization of wood 49–50
- by mineralization of wood 49–50, *51*, 56

Fourier transformed infrared (FTIR) spectroscopy 35, 194–195, 238, **267**, 269, 275

fractometer *273*, **274**

freezing of wood 279, 284

fungi
- ambrosia 119
- brown-rot 55, **56**, 66, 71, **77–78**, 83–87, 112–113, 118, 147
- cytology of 66
- decay-causing **9**, **13**, **15**, **20**, 62, **63**, 66, 75–88, **127**, 261–262
- decaying in exteriors 80, **83–84**
- decaying in interiors 85, **86**, 87–88
- enzymatic activity of 108–117
- fruiting bodies of *67*, 68, *69*, **83–84**, **86**, 87
- genes of 66
- hyphae of, *see* hyphae
- imperfecti, *see* Deuteromycota
- microscopic, *see* moulds
- mycelium of 68, 70, 87, 262
- parasitic 65, 71
- physiology of 70
- reproduction of 66, 68
- saprophytic 65, 80, 85
- soft-rot 66, **82**, 111
- spores of *67*, 68, *69*, 87–88
- staining, *see* staining fungi

symbiotic mycorrhizal **71**
taxonomic groups of 68, 70
white-rot **56**, 66, **79–80**, 108, 111, 115, 117, 147
wood-damaging 14, 65–66, 70–76, *128*
wood-decaying, *see* fungi decay causing
wood-destroying, *see* fungi decay causing
fungi activity influenced by
abiotic factors 72–76
acidity of wood substrate 72
anthropogenic factors 76
biological factors 76
carbon dioxide 72–73
fungicides 154–163
limit and optimum moisture of wood 72, *73*, **74**, 87–88, 131, 262
limit and optimum temperature **74**, 87–88, 281
mineral substances 71–72
organic extractives 72
solar radiation 75–76
trophic factors 70–72
fungicides **19**, 145, 172
creosotes 153, 156–157
fixable in wood *23*
inorganic 48, 153–155
organic natural 153, 161–163
organic synthetized 153, 157–161
requirements for use of **148**
furfuryl-alcohol resins 221, 232, 238–239, 244–245
furfurylation **20**
furniture
damaging of 80, 100
protection of garden 126, 187, **225**, 231, 245, 247, 261
renovation of 308
resistance to damage of 52

galleries, causing of
beetles 97–102
hymenoptera insects 102–104
marine borers 106–108
termites 104–106
galleries in wood, effect of
on acoustic properties 119
on density 119
on physical properties 119
on mechanical properties 119
gas
aggressive 45–46
flammable 34–36, 38, *39*, 41–42, 44, 168–170
inflammable 168–169, 198
nontoxic 181, 286, 288–289
toxic **45**, 181, 286–288
gas chromatography/mass spectrometry (GC/MS) 35
gene engineering **19**, 249–250
glass point (T_g) 291, **296**, 297
glazing-latex coatings 170–171
Gloeophyllum abietinum **74**, **83**
Gloeophyllum sepiarium **74**, 75–76, **83**
Gloeophyllum trabeum 154, 163, 173, 243, 244
glucanases 65, **77**
1,4-β-D-glucopyranose unit **9**
glues
degradation of 1, 176
durability of **19**
glulam (glued lamellae) *2*, **24**, 41, 131, 196
chemical protection of **200**, *201*
degradation of 198
glyoxal 236
grooves in wooden products 131, *133*

Hadrobregmus pertinax 94, **95**, 100, *101*
heavy metals 151
hemicelluloses **7–9**
degradation of **13**, 31, 35 *34*, 53, **77**, **79**, **109**, 113–114, 223, 228–229
hemimetabolism 91, 104
hindered amine light stabilizers 171
holometabolism 91, 97, 102
human health 25, **45**, 157, 195, 221, 249
humidity
relative of air **11**, 28, 140, 142, 294, 306, 326
of wood, *see* moisture of wood
hydrated aluminium sulphate 169
hydrogen bonds, *see* bonds
hydrogen cyanide **45**, 286–287
hydrogen peroxide *47*, **55–56**, **77**, 108, 112, 115
hydrolysis **9**, *30*, 77, **109**, 110, *111*

hydrolytic reaction **13**, 50
hydrophobic effect **19–20**, 175
hydrophobic material 220, 223, 226
hydrothermal
effects **13**
modification **56**, 219
hydrothermolysis 224
2-hydroxyethylmethacrylate (HEMA) 291, 307
hydroxyl
group **8**, 156, 159, 175, 182, **183**, 221, 223, 290, 302
groups of wood at its chemical modification 231–238, 242–243, 246
radical 112, 115
hydroxyphenyl-benzotriazoles 170
hydroxyphenyl-*s*-triazines 170
hygroscopicity
of damaged wood 13, 52, 117–119
of modified wood 228, 232–234, 238, 245, 294
Hylotrupes bajulus 15, 91, 93, **95**, *96*, 98, *99*, 164, 166, 263, 281, 283
hymenium 66, **86**
Hymenochaete rubiginosum **80**
hymenophore 66, *69*, **86**
hymenoptera (*Hymenoptera*) 93, 102–104
hyphae **15**, *67*
cytology of 68
micro 68, 88
rhizomorphs 68, 85, 87–88, 262
substrate 72, **77**, **80**
vegetative 66, **68**

ignition 34, 37–38, 40–41, 44
imidacloprid **150**, 165, **200**
imidized nanoparticles 170
impregnability of wood
classes of 146, *147*, 194
improvement of 148, 174–176
impregnated wood, *see* autoclave impregnations of wood
impulse hammer 270
incandescence *35*, 38, *39*, 41
index of swelling (IS) 290
inflammation 37–38
inhibitors of chitin production *166*, 167
in-process treatment (IPT) 200–205
input energy (IE) 38
insect growth regulators (IGR) **19**, 165, *166*
insect hormones
anti-juvenile 165, *166*
juvenile 165, *166*
skin-sloughing 165, *166*
insecticides **19**, 145, 172
creosotes 157, 163
inorganic 163–164
nontoxic hormonal 163, 165–167, 264
organic natural 163–164
organic toxic synthetized 163–165
pheromones 163, 167
repellents 163, 167
insects
adults of 91–93, 98–106
eggs of 91–93, 98
galleries of 92, 97–106, *305*
generation of 93, 98, 100–101, 103
imago of, *see* insects, adults of
larvae of 91–93, *96*, 98–104
pupa of 91–93
wood-boring, *see* wood-damaging
wood-damaging **10**, **13**, 15, 91–106, **127**, *128*, 263
xylophagous 94, 98, 100
insects activity influenced by
abiotic factors 94, **95**, 96, 288
anthropogenic factors 97
biological factors 96–97
limit and optimum moisture of wood **95**, *96*, 98, 100, 102, 104, 131, 263
limit and optimum temperature **95**, 98, 100, 104, 280–284
trophic factors 94
inspection of wooden products **24–25**, 117, 194–196, 261
instar of insect larvae 92
insulation
acoustic 20
air 137, 140
air gap *138–139*, *141*
hydro *136*, 137–138, *139*, 262
mass 167–168

thermal **12**, 20, 42, *139*, 140–141, 167–168
waterproofing 136
International Council-Evaluation Service (ICC/ES) 151
intumescent paints, expansion factor of 168
3-iodo-2-propynyl-butyl-carbamate (IPBC) *150*, 151, 159–160, **200**
Ips typographus 91, 93, 96–97, 167
isocyanates 236, 243, 286, **295**
isolation, *see* insulation
Isoptera, *see* termites
isothiazolones (ITA) 160–161

joint
strength of glued 176
tongue and grove *133–135*

Kalotermes flavicollis 106
Kebony wood 221, 239, **244**, 247
knots **4**
Krebs cycle 115, *116*

laminated veneer lumber (LVL) 32, **43**, 196
laser 219–220, 278, 285
Lentinus lepideus **74**, 76, **83**, 244
lethal dose (LD) **150**
levoglucosan (1,6-anhydro-β-D-glucopyranose) 35
life cycle assessment (LCA) 22, 25
lifetime 1, 20–22
economical 20
estimated end of wooden product **24–25**, 194
ethical 20
physical 20–21
lignamon 219
lignostone 219
lignin **7–10**
biodegradation of **79**, **82**, **109**, 114–117
chemical degradation of 46
coniferyl phenyl-propane units **9**
depolymerized fractions of 32
guaiacyl-type **9**, **183**
phenyl-propane units **9**, 35, **80**
photodegradation of 30–31
photo-oxidized **29**, 30–31, 51
plasticizing of **13**
synapyl phenyl-propane units **9**
syringyl-type **9**
thermal decomposition of *34*, 35, *36*
Limnoria, *see* sea crustaceans
Limnoria lignorum 107–108
log-cabins 1
damaging of 80, 98, 100
diagnosis of 265
lifetime of 21
protection of 126, *136*, 182–183, *185*
renovation of 308, *312*, 313
low-molecular weight coatings 170, 172
lumens 50, *51*, *65*
filled with modifying substance 220–221, *231*, 232–233
Lyctidae 93, 95, 102
Lyctus brunneus **95**, *96*, 102, 281
Lyctus linearis *102*
Lyctus pubescens 102

maintenance
cost of 20
level of 22, **24–25**, 309
neglected 311
principles of 260–261
mannanes **9**, **109**, 113
marine borers, *see* marine organisms
marine organisms **13**, **19**, 49, **63**, 106–108, **127**, 148
mechanical property increase (MPI) 222, **244**
melamine-formaldehyde (MF) resin 239, 245, *246*, 302
melting point (T_m) **296**
metamorphosis, *see* hemimetabolism; holometabolism
methyl bromide 286–287
microorganisms **13**, 28, *30*, 247, 286
microwave
heating 239
sterilization 263, 280–283
model of cylindrical capillaries 180–181
active area in 180–181
modernization 309

modification of wood, methods of
biological **20**, **219**, 221, 247–250, 275
chemical **20**, **219**, 220–221, 225, 231–233
enzymatic **20**, 221
hydrothermal **56**, 219–220, 224
laser **219**, 220
plasma **219**, 220
mechanical **219**, 220
thermal **20**, **219**, 220, 223–224
modifying protection of wood
ecology of 221, 235, 249
effectiveness of 221–222
methods of 219–221, 223, 232–233, 247–250
technologies of, *see* "individual indexes" for chemically, or thermally modified wood
modification substances, application properties of, *see* preservatives, application properties of
modulus of elasticity (MOE) 53, **54**, 229, *230*, 242, 270, 317
moisture
penetration to buildings *128*
properties of wood **11–12**
stresses in wood 13, 28, 170
transport in wood **19**
of wood **11**, 20, 38, 42, *57*, **127**
of wood, estimated by calculation 129
of wood, its limits and optimum for biological pests 72, *73*, **74**, 75, 87, **95**, *96*, 98, 100, 104, **127**, 131, **148**
of wood, its optimization before chemical protection and modification 148
of wood, its reduction by shape optimizations **129**, *131–136*
in wooden products 129, *130*
moulds
attack of wood by **13**, 62, **63**, 262
bio-control of wood by **20**, 221, 248–249
conditions for activity of 75, **127**
durability of modified wood against 226–227, 242
ecological risk of 91
preservatives against 148, 159, 163
reproduction of 66, *70*
species of *90*
musical instruments 229, 242, 283
mycelium 68, 70, 75
mycotoxins 76, 91, 221

nano-biocides 192–193, 199
nano-preservatives 191–193
natural durability of wood
classes of 15, **16–17**, **24**, 146, *147*, **148**
improving of **19–20**
predetermined by its structure 10, **15**
related to biological pests 14, 17–18, 128
testing of 14
near-infrared (NIR) spectroscopy 195, **267**, 275
nuclear-magnetic-resonance (C^{13} NMR) 35, 238, **267**, 269, 272

OHT-Wood 52
process of 224–226
properties of 227–229
oidia **68**
oil heat treated wood (OHT-Wood) 52
oil-based semi-transparent stains 172
Oligoporus placenta 113, 154
N-organodiazeniumdioxy-metals 157–158
organometal fungicides 158
organosilanes 241
oriented strand board (OSB) **19**, 196–198, **200**, 203–204
Oudemansiella mucida 248
oxidation **9**
bio **109–110**
photo **13**, 30
thermo **13**
oxidising wood attacking substance *3*

Paecilomyces sp. 79, 90
Paecilomyces variotii *70*, **90**
Paxillus panuoides **74**, **83**
paints, *see* coatings
Paints Directive 2004/42/EC 151
particleboard (PB) *2*, 196
degradations of **43**, 85, 197–198

wood-cement (cement chip board – CCB) **43**, 128, 196–198
chemical protection of **200**, *201*, *203*
penetration
into cell walls of wood 172, 232–234, 237, 240, 243, 245, 266, 289–290, 294, 297, 300, *303*, 307
class of preservative *147*, 194
depth of 148–149, 170, 173, *181*, 187, 189, 194, 289, 306
into lumens of wood *179*, 180–181, 232, 266, 289, 294, 297, *303*
Penicillium brevicompactum **90**, 286
Penicillium cyclopium *70*, **90**
pentachlorophenol (PCP) **150**, 151, 157
permeability
coefficient of wood 174, 180
of damaged wood 117, 119
of wood **8**, 41, 46, 149
permeability, improvement of refractory woods 174–176
biological, with bacteria and fungi 63–64, *88*, 175–176
chemical, with acetic acid, etc. 175
mechanical, by cutting and puncturing 175
physical, with laser, ultrasonic, etc. 175
permethrin **150**, 164
peroxidases, *see* enzymes
Phanerochaete chrysosporium 117
Phellinus pini 66
phenol-formaldehyde (PF) 205, 239–240, 302
phenolic resins 41, 48, 198, 221, 232, 238–240, 245–246, 292, **295**, 301
pheromones **19**, 94, 96–97, 163, 167
aggregation 97, 167
alarm 167
defensive 97
dispersion 97, 167
identification 97
sexual 96, 167
Pholiata carbonica 248
phosphine 286, 288–289
photo-oxidation **29**, 51, 172
photostability of wood 193
pH value of wood
change at fixation of preservative 158, **183**
limits and optimum for wood-damaging fungi 72, **74**
physical status
glassy **8**
plastic **8**
viscoelastic **8**
physiology
of wood-damaging fungi 66, 70–76
of wood-damaging insects 94–97
Physisporinus vitreus 176
pigments 170–173
Piptoporus betulinus 72
plasma 219–220, 285–286
plastic texture 32, 51
PlatoWood 52, 226
process of 224–225
properties of 226, *227*, 228, **230**
plywood *2*, **19**, 32, *33*, 221
chemical protection of 165, **200**, *201*, *203*
degradations of 41, **43**, 85, 197–198
Poiseuille equation 180
polychrome 281, 283, *305*, 318
polyesters 294, **295**, 298
polyethylene glycols (PEGs) 56, 294, **295**, 300–302, 307–308
polymer concrete 315, *316*, 318, *320–321*
polymerase chain reaction (PCR) 276, *277*
pore-filling-ratio (PFR) 173, 222
Poria monticola 113
Poria placenta 114, 163, 173, 245
post-manufacture treatment (PMT) 200, 204–205
preservatives **19**, 145, 149–173
accompanying compounds of 152
active substances of 151–172, **183**
diffusion in wood **10**
ecotoxicity of 150, 152, 161
efficiency of 145–146, 193, 195–196
evaluation of 172–173
fixation in wood **10**, 149, 182, **183**, 195
nano-compounds 191–193
notification price of 152
regulation of 151–152, 177

preservatives (*Continued*)
  stability in wood 145, 195
  toxicity of 149, **150**, 152, 160–161
  transport in wood **5**, 174, 177–182
preservatives, application properties of
  chemical stability 176–177
  diffusion coefficient 177, 182
  leachability 176, 201, 241
  penetration ability 233
  pH value 177, 202
  polarity 174, 177
  reactivity with wood 234–238
  solubility 177
  surface tension 174, 202
  thermal stability 177
  viscosity 174, 177, 180–181, 192, 202
  volatility 176, 201
  wide sphere of use 177
preservatives, evaluation of
  corrosiveness for wood 173
  health risk 173
  intended efficiency 172
  screening and standardization tests 172–173
  stability in wood 173
  treatability 172
pressure
  in autoclave 177, 179–180
  capillary, *see* capillary pressure
  driving forces 174, 177, **178**, 179–180, 189
  dynamic 177, 179
  hydrostatic 177, 179–180
  local mechanical 177, 179
properties of damaged wood 50–57, 117–119
propiconazole 150–151, 158–160, 165, **200**
protection of wood
  biological 76
  chemical 1, 18, 22, 145–206
  methods of 17–18
  modifying 1, 18, 218–250
  preventive **19**, 263
  structural 1, 14, 128–143
Pseudomonans spp. 63, 150, 176
Ptilinus pectinicornis *101*
putties 304
pyrethrins 164
pyrethroids **19**, 163–165, 199, **200**
pyrolysis *35–36*, 44, 169, 225

quality control of chemically protected wood
  certificate from 193
  penetration requirements at 194
  retention requirements at 194
quaternary ammonium compound (QAC) **19**, **150**, 152, *153*, 159, 164, 246

radiation
  infrared (IR) **29**, 30
  gamma (γ) 279, 270–271, 284–285, 291, 298
  microwave 52, 281
  photo *30*
  solar 30–32, 75–76
  thermal *30*
  ultraviolet (UV) **10**, **29**, 30–31
reactions
  biochemical 1, **13**, 108–117
  condensation 50, 223
  cross-linking 220, 223, 237–239, 298, 302
  dehydration **13**, 34, 159, 168, 220
  depolymerization 34, 113, 115
  endothermic 34, 37, 168
  exothermic 34–38, 40–41, 168–169
  hydrolytic **13**, 30, 50, **109**, 220
  ion-exchange **183**
  oxidation **13**, 30, 41, 108, **109–110**, *112*, 168, 226, 269
  oxidation–reduction **183**
  photolytic 30
  photo-oxidation **13**, **29**
  polyaddition 238, 292, **295**, 299
  polycondensation 238, 240, 292, **295**, 298
  polymerization 238, 291, **295**
  thermo-oxidation **13**, 36, 38, 168
reconstruction 18, 22, 309, 311
rectification process (NOW) 226, 228
recycling 25
reinforcement of damaged wood
  with conservation agents 289, 291–292, 298–301, 304, 306

in practice **24**, 48, 237, *314*, *316*
using renovation methods 308–327
reinforcing roads *320–321*
renovation of greater parts of wooden constructions
interlinking 308, 326, *327*
spatial bracing 308, *325*
stretching *326*
tightening 325, *326*
renovation of individual wooden elements
anchoring into steel brackets 308, **313**, 315, *316–317*
bracing 308, 323, *324*
height adjustment 308, **313**, 315, *316*
insertion of carbon lamella **313**, 315, 317, *318*
prosthesis 308, **313**, 314, 317–318
of "beta" system 318, *320–321*, 322
of carpentry types 318, *319*
sealing 308, 313–314, 321, *322*
splicing 308, **313**, 314–315
supporting 308, 322, *323*
suspending *324*
unloading 308
renovation project 309–311
repair 22, 300, 309, 311–312, *317*
repellents of pests **19**, 163, 167
reproduction
of wood-damaging fungi 66–70
of wood-damaging insects 91–93
resin canal **4–5**, 62
resistance of wood against
biological pests 15, 52, 72
fire **15**
fungi **15**, 72
ignition **15**
insects **15**
UV radiation **15**
resistograph **268**, *272*, 273, *274*
restoration 1, 260
retention
class of preservative *147*, 194
of conservation agent 291
of creosote 157, 189
of preservatives 173, *181*, 187, 194–195
Reticulitermes lucifugus 105
Rhinotermitidae, *see* subterranean termites
roof overlap 134, *135*
rot
principle of wood 77–82, 106–117
structure of wood damaged by 77–82
rothounds dog **86**, 266
rotten wood
acoustic properties of 118, **267**, 269–270
density of 117
hardness of **113**, 118
hygroscopicity of 117
impact bending strength of **113**, 118
mechanical properties of **113**, 118, **268**, 273–275
permeability of 117
physical properties of 117–118
structure of **77–78**, **80**, **82**

saccharose 56, 232, 292–294, 302, 307–308
Saperda carcharias 119
scanning electron microscope 181, 195, **267**, 268
Schizophyllum commune *73*, **74**, **84**
Scolytidae 93, 97, 263
Scolytus intricatus 119
sea crustaceans 107
Serpula himantioides 154, 276
Serpula lacrymans 66, 71, **74**, 75, *78*, 85, **86**, 87, 113, 118, 154, 159–160, 248, 262–263, 266, 276, 281, 284–285
service life 1, 21–22, 151, 311
estimated (ESL) 21–22, **23–24**
factor method for determination of 21–22
prediction of (SLP) 20–21
prolongation of 126, 221, 250
reference (RSL) 21–22
of wooden products 18, 21–22, 25, 157, 162, 172, 176, 205, 221
shape
deformation **12**, 51
optimization **129**, 131, *132–135*, 136
shellac 292–293
shingles 22, 131, 187, 189, 247
shipworms 106
silicates 49, 240, 246
silicon compounds 232, 246–247, 240–241
silicones **20**, 240–241

silver 152, 154
  nano-compounds 192–193
silver chloride 152, 154
Sirex juvencus 103
Siricidae, see woodwasps
sleepers
  damaging of 63, 80
  life time of 21
  protection of 145, 156, 186, 189
  resistance to damage 52
smoking 41–42, 45
sodium octaborate 155, 164
sodium tetraborate **46**, 155, 164, 168, 170, 308
soft rot **13**
  fungi causing 66, **82**, 111, 148
  mechanisms of **82**
staining fungi **13**, 62, **63**, 66, 71, 75, *88*, **89**, **127**, 148
state of wood in construction (V) 264
static of construction 309, 311–312
Staypak/Staywood 219
Stereum hirsutum **74**, **84**
sterilization of wood 145, 263, 279–280
  chemical 181, 286–289
  laser 220
  plasma 220
  radiant 284–286
  thermal 87, 280–284
stilbenes **10**, 17, 72, 162
strength
  of chemically protected composites 205
  of degraded wood 13, **29**, 48, 52–53, **55–56**, *57*, 62, **113**, 118–119, 280
  of renovated wood 308
  of treated wood **10**, 220, 229, *230*, 232, 238, **244**, 245, *246*, 280, 285, 292, **296**, 298–302
strengthening of damaged wood *246*, 262, *305*
structural protection of wood
  design proposals for 129, 132–141
  fire sections for 142–143
  methodology of 126
  methods of **129**
  by selection of suitable wood materials 126–128
styrene 232, 298
sulphur dioxide 286
sulphuryl fluoride 286–289
surface of wood
  damage of 28, 32, *33*
  flame spreading on 37, 41
  roughness of **4**, *31*, **15**, 40, 51, 220
swelling of wood, *see* wood swelling
  inhibition of 134, 228, 232
symbiosis 65, 95
synergy 30, 76, 111, 113, 160, 197, 225, 247, 288

tannins **9**, 15, 17, **19**, 72, 85, 100, 176, 248
  as fixators of boron preservatives 155
taxonomic groups of biological pests 68, 93
tebuconazole 150–151, 158, 160, 165, **200**
technologies of wood treatment with chemicals
  application of pastes **178**
  application of solutions **178**
  bandages **178**, 179, 183–184, *185*, 261, 306–307
  cartridges **24**, 164, 184, *186*
  cyclic impregnation **178**
  dipping **24**, 172, **178**, 179, 182–183, *184*, 204, 233, 261, 306–308
  dipping with preheating **178**
  dipping after vacuum **178**
  immersion **178**, 179, 182–183, *184*
  injecting **178**, 179, 186, 261, 264, 306
  nanotechnologies 191–193
  painting 149, 157, 172, **178**, 179, 182–183, 233
  panel methods **178**, 179, 183, *185*, 307
  pressure-diffusion **178**, 187, 189–191, 300, 307–308
  pulse impregnation **178**
  spraying 149, 172, **178**, 179, 182–183, *184*, 204, 233, 264, 306–307
  supercritical carbon dioxide (SC-CO_2) 200, 204–205
  vacuum pressure “individual technologies”, *see* Autoclave wood impregnation
  vacuum-pressure **24**, 149, 156–157, 172, **178**, 179, 186–189, *190*, 204, 233, 261, 306

Teredinidae, *see* shipworms
Teredo navalis 106, *107*
termites 93, 104–106, **127**, 165, 198
  nonsubterranean 94, 105
  subterranean 93, 105
  tunnel damaging of wood with 104, *105*
terpenes **9–10**
  biologically effective **9–10**, 161–162, 275
  influencing activity of insects 98
terpenoids **19**, 161–162, 223, 263
Tetropium castaneum 93, *99*
thermal conductivity 14, 38, 40–41, 140, 283
thermal decomposition *34*, 35, *36*, 37, *39*, 41, 168–169, 198
thermal properties of wood **12**
thermally modified wood
  biological durability of 199, 226–228
  changes in structure of 223
  dimensional stability of 228
  fire resistance of 229
  technologies, *see* processes of ThermoWood, PlatoWood, or OHT-Wood
  weather resistance of 228
thermograph 140, **267**
thermo-gravimetric (TG) 35
ThermoWood 52, 226
  processes of *224*
  properties of **225**, 226, *227*
thujaplicins 162
titanium dioxide 152, 171, 193
tolylfluanid **150**
torus *64*
Trametes versicolor *73*, **74**, **80**, *81*, 113, 163, 173, 192, 244, *277*, 278
tri-arylphosphates 170, 202
triazoles **19**, *153*, 159–160, 165, 199–200, 202
tri-butyl-tin naphtenate (TBTN) 157
Trichoderma gliocladium 176
Trichoderma harzianum 163, 248
Trichoderma viride *70*, 76, **90**, 176, 248–249
tri-methyl borate (TMB) 155, 204–205
trophic factors
  needed for activity of wood-damaging fungi 70–72
  needed for activity of wood-damaging insects 94
trusses 1, 22
  damaging of *40*, 85, 98, 100, 103
  protection of 126, 145, 182
  renovation of 308, 313, *323–326*
Trypodendron lineatum 93–94, 97, 119
tylose in vessel **6**

urea-formaldehyde (UF) 205, 239, 245, *246*, 302, *303*
urethanization 234, 236
Urocerus gigas *103*
utility poles
  damaging of 63, 80, 103
  diagnosis of 266, *272*
  protection of 145, 156, 164, *185*, 186, 189–190
  resistance to damage 45, 52, 131
UV radiation **13**, 14, 28
  increasing of resistance to 170–172, 220
  reflectors of 170
UV stabilizers **20**, 170–172, 193

vacuum sublimation 301, 306–307
vapour-impermeable 126, 130, 170, *185*
vegetable extracts **19**
vegetable oils 161–162, 172, 225–227, 232, 261, 292
veneers *33*, 221, 239–240, 269
ventilation
  gap *141*
  of interior 142
  of roof *139*
cis-verbenol 167
volatile organic compounds (VOC) 151, 172, 205, 245, 249, 275

water
  absorptivity of 88, 117, 119
  capillary 20, *128*, 138
  condensation of 134, *137*, 140, 324
  condensed *128*, 136, 140–142
  evaporation of 134

water (*Continued*)
hygroscopic *128*
isolation **129**
precipitations **25**
rainfall **24**, 28, 31, 51, 126, *128*, 131, *132–135*, 137–138
splashing *128*
transport of 87, 134
vapour 38
water repellents **20**, **24**, 42, 172, 197
water resistance of wooden composites 198, 205
water-based semi-transparent stains 172
waxes 3, 172, 201, 232, 292
weather factors 28, 30, 51–52
weathering 14, 30
accelerated 32, 171, 195
resistance to 52, 198
natural (field) 28, 32, 171, 195
protection against 170–172, 192, 220
weight-percent-gain (WPG)
of modifying substance 222, 242–245
of preservative 156, 173
West European Institute for Wood Impregnation (WEI) 156
white rot **13**, *81*
of delignification type **79–80**, *81*
of erosion type **79–80**, *81*
mechanisms of **79–80**, 111–112
wood properties with **113**, **274**
windows
damaging of 80, 85, 100, 103
lifetime of 21–22, 25
protection of 126, *132*, 134–136, *137*, 142, 160, 164, 183, *186*, **225**, 230, 247, 261
renovation of 308, 314
resistance to damage 52
venting gaps in 137
wood
annual rings of **4**, **6**, *33*
archaeological, *see* archaeological wood
broadleaved **5–6**, **9–11**, **16–17**, 45–46, **74**, **84**, **95**, **225**
cell elements of **4–6**
cell walls of **4–7**
compression **4**
coniferous **5–6**, **9**, **11**, **16–17**, 32, 45–46, **74**, **83**, **95**, 98, 103, *104*, **225**
density **10–11**, **16**, 40–41, 53, **56**
durable species of **9**, **16**, **19**, 126
early **4**, **15**, 32, 103, 278
heart **4**, **16–17**, **174**
juvenile **4**, 14
late **4**, 32, 278
mature **4**, 98, **174**
moisture, *see* moisture of wood
plastic texture of 32, *33*, 51
polymers of **8–10**
porosity of **11**, 40, 72
sap **4**, **15**, *64*, 88, **174**
shrinkage of **12**, **24**, 53, **55–56**, 228
strength of **11**, **55**, *230*, *246*
subfossil (fossil) **11**, 17, 53, 55, *57*
swelling of **12**, **24**, 134, 228, 232
tension **4**
tissues in **5**
wood for constructions, selection of 126, 128
wood degradation, types of **13**
wood structure
anatomical *2*, 3, **7–8**, 13, **15**, 42, 108, 117, **219**, 266
geometric *2*, 3, **4**, 13, **15**, 28, 42, 117, **219**, 266
molecular *2*, 3, **8–10**, 12, **15**, 42, **56**, 108, 117, **219**, 233, 244, 266, 269, 275
morphological *2*, 3, **4–6**, 13, **15**, 42, 117, **219**, 266
wood-damaging insects, *see* insects wood-damaging
wood-decaying fungi, *see* fungi, decay-causing
wooden composites
chemically protected, *see* chemical protection of wooden composites
degradation of 32, 42, **43**, 85, 90, 104
durability of **19**, **43**
modified 221
structure of 1, *2*
susceptibility to damage of 197–199
types of *2*, **43**, 149, 196–197

wooden constructions
  faces of *132–133*
  life time of 20
  protection of **128**
wooden products
  life time of 20
  protection of **128**
wood-plastic composite (WPC) 196, 198, **200**, 241
woodwasps 102–103

Xestobium rufovillosum *101*
X-ray
  diagnostic 194–195, **267**, 270–272
  sterilization 284
xylanes **9**, **109**, 113
Xyleborus dispar 97

zinc borate 154–155, 193, **200**, 204
zinc oxide 154, 171, 192–193, 202, 204